T0343349

Human Herpesviruses HHV-6A, HHV-6B, and HHV-7

Diagnosis and Clinical Management

Human Herpesviruses HHV-6A, HHV-6B, and HHV-7

Diagnosis and Clinical Management

Louis Flamand
Quebec, PQ, Canada

Irmeli Lautenschlager
Helsinki, Finland

Gerhard R.F. Krueger
Houston, TX, USA

Dharam V. Ablashi
Santa Barbara, CA, USA

ELSEVIER

AMSTERDAM • BOSTON • HEIDELBERG • LONDON • NEW YORK • OXFORD
PARIS • SAN DIEGO • SAN FRANCISCO • SINGAPORE • SYDNEY • TOKYO

Elsevier
The Boulevard, Langford Lane, Kidlington, Oxford OX5 1GB, UK
Radarweg 29, PO Box 211, 1000 AE Amsterdam, The Netherlands

First edition 1992
Second edition 2006
Third edition 2014

British Library Cataloguing-in-Publication Data
A catalogue record for this book is available from the British Library

Library of Congress Cataloging-in-Publication Data
A catalog record for this book is available from the Library of Congress

ISBN: 978-0-444-62703-2

For information on all Elsevier Science publications visit our website at http://store.elsevier.com

Typeset by MPS Limited, Chennai, India
www.adi-mps.com

Cover images courtesy of Dr. Bhupesh K Prusty, University of Wuerzburg, Germany.

Co-fluorescent *in-situ* hybridization (Co-FISH) showing localization of ciHHV-6 within human telomeric region. HHV-6 (Green), telomeric probe (Red) and DNA (Blue).

Contents

List of Contributors

Dharam V. Ablashi The HHV-6 Foundation, Santa Barbara, CA, USA

Henri Agut AP-HP Groupe hopital Pitié-Salpêtrière, Paris, France; Université Pierre-et-Marie-Curie, Paris, France

Bridgette Jeanne Billioux The Johns Hopkins University, Baltimore, MD, USA

Francesco Broccolo University of Milano–Bicocca, Monza, Italy

L. Maximilian Buja The University of Texas Health Science Center at Houston and Texas Heart Institute, St Luke's Episcopal Hospital, Texas Medical Center, Houston, TX, USA

J. Mauricio Calvo-Calle University of Massachusetts Medical School, Worcester, MA, USA

John R. Crawford University of California and Rady Children's Hospital, San Diego, CA, USA

Leon G. Epstein Ann & Robert Lurie H. Children's Hospital of Chicago, Northwestern University Feinberg School of Medicine, Chicago, IL, USA

Louis Flamand Axe maladies infectieuses et immunitaires, Centre de recherche du CHU de Québec and Laval University, Québec, Canada

Agnès Gautheret-Dejean AP-HP Groupe hopital Pitié-Salpêtrière, Paris, France; Université Pierre-et-Marie-Curie, Paris, France; Université René Descartes, Paris, France

Steven Jacobson National Institute of Neurological Disorders and Stroke, National Institute of Health, Bethesda, MD, USA

Yoko Kano Kyorin University School of Medicine, Shinkawa Mitaka, Tokyo, Japan

Anthony L. Komaroff Harvard Medical School, Boston, MA, USA

Gerhard R.F. Krueger The University of Texas Medical School at Houston, Houston, TX, USA

Uwe Kuehl Charité—Universitätsmedizin Berlin, Campus Benjamin Franklin, Berlin, Germany

Roberto Alvarez Lafuente Hospital Clínico San Carlos, Madrid, Spain

Dirk Lassner Institute Cardiac Diagnostics and Therapy, Berlin, Germany

Irmeli Lautenschlager Helsinki University Hospital and University of Helsinki, Helsinki, Finland

Jin-Mei Li West China Hospital of Sichuan University, Sichuan, China

Mario Luppi Università degli Studi di Modena e Reggio Emilia, Modena, Italy

John J. Millichap Ann & Robert Lurie H. Children's Hospital of Chicago, Northwestern University Feinberg School of Medicine, Chicago, IL, USA

Jose Montoya Stanford University School of Medicine, Stanford, CA, USA

Guillaume Morissette Axe maladies infectieuses et immunitaires, Centre de recherche du CHU de Québec, Québec, Canada

Lieve Naesens Rega Institute, Katholieke Universiteit Leuven, Leuven, Belgium

Pitt Niehusmann University of Bonn Medical Center, Bonn, Germany

Masao Ogata Oita University Hospital, Hasama-machi, Yufu-city, Oita, Japan

Joseph Ongrádi Institute of Medical Microbiology, Semmelweis University, Budapest, Hungary

Leonardo Potenza Università degli Studi di Modena e Reggio Emilia, Modena, Italy

Joshua C. Pritchett The HHV-6 Foundation, Santa Barbara, CA, USA

Sylvie Ranger-Rogez Dupuytren University Teaching Hospital and Faculty of Pharmacy, Limoges, France

Raymund R. Razonable College of Medicine, Mayo Clinic, Rochester, MN, USA

Tetsuo Shiohara Kyorin University School of Medicine, Shinkawa Mitaka, Tokyo, Japan

Balázs Stercz Institute of Medical Microbiology, Semmelweis University, Budapest, Hungary

Lawrence J. Stern University of Massachusetts Medical School, Worcester, MA, USA

Mikoko Tohyama Ehime University Graduate School of Medicine, Shitsukawa, Toon-city, Ehime, Japan

Vincent Descamps Hôpital Bichat, Université Paris-Diderot, Paris, France

Tetsushi Yoshikawa Fujita Health University School of Medicine, Toyoake, Aichi, Japan

Danielle M. Zerr University of Washington, Seattle Children's Research Institute, Seattle, WA, USA

Dong Zhou West China Hospital of Sichuan University, Sichuan, China

Foreword

We have come a long way.

Named human B lymphotropic virus (HBLV) at the time of discovery in 1986, and then renamed human herpesvirus 6 (HHV-6) shortly thereafter, the viruses subsequently known for over 20 years as human herpesvirus 6 variants A and B are now recognized by the international virology community as two closely related but distinct viruses, human herpesvirus 6A and human herpesvirus 6B (HHV-6A and HHV-6B). A somewhat more distantly related virus, human herpesvirus 7 (HHV-7) was discovered in 1990. Because of their shared genetic and biological properties, these three viruses have been assigned to species in the *Roseolovirus* genus of the Betaherpesvirus subfamily in the virus family Herpesviridae. The nearly coincident timing of their discoveries, their close phylogenetic relationships, their many shared and related properties, and an international cadre of biomedical researchers who have studied more than one of them (sometimes simultaneously) have resulted in highly intertwined, interwoven, and fused scientific histories.

We have come a long way.

Since their discoveries, connections have been sought between these viruses and human disease. In 1988, HHV-6B was identified as the cause of most cases of exanthem subitum (also known as roseola, roseola infantum, and sixth disease). As might be expected, some initially promising etiologic associations faded after further study, while other stories have percolated along for 20 or so years without resolution. Importantly, diagnostic methods have improved and the epidemiologic and clinical questions have become more sharply focused. We are in the midst of a period of discovery where significant progress is being made in identifying new clinical associations and more completely defining the spectrum of disease associated with roseoloviruses.

We have come a long way.

All three human roseoloviruses are T lymphotropic. Nonetheless, all three have been associated with neurologic diseases, and the stage is nearly set for controlled clinical trials of treatments for some of these diseases.

We have come a long way.

Approximately 1% of humans harbor chromosomally integrated HHV-6 genomes that were transmitted via the germ line and are thus present in essentially every cell in the body. Chromosomally integrated HHV-6 has plausible but unproven clinical significance and an intriguing biology that almost certainly extends to vertically transmitted HHV-6 infections.

We have indeed come a long way.

This book, the third edition of an important series, is a timely encapsulation of the state of the art regarding *Roseolovirus* molecular and cellular biology, immunobiology, diagnosis, pathogenesis, disease associations, and therapeutic approaches. Every chapter is written by active, internationally respected experts. The editors are to be commended for assembling such a stellar panel of authors and for ensuring such comprehensive coverage.

Finally, I note the recent passing of Dr. Caroline Breese Hall, known to many of us as Caren. For 20 years, Dr. Hall led clinical studies of roseoloviruses in young children at the University of Rochester. Her work was at an extraordinary quality level and scale (over 10,000 children were studied), and provided many observations, insights, and tools that guide ongoing research into these viruses. Dr. Hall's trainees and the other young (and not-so-young) scientists she encouraged in so many ways stand as an enduring legacy and inspiration.

We have much to live up to and much to do.

Philip E. Pellett
Wayne State University
Detroit, Michigan, USA
November 2013

Foreword

Human herpesvirus-6 (HHV-6) was first discovered from the immunocompromised patients in 1986 in the US. Soon after discovery of the virus, it has been demonstrated in Japan that primary viral infection was associated with exanthem subitum that is common febrile exanthematous illness in infants. Although the disease had been considered to be benign disease, it has been demonstrated that several severe complications including encephalitis, fulminant hepatitis, and hemophagocytic syndrome occur in some patients with primary HHV-6 infection. In addition to primary viral infection, HHV-6 reactivation has been shown to be implicated in several severe diseases in immunocompromised patients. Among the suggested diseases caused by HHV-6 reactivation in transplant recipients, post-transplant acute limbic encephalitis is the most important disease because of strong correlation between the disease and virus infection and disease severity. Therefore, neuropathogenicity of HHV-6 is considered to be particularly important research field of the virus infection.

HHV-6 was initially divided into two variants; HHV-6 variant A and B, based on genetic differences and differences of reaction against monoclonal antibodies. Recently, HHV-6 was divided into two distinct species; HHV-6A and B. Although seroepidemiology and primary infection of HHV-6B have been well understood, those of HHV-6A remain to be unclear. Limited molecular epidemiological studies have suggested that there was geographical difference in HHV-6A infection; low endemic in Europe, United States, and Japan vs. high endemic in Zambia. No specific serological assay system discriminating between HHV-6A and B antibodies has hampered to elucidate seroepidemiological difference between the two viruses. In order to clarify the important issues, it is necessary to develop the specific serological assay and to prepare the international collaborative study for performing international seroepidemiological analysis.

The most interesting characteristic of HHV-6 is integration of viral genome into human chromosome (chromosomally integrate HHV-6, CIHHV-6). It has been demonstrated that human telomere like repeat plays an important role in integration of viral genome. Although CIHHV-6 has been suggested to have pathogenicity such as immunosuppression in CIHHV-6 transplant recipient, it is considered that further study is necessary to confirm pathogenic role of CIHHV-6. Moreover, it has been demonstrated that infectious virus could be replicated from integrated viral genome by chemical stimulation in-vitro. Although several investigators have demonstrated that in-vivo HHV-6 reactivation from CIHHV-6 patients, I think that we have to continue to performing careful experiments to determine whether viral reactivation occur really from CIHHV-6 patient.

Since various important findings such as the mechanism of CIHHV-6 have been demonstrated in the last couples of years, numerous chapters have been updated to include the new information with regard to basic virology and clinical virology. Additionally, HHV-6 has been divided recently to the two different species; HHV-6A and HHV-6B. Therefore, difference of the two species was intended to be clear in this edition. New important questions with regard to HHV-6 infection appear one after another. I think that this book may provide answers of those questions.

Yoshizo Asano

Dedication

Since its first isolation in 1986 to the late 1990s the Human Herpesvirus 6 (HHV-6) field of research was prosperous, attracting numbers of excellent fundamental scientists and clinicians. Findings were numerous as often the case with isolation of a new pathogen, and the HHV-6 research community was slowly building with the first HHV-6 international conference held in Atlanta in 1995. Although many important biological discoveries were made during this time, reductions in research funding coupled with the lack of clear-cut disease associations (other than roseola) have gradually eroded the number of scientists working on HHV-6/HHV-7. By the mid-2000s the HHV-6 research community reached a minimum.

In 2004 the HHV-6 Foundation was created to promote HHV-6 awareness in clinical and basic research. Mrs. Kristin Loomis, the HHV-6 Foundation President and co-founder along with its Scientific Director Dr. Dharam Ablashi, reached out to the HHV-6 scientists and encouraged them to actively collaborate and renew their research interest on HHV-6. One of the Foundation's first initiatives was to relaunch the HHV-6/HHV-7 international conference (Barcelona in 2006). Another major initiative was the creation of a repository for HHV-6 reagents. With the generosity of many scientists, key reagents are now available to those interested in conducting HHV-6 research. In addition, a generous pilot grant program enables many investigators to initiate and conduct promising studies and generate essential preliminary results required by conventional funding agencies. Lastly, the HHV-6 Foundation played a central role convincing the ICTV that HHV-6A and HHV-6B should be recognized as distinct viruses. Without their constant and relentless efforts, research on HHV-6 would not have progressed at the same pace nor would the field be as healthy as it is today. We would also like to acknowledge past and current Scientific Advisory Board members for their involvement in many of the projects carried out by the HHV-6 Foundation. Finally, the HHV-6 Foundation contributed in several other ways, including the enticement of young researchers to the HHV-6 field as well as the dissemination of relevant information to patients, clinicians, and researchers by the publication of a monthly newsletter. It is our great pleasure to dedicate this book to the entire staff of the HHV-6 Foundation for their constant support and efforts over that past decade.

Louis Flamand
Irmeli Lautenschlager
Gerhard R.F. Krueger
Dharam V. Ablashi
Editors

Chapter 1

HHV-6A and HHV-6B Officially Classified as Distinct Viruses

Joshua C. Pritchett and Dharam V. Ablashi
The HHV-6 Foundation, Santa Barbara, California

INTRODUCTION AND CLASSIFICATION HISTORY

In 1986, a new virus was isolated from patients in the United States with AIDS and other lymphoproliferative disorders.[1] Initially called human B-lymphotrophic virus (HBLV), the virus was renamed human herpesvirus 6 (HHV-6) (GS strain) soon thereafter. In 1987 and 1988, independent isolates were obtained from AIDS patients, designated U1102 (from Uganda)[3] and Z-29 (from Zaire).[4,5] As other strains were isolated from various geographic regions and clinical settings, it became gradually apparent that all HHV-6 isolates could be unambiguously included in one of two well-defined groups, differing in molecular, epidemiological, and biological properties.[6–10] The two groups showed different *in vitro* tropism for selected T cell lines, specific immunological reactions to monoclonal antibodies, distinct patterns of restriction endonuclease sites, and specific and conserved interstrain variations in their DNA sequences.[11]

In the early 1990s, the scientific community debated whether the two groups reflected a normal heterogeneity of the virus population,[6] and in 1992 an initial consensus was reached to designate the groups as two variants of the same virus: HHV-6A and HHV-6B.[12] This decision was based on two main factors: (1) the interspecific divergence of nucleic acids was remarkably low, and (2) differential epidemiology and pathogenic potential were still unknown.[12] However, as new evidence was provided, several authors began to suggest that the two variants should be designated as different viruses.[13–16]

Recently, genomic sequencing has confirmed the indisputable distinction between HHV-6A and HHV-6B. The genomes of the two HHV-6 variants are colinear and share an overall identity of 90%, but intervariant divergence of

Human Herpesviruses HHV-6A, HHV-6B & HHV-7.

specific sequences (i.e., the immediate-early locus) is higher than 30%,[17,18] and there are differences in function between IE1A and IE1B.[19] Remarkably, even though the IE1 gene differs substantially between HHV-6A and HHV-6B, this region is highly conserved (>95%) within clinical and laboratory isolates of the same group.[20] Analysis of different viral strains shows that even highly conserved sequences with homology higher than 95%, such as gH, gB, and U94, are characterized by specific amino acid signatures, permitting unambiguous distinction between the two variants.[21,22] Furthermore, several reports describe that the splicing pattern and temporal regulation of transcription of selected genes are different.[17,18,23–25] The lack of a genetic gradient and the absence of evidence of intervariant recombination suggest that the two groups do occupy different ecological niches *in vivo*.[13]

An Ad Hoc Committee on HHV-6A & HHV-6B Genomic Divergence was formed in 2009 by the HHV-6 Foundation to generate an official proposal to recognize HHV-6A and HHV-6B as distinct viruses; the proposal was submitted to the International Committee on Taxonomy of Viruses (ICTV) in 2010. In 2012, the ICTV officially ratified the classification of HHV-6A and HHV-6B as viruses species, replacing human herpesvirus 6 with human herpesvirus 6A and human herpesvirus 6B in the genus *Roseolovirus*, subfamily Betaherpesvirinae, family Herpesviridae, order Herpesvirales. HHV-6A has been designated as the type species in this genus.[26]

In spite of suggestions originally advanced by Braun et al.,[15] authors of many HHV-6-related papers have not investigated or specified which virus was studied. The lack of clear distinction between HHV-6A and HHV-6B in the literature makes it difficult to properly assess epidemiological differences and etiologic associations. In light of the ICTV's official reclassification, the utilization of virus-specific clinical and laboratory assays for HHV-6A and HHV-6B is especially crucial.[27–30] However, because HHV-6A is usually present at lower copy numbers than HHV-6B, assays that rely strictly on melting point analysis for differentiation may be biased toward the detection of HHV-6B, resulting in further confusion.[31] To avoid this complication, the utilization of comprehensive virus-specific assays is preferred.[30,32–34] In an effort to bring additional clarity to the important biological and clinical distinctions between HHV-6A and HHV-6B, we herein urge scientists and physicians to carefully differentiate, whenever possible, between HHV-6A and HHV-6B in all future publications.

DISTINCT EPIDEMIOLOGY AND DISEASE ASSOCIATIONS

In the United States and Japan, 97 to 100% of primary infections are from HHV-6B and occur between the ages of 6 and 12 months.[35–39] Less is known about the epidemiology of HHV-6A infection. One report has indicated that HHV-6A infection is acquired later in life and primary infection is typically without clinical symptoms.[40] However, several groups have now documented symptomatic HHV-6A primary infections among children from both the United States and Africa.[41,42] In addition, HHV-6A was found to be the predominant viruses associated with viremic infection in a pediatric population of Sub-Saharan Africa,[41] and it has also been shown to cause roseola and febrile disease in this population. In one study, HHV-6A was detected in seven out of nine hospitalized human immunodeficiency virus (HIV)-positive children from this geographic population.[41,43] Although this specific correlation awaits further confirmation, this finding is potentially significant because HHV-6 has been proposed as a potential accelerating factor in HIV infection, as corroborated by the results of *in vivo* studies in macaques.[44,45]

Human herpesvirus 6A and HHV-6B have differential distributions in human tissues. HHV-6B is the dominant virus present in the peripheral blood mononuclear cells (PBMCs) of healthy adults, at least in industrialized countries, and it is also the virus that reactivates in a significant majority of both solid organ and stem cell transplant cases,[14,34,46–53] while both HHV-6A and HHV6B are detected with similar frequency in the plasma of bone marrow transplant patients.[54,55] HHV-6B is also frequently detected in the gastrointestinal tract of solid organ transplant patients,[56] has been identified in endodontic abscesses,[57] and is the virus found in adenoids and tonsils, particularly in children affected by upper airway infections.[58] It has been reported that 54% of healthy adults harbor HHV-6A and HHV-6B coinfection in the lungs.[59] Both HHV-6A and HHV-6B have been identified in vitreous fluid samples and implicated in ocular inflammatory diseases;[60,61] however, it must be noted that these observed differential distribution patterns in human tissues may reflect, at least in part, the differing prevalence of the two viruses among separate geographic regions.

Although HHV-6A and HHV-6B are both neurotropic, there is evidence suggesting an increased severity of HHV-6A over HHV-6B in cases of clinical neurological disease.[40,50,62,63] In addition, although an overwhelming majority of posttransplant reactivation occurs with HHV-6B,[51,52,64] HHV-6A DNA and mRNA are found more frequently than those of HHV-6B in patients with neuroinflammatory diseases such as multiple sclerosis (MS)[65–68] and rhomboencephalitis.[62] HHV-6A has been found predominantly in the central nervous system (CNS) of a subset of patients with MS, and active HHV-6A infection has been detected in the blood[65,69,70] and cerebrospinal fluid (CSF)[71] of patients with relapsing/remitting MS.[65,66,68,69,71–74] Although the finding has not yet been confirmed in humans, marmosets intravenously inoculated with HHV-6A intravenously exhibited neurological symptoms while those inoculated with HHV-6B were asymptomatic.[75] A strain of HHV-6A has also recently been isolated from the fluid specimens from a glioma cyst.[76] HHV-6B, but not HHV-6A, has been associated with mesial temporal lobe epilepsy and status epilepticus.[77–79] HHV-6A, but not HHV-6B, has been associated with Hashimoto's thyroiditis[80] as well as syncytial giant cell-hepatitis in liver transplant patients.[81]

ACKNOWLEDGMENT

We would like to thank Kristin Loomis, president and executive director of the HHV-6 Foundation, for supporting and organizing the Ad Hoc Committee on HHV-6A & HHV-6B Genomic Divergence, which submitted the original reclassification proposal document to the ICTV, leading to the official recognition of HHV-6A and HHV-6B as two distinct viruses of human betaherpesvirus. We would also like to thank Reed Hardin of the HHV-6 Foundation, and Drs. Dario Di Luca, Paolo Lusso, Louis Flamand, Phil Pellett, and Yasuko Mori for their assistance in developing this chapter.

REFERENCES

1. Salahuddin SZ, Ablashi DV, Markham PD, et al. Isolation of a new virus, HBLV, in patients with lymphoproliferative disorders. *Science* 1986;**234**(4776):596–601.
2. Ablashi DV, Salahuddin SZ, Josephs SF, et al. HBLV (or HHV-6) in human cell lines. *Nature* 1987;**329**(6136):207.
3. Downing RG, Sewankambo N, Serwadda D, et al. Isolation of human lymphotropic herpesviruses from Uganda. *Lancet* 1987;**2**(8555):390.
4. Lopez C, Pellett P, Stewart J, et al. Characteristics of human herpesvirus-6. *J Infect Dis* 1988;**157**(6):1271–3.
5. Tedder RS, Briggs M, Cameron CH, Honess R, Robertson D, Whittle H. A novel lymphotropic herpesvirus. *Lancet* 1987;**2**(8555):390–2.
6. Schirmer EC, Wyatt LS, Yamanishi K, Rodriguez WJ, Frenkel N. Differentiation between two distinct classes of viruses now classified as human herpesvirus 6. *Proc Natl Acad Sci USA* 1991;**88**(13):5922–6.
7. Aubin JT, Collandre H, Candotti D, et al. Several groups among human herpesvirus 6 strains can be distinguished by Southern blotting and polymerase chain reaction. *J Clin Microbiol* 1991;**29**(2):367–72.
8. Ablashi DV, Balachandran N, Josephs SF, et al. Genomic polymorphism, growth properties, and immunologic variations in human herpesvirus-6 isolates. *Virology* 1991;**184**(2):545–52.
9. Wyatt LS, Balachandran N, Frenkel N. Variations in the replication and antigenic properties of human herpesvirus 6 strains. *J Infect Dis* 1990;**162**(4):852–7.
10. Josephs SF, Ablashi DV, Salahuddin SZ, et al. Molecular studies of HHV-6. *J Virol Methods* 1988;**21**(1–4):179–90.
11. Iyengar S, Levine PH, Ablashi D, Neequaye J, Pearson GR. Sero-epidemiological investigations on human herpesvirus 6 (HHV-6) infections using a newly developed early antigen assay. *Int J Cancer* 1991;**49**(4):551–7.
12. Ablashi D, Agut H, Berneman Z, et al. Human herpesvirus-6 strain groups: a nomenclature. *Arch Virol* 1993;**129**(1):363–6.
13. Campadelli-Fiume G, Mirandola P, Menotti L. Human herpesvirus 6: an emerging pathogen. *Emerg Infect Dis* 1999;**5**(3):353–66.
14. Di Luca D, Dolcetti R, Mirandola P, et al. Human herpesvirus 6: a survey of presence and variant distribution in normal peripheral lymphocytes and lymphoproliferative disorders. *J Infect Dis* 1994;**170**(1):211–5.
15. Braun DK, Dominguez G, Pellett PE. Human herpesvirus 6. *Clin Microbiol Rev* 1997;**10**(3):521–67.
16. Clark DA. Human herpesvirus 6. *Rev Med Virol* 2000;**10**(3):155–73.
17. Isegawa Y, Mukai T, Nakano K, et al. Comparison of the complete DNA sequences of human herpesvirus 6 variants A and B. *J Virol* 1999;**73**(10):8053–63.

18. Dominguez G, Dambaugh TR, Stamey FR, Dewhurst S, Inoue N, Pellett PE. Human herpesvirus 6B genome sequence: coding content and comparison with human herpesvirus 6A. *J Virol* 1999;**73**(10):8040–52.
19. Gravel A, Gosselin J, Flamand L. Human herpesvirus 6 immediate-early 1 protein is a sumoylated nuclear phosphoprotein colocalizing with promyelocytic leukemia protein-associated nuclear bodies. *J Biol Chem* 2002;**277**(22):19679–87.
20. Stanton R, Wilkinson GW, Fox JD. Analysis of human herpesvirus-6 IE1 sequence variation in clinical samples. *J Med Virol* 2003;**71**(4):578–84.
21. Achour A, Malet I, Le Gal F, et al. Variability of gB and gH genes of human herpesvirus-6 among clinical specimens. *J Med Virol* 2008;**80**(7):1211–21.
22. Rapp JC, Krug LT, Inoue N, Dambaugh TR, Pellett PE. U94, the human herpesvirus 6 homolog of the parvovirus nonstructural gene, is highly conserved among isolates and is expressed at low mRNA levels as a spliced transcript. *Virology* 2000;**268**(2):504–16.
23. Pfeiffer B, Berneman ZN, Neipel F, Chang CK, Tirwatnapong S, Chandran B. Identification and mapping of the gene encoding the glycoprotein complex gp82–gp105 of human herpesvirus 6 and mapping of the neutralizing epitope recognized by monoclonal antibodies. *J Virol* 1993;**67**(8):4611–20.
24. Pfeiffer B, Thomson B, Chandran B. Identification and characterization of a cDNA derived from multiple splicing that encodes envelope glycoprotein gp105 of human herpesvirus 6. *J Virol* 1995;**69**(6):3490–500.
25. Mirandola P, Menegazzi P, Merighi S, Ravaioli T, Cassai E, Di Luca D. Temporal mapping of transcripts in herpesvirus 6 variants. *J Virol* 1998;**72**(5):3837–44.
26. Adams MJ, Carstens EB. Ratification vote on taxonomic proposals to the International Committee on Taxonomy of Viruses. *Arch Virol* 2012;**157**(7):1411–22.
27. Burbelo PD, Bayat A, Wagner J, Nutman TB, Baraniuk JN, Iadarola MJ. No serological evidence for a role of HHV-6 infection in chronic fatigue syndrome. *Am J Transl Res* 2012;**4**(4):443–51.
28. Higashimoto Y, Ohta A, Nishiyama Y, et al. Development of a human herpesvirus 6 species-specific immunoblotting assay. *J Clin Microbiol* 2012;**50**(4):1245–51.
29. Ihira M, Enomoto Y, Kawamura Y, et al. Development of quantitative RT-PCR assays for detection of three classes of HHV-6B gene transcripts. *J Med Virol* 2012;**84**(9):1388–95.
30. Cassina G, Russo D, De Battista D, Broccolo F, Lusso P, Malnati MS. Calibrated real-time polymerase chain reaction for specific quantitation of HHV-6A and HHV-6B in clinical samples. *J Virol Methods* 2013;**189**(1):172–9.
31. Lou J, Wu Y, Cai M, Wu X, Shang S. Subtype-specific, probe-based, real-time PCR for detection and typing of human herpesvirus-6 encephalitis from pediatric patients under the age of 2 years. *Diagn Microbiol Infect Dis* 2011;**70**(2):223–9.
32. de Pagter PJ, Schuurman R, de Vos NM, Mackay W, van Loon AM. Multicenter external quality assessment of molecular methods for detection of human herpesvirus 6. *J Clin Microbiol* 2010;**48**(7):2536–40.
33. Flamand L, Gravel A, Boutolleau D, et al. Multicenter comparison of PCR assays for detection of human herpesvirus 6 DNA in serum. *J Clin Microbiol* 2008;**46**(8):2700–6.
34. Geraudie B, Charrier M, Bonnafous P, et al. Quantitation of human herpesvirus-6A, -6B and -7 DNAs in whole blood, mononuclear and polymorphonuclear cell fractions from healthy blood donors. *J Clin Virol* 2012;**53**(2):151–5.
35. Yamamoto T, Mukai T, Kondo K, Yamanishi K. Variation of DNA sequence in immediate-early gene of human herpesvirus 6 and variant identification by PCR. *J Clin Microbiol* 1994;**32**(2):473–6.
36. Zerr DM, Meier AS, Selke SS, et al. A population-based study of primary human herpesvirus 6 infection. *N Engl J Med* 2005;**352**(8):768–76.

37. Enders G, Biber M, Meyer G, Helftenbein E. Prevalence of antibodies to human herpesvirus 6 in different age groups, in children with exanthema subitum, other acute exanthematous childhood diseases, Kawasaki syndrome, and acute infections with other herpesviruses and HIV. *Infection* 1990;**18**(1):12–15.
38. Dewhurst S, McIntyre K, Schnabel K, Hall CB. Human herpesvirus 6 (HHV-6) variant B accounts for the majority of symptomatic primary HHV-6 infections in a population of U.S. infants. *J Clin Microbiol* 1993;**31**(2):416–8.
39. Thader-Voigt A, Jacobs E, Lehmann W, Bandt D. Development of a microwell adapted immunoblot system with recombinant antigens for distinguishing human herpesvirus (HHV)6A and HHV6B and detection of human cytomegalovirus. *Clin Chem Lab Med* 2011;**49**(11):1891–8.
40. De Bolle L, Naesens L, De Clercq E. Update on human herpesvirus 6 biology, clinical features, and therapy. *Clin Microbiol Rev* 2005;**18**(1):217–45.
41. Bates M, Monze M, Bima H, et al. Predominant human herpesvirus 6 variant A infant infections in an HIV-1 endemic region of Sub-Saharan Africa. *J Med Virol* 2009;**81**(5):779–89.
42. Hall CB, Caserta MT, Schnabel KC, et al. Characteristics and acquisition of human herpesvirus (HHV) 7 infections in relation to infection with HHV-6. *J Infect Dis* 2006;**193**(8):1063–9.
43. Kasolo FC, Mpabalwani E, Gompels UA. Infection with AIDS-related herpesviruses in human immunodeficiency virus-negative infants and endemic childhood Kaposi's sarcoma in Africa. *J Gen Virol* 1997;**78**(Pt 4):847–55.
44. Lusso P, Crowley RW, Malnati MS, et al. Human herpesvirus 6A accelerates AIDS progression in macaques. *Proc Natl Acad Sci USA* 2007;**104**(12):5067–72.
45. Biancotto A, Grivel JC, Lisco A, et al. Evolution of SIV toward RANTES resistance in macaques rapidly progressing to AIDS upon coinfection with HHV-6A. *Retrovirology* 2009;**6**:61.
46. Hudnall SD, Chen T, Allison P, Tyring SK, Heath A. Herpesvirus prevalence and viral load in healthy blood donors by quantitative real-time polymerase chain reaction. *Transfusion* 2008;**48**(6):1180–7.
47. Reddy S, Manna P. Quantitative detection and differentiation of human herpesvirus 6 subtypes in bone marrow transplant patients by using a single real-time polymerase chain reaction assay. *Biol Blood Marrow Transplant* 2005;**11**(7):530–41.
48. Pellett PE, Lindquester GJ, Feorino P, Lopez C. Genomic heterogeneity of human herpesvirus 6 isolates. *Adv Exp Med Biol* 1990;**278**:9–18.
49. Wang LR, Dong LJ, Zhang MJ, Lu DP. The impact of human herpesvirus 6B reactivation on early complications following allogeneic hematopoietic stem cell transplantation. *Biol Blood Marrow Transplant* 2006;**12**(10):1031–7.
50. Boutolleau D, Duros C, Bonnafous P, et al. Identification of human herpesvirus 6 variants A and B by primer-specific real-time PCR may help to revisit their respective role in pathology. *J Clin Virol* 2006;**35**(3):257–63.
51. Chapenko S, Trociukas I, Donina S, et al. Relationship between beta-herpesviruses reactivation and development of complications after autologous peripheral blood stem cell transplantation. *J Med Virol* 2012;**84**(12):1953–60.
52. Faten N, Agnes GD, Nadia BF, et al. Quantitative analysis of human herpesvirus-6 genome in blood and bone marrow samples from Tunisian patients with acute leukemia: a follow-up study. *Infect Agent Cancer* 2012;**7**(1):31.
53. Lautenschlager I, Razonable RR. Human herpesvirus-6 infections in kidney, liver, lung, and heart transplantation: review. *Transpl Int* 2012;**25**(5):493–502.

54. Nitsche A, Muller CW, Radonic A, et al. Human herpesvirus 6A DNA is detected frequently in plasma but rarely in peripheral blood leukocytes of patients after bone marrow transplantation. *J Infect Dis* 2001;**183**(1):130–3.
55. Secchiero P, Carrigan DR, Asano Y, et al. Detection of human herpesvirus 6 in plasma of children with primary infection and immunosuppressed patients by polymerase chain reaction. *J Infect Dis* 1995;**171**(2):273–80.
56. Lempinen M, Halme L, Arola J, Honkanen E, Salmela K, Lautenschlager I. HHV-6B is frequently found in the gastrointestinal tract in kidney transplantation patients. *Transpl Int* 2012;**25**(7):776–82.
57. Ferreira DC, Paiva SS, Carmo FL, et al. Identification of herpesviruses types 1 to 8 and human papillomavirus in acute apical abscesses. *J Endod* 2011;**37**(1):10–16.
58. Comar M, Grasso D, dal Molin G, Zocconi E, Campello C. HHV-6 infection of tonsils and adenoids in children with hypertrophy and upper airway recurrent infections. *Int J Pediatr Otorhinolaryngol* 2010;**74**(1):47–9.
59. Cone RW, Huang ML, Hackman RC, Corey L. Coinfection with human herpesvirus 6 variants A and B in lung tissue. *J Clin Microbiol* 1996;**34**(4):877–81.
60. Sugita S, Shimizu N, Watanabe K, et al. Virological analysis in patients with human herpes virus 6-associated ocular inflammatory disorders. *Invest Ophthalmol Vis Sci* 2012;**53**(8):4692–8.
61. Cohen JI, Fahle G, Kemp MA, Apakupakul K, Margolis TP. Human herpesvirus 6-A, 6-B, and 7 in vitreous fluid samples. *J Med Virol* 2010;**82**(6):996–9.
62. Crawford JR, Kadom N, Santi MR, Mariani B, Lavenstein BL. Human herpesvirus 6 rhombencephalitis in immunocompetent children. *J Child Neurol* 2007;**22**(11):1260–8.
63. Hall CB, Caserta MT, Schnabel KC, et al. Persistence of human herpesvirus 6 according to site and variant: possible greater neurotropism of variant A. *Clin Infect Dis* 1998;**26**(1):132–7.
64. Cheng FW, Lee V, Leung WK, et al. HHV-6 encephalitis in pediatric unrelated umbilical cord transplantation: a role for ganciclovir prophylaxis? *Pediatr Transplant* 2010;**14**(4):483–7.
65. Alvarez-Lafuente R, De Las Heras V, Bartolome M, Picazo JJ, Arroyo R. Relapsing-remitting multiple sclerosis and human herpesvirus 6 active infection. *Arch Neurol* 2004;**61**(10):1523–7.
66. Soldan SS, Berti R, Salem N, et al. Association of human herpes virus 6 (HHV-6) with multiple sclerosis: increased IgM response to HHV-6 early antigen and detection of serum HHV-6 DNA. *Nat Med* 1997;**3**(12):1394–7.
67. Dominguez-Mozo MI, Garcia-Montojo M, De Las Heras V, et al. MHC2TA mRNA levels and human herpesvirus 6 in multiple sclerosis patients treated with interferon beta along two-year follow-up. *BMC neurology* 2012;**12**(1):107.
68. Alvarez-Lafuente R, Martinez A, Garcia-Montojo M, et al. MHC2TA rs4774C and HHV-6A active replication in multiple sclerosis patients. *Eur J Neurol* 2010;**17**(1):129–35.
69. Akhyani N, Berti R, Brennan MB, et al. Tissue distribution and variant characterization of human herpesvirus (HHV)-6: increased prevalence of HHV-6A in patients with multiple sclerosis. *J Infect Dis* 2000;**182**(5):1321–5.
70. Alvarez-Lafuente R, Garcia-Montojo M, De las Heras V, Bartolome M, Arroyo R. Clinical parameters and HHV-6 active replication in relapsing-remitting multiple sclerosis patients. *J Clin Virol* 2006;**37**(Suppl. 1):S24–6.
71. Rotola A, Merlotti I, Caniatti L, et al. Human herpesvirus 6 infects the central nervous system of multiple sclerosis patients in the early stages of the disease. *Mult Scler* 2004;**10**(4):348–54.

72. Alvarez-Lafuente R, De Las Heras V, Bartolome M, Garcia-Montojo M, Arroyo R. Human herpesvirus 6 and multiple sclerosis: a one-year follow-up study. *Brain Pathol* 2006;**16**(1):20–7.
73. Yao K, Gagnon S, Akhyani N, et al. Reactivation of human herpesvirus-6 in natalizumab treated multiple sclerosis patients. *PLoS One* 2008;**3**(4):e2028.
74. Blanco-Kelly F, Alvarez-Lafuente R, Alcina A, et al. Members 6B and 14 of the TNF receptor superfamily in multiple sclerosis predisposition. *Genes Immun* 2011;**12**(2):145–8.
75. Leibovitch E, Wohler JE, Cummings Macri SM, et al. Novel marmoset (*Callithrix jacchus*) model of human herpesvirus 6A and 6B infections: immunologic, virologic and radiologic characterization. *PLoS Pathog* 2013;**9**(1):e1003138.
76. Chi J, Gu B, Zhang C, et al. Human herpesvirus 6 latent infection in patients with glioma. *J Infect Dis* 2012;**206**(9):1394–8.
77. Theodore WH, Epstein L, Gaillard WD, Shinnar S, Wainwright MS, Jacobson S. Human herpes virus 6B: a possible role in epilepsy? *Epilepsia* 2008;**49**(11):1828–37.
78. Li JM, Lei D, Peng F, et al. Detection of human herpes virus 6B in patients with mesial temporal lobe epilepsy in West China and the possible association with elevated NF-κB expression. *Epilepsy Res* 2011;**94**(1–2):1–9.
79. Epstein LG, Shinnar S, Hesdorffer DC, et al. Human herpesvirus 6 and 7 in febrile status epilepticus: the FEBSTAT study. *Epilepsia* 2012;**53**(9):1481–8.
80. Caselli E, Zatelli MC, Rizzo R, et al. Virologic and immunologic evidence supporting an association between HHV-6 and Hashimoto's thyroiditis. *PLoS Pathog* 2012;**8**(10):e1002951.
81. Potenza L, Luppi M, Barozzi P, et al. HHV-6A in syncytial giant-cell hepatitis. *N Engl J Med* 2008;**359**(6):593–602.

Chapter 2

Practical Diagnostic Procedures for HHV-6A, HHV-6B, and HHV-7

Agnès Gautheret-Dejean[a,b,c] and Henri Agut[a,b]
[a]AP-HP Groupe hospitalier Pitié-Salpêtrière, Paris, France, [b]Université Pierre-et-Marie-Curie, Paris, France, [c]Université Paris Descartes, Paris, France

AN OVERVIEW OF DIAGNOSTIC STRATEGIES

Objectives

The goal of diagnostic approaches is to provide the proof of a human herpesvirus 6A (HHV-6A), HHV-6B, or HHV-7 infection and to precisely define the status of this infection in terms of distinction between latent and active form, quantification of spread within the infected organism, and causative relationship with the concomitant clinical symptoms. A specific concern is that of chromosomal integration of HHV-6 (ciHHV-6), which raises numerous novel questions regarding both pathophysiology and diagnosis. Diagnostic procedures are undertaken either in the context of medical management of a given patient or for a scientific purpose within the general framework of a planned study involving a cohort of human subjects. Beyond the recognition of viral infection, the objective is to use the results of virological diagnosis to decide whether or not the current infection has to be treated, using the limited means at our disposal. In addition, it may be necessary to characterize the causative virus strain more precisely by means of molecular studies if its interhuman transmission or phylogeny have to be investigated. This is also necessary if the infection exhibits an unusual level of complexity such as in the case of a mixed infection involving distinct viral variants or viruses. Of note, the fact that HHV-6A and HHV-6B are now defined as distinct viruses (see Chapter 1) makes their formal differentiation relevant early in any HHV-6 infection diagnosis procedure and strictly necessary for the publication of HHV-6-related scientific articles. Finally, the diagnosis procedures are not dispensable for the follow-up of infection, once diagnosed, and specific treatment, if initiated.

This implies not only the serial quantification of virus replication but also the detection of a putative resistance to antivirals in case of therapeutic failure.

Direct and Indirect Approaches

Two general complementary approaches can be used.[1] Direct diagnosis is based on the detection and characterization of whole virions or some of their components, the most convenient ones currently being nucleic acids. The detected viral components may come from either cell-free virions or infected cells, the virus factories, which provide an even larger set of viral targets than virus particles. Indeed, infected cells synthesize not only the components of released viral particles but also numerous virus-encoded proteins and nucleic acids, not structural, which are necessary for the sequence of intracellular viral life cycle but are absent from virions. A wide set of methods can be used for direct diagnosis depending on the target selected for detection: infectious viral particles, genomic viral DNA, viral messenger RNAs, and viral proteins. As a consequence, the constraints of getting appropriate human specimens, the time necessary for obtaining a contributive result, and the cost of reagents, as well as the relevance of information provided, may vary significantly according to the method applied for diagnosis. The rapidity, accessibility, sensitivity, and specificity of these techniques thus remain crucial factors that will essentially determine the final choice of strategy.

The indirect approach, also known as *serology*, is based on the detection and characterization of virus-specific antibodies in a body fluid, mainly serum or plasma (Fig. 2.1). The presence of antibodies is established through their reactivity against reference viral antigens. The antigen–antibody complexes are detected and, when necessary, quantified thanks to a limited number of methods, some of them automated. Antibodies are stable proteins that can keep their immunologic properties after prolonged storage of body fluids, which confers a good reproducibility to serologic assays and provides convenient opportunities for retrospective studies. Apparently, the specimens and techniques used in serology are much easier to manage than those needed for direct diagnosis; however, the interpretation of serologic results may be equivocal, for many different reasons. The most basic one is that HHV-6A, HHV-6B, and HHV-7 infections are widespread in the general population; they are acquired primarily within the first years of life and definitely persist as lifelong latent infections. For this reason, seropositivity to any of these three viruses is a highly frequent finding of limited medical interest. In addition, the kinetics of antibody response and the presence of immunoglobulin M (IgM) antibodies that are useful to identify a primary infection often fail to recognize acute infections related to virus reactivations. The serologic profile is even more complex in case of immune suppression or ciHHV-6, two situations in which the capacity to mount or maintain a humoral immune response may be deeply altered. Ultimately, a cross-reactivity of antibodies directed against the

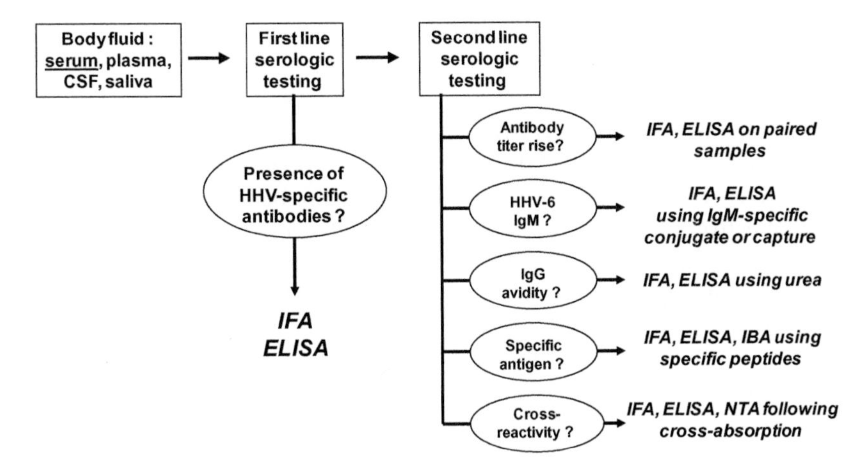

FIGURE 2.1 Overview on the current approaches for HHV-6A, HHV-6B, and HHV-7 serologic testing. Serum (or another body fluid) is assayed for HHV-6-specific antibodies according to a theoretical two-step procedure: first-line analysis is dedicated to the detection of infection, and second-line analysis is dedicated to the detailed characterization of the current stage of infection. *Abbreviations:* ELISA, enzyme-linked immunosorbent assay; IBA, immunoblot assay; IFA, immunofluorescence antibody assay (includes anticomplement immunofluorescence assay, ACIF); NTA, neutralization assay.

four human betaherpesviruses has been reported, and currently available serologic assays cannot discriminate HHV-6A from HHV-6B infections.[2,3]

Specimens

A wide range of human specimens can support virological diagnosis procedures. The most common ones are whole blood, plasma, and serum. The relevance of saliva samples for diagnosis has to be established but this body fluid has the advantages of being readily obtained and stored and containing virions (and/or viral DNA), infected cells, and antibodies.[4] Cerebrospinal fluid is mandatory for the diagnosis of central nervous system infections. Bronchoalveolar lavages allow exploration of lung infections. Finally, any cell fraction obtained from a body fluid, as exemplified by peripheral blood mononuclear cells (PBMCs), cell smear, or tissue biopsy, can be tested for the presence of virus components by means of direct diagnosis assays. Of particular interest in this domain are the molecular biology techniques, particularly the most recent refinements of PCR, such as digital PCR, which allows clonal characterization of infecting viruses in human samples and will be a prominent tool in future pathophysiology investigations. Whatever the sample and assay used, the quality of sample collection and appropriate conditions for their transport and storage remain key factors for the success of diagnostic procedures.

METHODS OF DIRECT DIAGNOSIS

Cell Culture

The isolation of HHV-6A, HHV-6B, and HHV-7 in cell cultures is the method that enabled their discovery.[5,6] This is also the only way to prove the presence of infectious viral particles in a sample, and it may be used for the diagnosis of an active infection. However, these methods are time consuming, laborious, and expensive, and they cannot be used for routine work in a clinical setting. All clinical HHV-6 isolates may grow on PBMCs or cord blood lymphocytes activated by phytohemagglutinin, in the presence of interleukin-2 and polybren. A cytopathic effect may appear within about 10 days, consisting of enlarged and refractive cells, but some strains do not exhibit any cytopathic effect. Isolation on other cell types such as fibroblasts is also possible, mainly for HHV-6A strains, but without high viral titers.[7,8] Some strains have been adapted to grow on cell lines (e.g., MT4 cells transformed by HTLV-I for HST strain or HSB2 for GS strain), which simplifies culture conditions.[9,10] These models are often used to evaluate the antiviral activity of new drugs.

Antigen Detection

The detection and quantitation of viral antigens in PBMCs (antigenemia; see Fig. 2.2) and tissue biopsies provide evidence of viral proteins produced at different stages of the multiplication cycle, according to the antibodies used. Some antibodies target proteins produced during the late phase, signing an active infection. The detection of antigens in tissue biopsies is very useful to precisely locate viral infection, either in inflammatory cells infiltrating the tissue or in the tissue cells: oligodendrocytes,[11] liver,[12] kidney,[13] lung,[14,15] and salivary glands.[16,17] Specific identification of HHV-6A or HHV-6B and quantitative antigenemia are also possible.[18,19] Antigen detection methods remain difficult to develop, due to a reduced number of available antibodies, the existence of cross-reactivities with viruses sharing antigenic determinants, nonspecific staining, and usually a low sensitivity.

Nucleic Acid Detection, Quantitation, and Sequencing

Detection of DNA by in situ *Hybridization*

The detection of HHV-6 DNA by *in situ* hybridization localizes the virus within a tissue, a cell, or a chromosome for ciHHV-6.[17,20–22] However, this does not indicate the type of viral infection, active or latent, and may then be associated, in some studies, to the detection of viral antigens.[22,23]

Detection and Quantitation of DNA by PCR and Real-Time PCR

Because of their sensitivity, specificity, and quickness, methods for the detection and quantitation of viral genomic DNA by means of polymerase chain

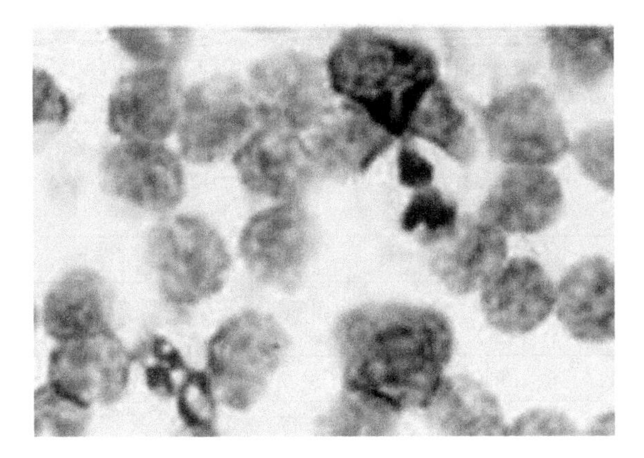

FIGURE 2.2 HHV-6B antigenemia. *(Kindly provided by Irmeli Lautenschlager, from Loginov R et al. J Clin Virol 2006; 37(Suppl. 1): S76–81).*

reaction (PCR) have become essential for routine diagnosis. However, as for human cytomegalovirus (HCMV) or Epstein–Barr virus (EBV), a qualitative assay applied to biological samples such as whole blood or plasma remains inadequate to assess the level of viral activation. For this reason, several authors have developed semi-quantitative or quantitative PCR assays. Semi-quantitative assays are based on either endpoint dilution amplification leading to time-consuming protocols and low reproducibility of results, or comparison of the signal obtained with the amplification of an external serially diluted standard.[24–26] Quantitative amplification methods were initially based on competitive PCR assays, but they had limitations, including a reduced range of linearity as well as possible competition between the patient's virus and the internal standard that required the use of different quantities of internal control.[27–30] Real-time PCR methods have replaced previous methods and offer several advantages, such as quickness (2 hours for combined amplification and read out), a wide range of signal linearity (fewer than 10 to 10^8 genomic copies), collection of all fluorescent signals during the amplification steps without opening the reaction tubes, stability of the fluorescent probe, and the 96-well format that allows performing large series. To date, the lack of an international standard is a problem, and comparison of results obtained using different methods remains difficult. Several methods have been developed for HHV-6A, HHV-6B, and HHV-7, and they are presented in Table 2.1. Differentiation of HHV-6A and HHV-6B viruses is sometimes possible, whereas the other methods equally quantify both viral genomes. Some authors have developed multiplex PCR for the concomitant detection or quantitation of several herpesviruses—for example, HCMV and HHV-6 by Pradeau et al.;[31] HSV, VZV, HCMV, EBV, and HHV-6 by Engelmann et al.;[32] and HSV, VZV, HCMV, EBV, HHV-6, and HHV-7 by Tanaka et al.[33]

TABLE 2.1 Real-Time Amplification Methods for Quantitation of HHV-6 or HHV-7 Genomic DNA

Virus	HHV-6 Virus Identification	Probe/ Amplification Technology	References
HHV-6A and B	No	TaqMan®	Locatelli et al.[122]
HHV-6A and B	No	TaqMan®	Gautheret-Dejean et al.[123]
HHV-6A and B	Yes (melting curves)	MGB	Hymas et al.[124]
HHV-6A and B	No	TaqMan®	Tanaka et al.[125]
HHV-6A and B	No	LAMP	Ihira et al.[126]
HHV-6A and B	No	TaqMan® (Qiagen RealArt™)	Flamand et al.[127]
HHV-6A and B	No	Real-time	Nagate et al.[128]
HHV-6A and B	No	TaqMan® (Nanogen Advanced Diagnostics)	Gaeta et al.[129]
HHV-6A or B	Yes (two specific systems)	TaqMan®	Nitsche et al.[130]
HHV-6A and B	Yes (melting curves)	NA (Roche molecular biochemicals)	Razonable et al.[131]
HHV-6A and B	Yes (melting curves)	MGB	Reddy et al.[132]
HHV-6A and B	No	TaqMan®	Zerr et al.[133]
HHV-6A, B and HHV-7	Yes (two specific probes for HHV-6 and HHV-7 and melting curves to differentiate HHV-6A and HHV-6B)	FRET	Safronetz et al.[134]
HHV-7	NA	Real-time, MGB probe	Fernandez et al.[135]
HHV-7	NA	NA (Roche molecular biochemicals)	Razonable et al.[131]
HHV-7	NA	TaqMan®	Broccolo et al.[136]
HHV-7	NA	TaqMan®	Zerr et al.[133]

Abbreviations: FRET, fluorescence resonance energy transfer; LAMP, loop-mediated isothermal amplification; MGB, minor groove binding; NA, not available.

Detection and Quantitation of Transcripts by RT-PCR

Viral transcripts detection is used to determine the stage of viral infection (α, β, or γ) and episomal latency characterized by the production of U94 transcripts.[34–39] However, the instability of mRNA and the method's low sensitivity are a hindrance to its routine use. The chronology of HHV-7 viral transcripts has been described but no clinical application has been reported to date.[40]

Nucleotide Sequencing

Initially used to differentiate HHV-6A from HHV-6B viruses, nucleotide sequencing is now primarily used to identify mutations associated with drug resistance and to study viral polymorphism.[41]

Differential Diagnosis of HHV-6A and HHV-6B Infection

Methods to differentiate HHV-6A from HHV-6B in biological samples were initially based on restriction fragment length analysis after enzyme digestion, amplification with variant-specific primers, or amplimer hybridization with variant-specific probes.[42,43] Nevertheless, those methods were time consuming and generally provided good results only when one variant was implicated. Similar to qualitative PCR methods that simultaneously detect and differentiate HHV-6A and HHV-6B,[44] real-time PCR assays are based on the analysis of melting curve features (see Table 2.1). However, the ability of those methods to detect both variants in the case of mixed infections, especially when one variant is a minority, has not been clearly evaluated. Reports on various real-time PCR assays based on the use of either virus-specific primers or probes have been published (see Table 2.1).

Antiviral Susceptibility Assays

In vitro HHV-6 susceptibility to antiviral compounds may be evaluated by phenotypic or genotypic approaches. Phenotypic tests investigate the replication capacity (fitness) of viruses in the presence of antiviral drugs. Viral multiplication is assessed by observing a cytopathic effect, by detecting viral antigens by immunofluorescence or immunohistochemistry assays, flow cytometry, DNA hybridization, or amplification.[9,45–48] After identification of a mutation associated with antiviral resistance by phenotypic analysis, genotypic analysis may be possible. It consists of identifying genetic modifications, mutations, or deletions through nucleotide sequencing of the gene coding for the antiviral target protein.[41,49,50]

SEROLOGIC METHODS

Techniques

Indirect immunofluorescence antibody assays (IFAs) were the first ones to be used and remain still widely employed.[5] In these tests, HHV-infected cells are fixed on a glass slide, a serum dilution is added, and a fluorochrome-conjugated anti-immunoglobulin antibody is then applied to detect the binding of serum antibodies to specific antigens. When illuminated with ultraviolet light and observed with a microscope, both the number of fluorescent foci and the characteristic patterns of cell staining allow conclusions regarding positivity or negativity. Infected cells consist of primary cells such as cord blood mononuclear cells or continuous cell lines previously inoculated with a reference viral strain and collected when a significant cytopathic effect (CPE) has been observed. Anticomplement immunofluorescence assay (ACIF) is a modification of conventional IFA in which human complement is applied to slides and the antigen–antibody–complement complex is detected by a fluorochrome-conjugated anti-C3 antibody.[51,52] It is essential to lower nonspecific fluorescent signals by means of extensive washing after each exposure to a specific reagent (serum, complement if used, conjugate). Nevertheless, this does not prevent the binding of cross-reactive antibodies, in particular those directed against other betaherpesviruses.[3,53]

Enzyme-linked immunosorbent assays (ELISAs)[54] have been developed using either crude lysate of infected cells or purified virus obtained from cell culture supernatant as antigens.[55,56] The antigen preparation is coated on polystyrene plates and the serum sample is then added, after which an enzyme-conjugated antihuman immunoglobulin is applied and the catalyzed chromogenic reaction is identified. Generally speaking, ELISA is highly sensitive, simple to perform, rather inexpensive, and capable of being automated; however, for HHV-6A, HHV-6B, and HHV-7, the number of commercialized ELISA kits is much lower and experience in their use is more limited than for other domains of clinical virology. The specificity of the HHV-6 ELISA has often been questioned, and this test has not produced better results than IFA, according to some authors.[57] In contrast, other authors have demonstrated that discrepancies between IFA and ELISA mainly correspond to either false-positive or false-negative results in IFA.[58] The use of synthetic peptides as antigens is expected to improve the specificity of future ELISAs. Promising results have also been obtained with immunoassays based on electrochemiluminescence (ECL) technology.[59]

Western blot and other immunoblot assays (IBAs) allow identification of antibodies to specific viral proteins with a higher specificity and usually a lower sensitivity than ELISA. Infected cell lysates or recombinant proteins are denatured, separated by means of gel electrophoresis or isolated deposition, transferred to a membrane, and finally allowed to react with the serum specimen, basically using the same indicator system as in ELISA.[60] A comparison of IFA, ELISA, and IBA in the case of HHV-7 has confirmed that IBA was the most specific, exhibiting a sensitivity lower than ELISA and higher than IFA.[61] Accordingly, the high specificity of an IBA designed to detect HHV-6-specific IgM antibodies was demonstrated in children under 2 years of age.[62] However, the availability of these assays, mainly developed for research purposes, is limited.

In a neutralization assay (NTA), the search for neutralizing antibodies is performed by making serial serum dilutions react with a standardized amount of infectious virus. The antibody titer is generally expressed as the highest serum dilution blocking viral infectivity. Virus multiplication is evaluated by observing the CPE, counting HHV-6-positive cells by means of IFA, or by measuring of viral antigen or DNA synthesis by dot blot assay or PCR.[63–65] NTA is generally considered to be highly specific, sensitive, and correlated with protective immunity, but it is labor intensive and expensive.

Viral Antigens Used for Serologic Testing

In general, IFA and ELISA based on infected cell lysate investigate the immune response directed against all viral proteins expressed in infected cells. In contrast with Epstein–Barr virus[66] and human herpesvirus 8 serologic assays, no discrimination is made between the proteins corresponding to either lytic cycle or latency. As a functional binding assay, NTA reflects the interaction between neutralizing antibodies and glycoproteins present on the viral envelope but little is known about the precise target of these antibodies.

The identification of target proteins has been undertaken by means of immunoblot or immunoprecipitation assays combined with the use of monoclonal antibodies.[67] This approach allowed identification of the tegument protein p100 (U11 gene product) of HHV-6A (U1102) as a major determinant of immune response.[68] Similarly, for HHV-6B (Z29), a 101-kDa protein (101 K coded by U11 ORF) has been identified as an immunodominant virion protein for both IgG and IgM reactivity.[62] This opens the door to the use of synthetic peptides for the development of highly specific assays. The early antigen p41/38 (U27 gene product) was proposed for development of a specific ELISA, but further results obtained with recombinant p41 as the antigen have been rather disappointing.[69,70] The recombinant protein REP (U94 gene product expressed during virus latency) has been used to develop a novel ELISA that would be suitable to explore the antiviral immune response in HHV-6-related chronic diseases.[71] Future experiments will complete the list of relevant HHV-6 epitopes for serology, but the challenge of finding protein sequences specific to HHV-6A and HHV-6B, respectively remains. Indeed, the absence of available specific assays for each of the two HHV-6 viruses is a major drawback of current serologic approaches. Another important requirement for the future is standardization of the different serologic assays to allow clear definition of antibody titer thresholds, comparisons among results obtained in different places, and development of multicentric studies.

Characterization of Specific Antibodies

The use of conjugated antihuman IgM antibody in IFA as well as ELISA and IBA allows the detection of HHV-specific IgM,[62,72] although a few important points must be kept in mind when interpreting the results. In many acute viral infections, the detection of IgM in a single serum specimen is sufficient to conclude that the infection is recent; however, in the case of herpesviruses, IgM may be also detected during viral reactivation from latency. In addition, the heterotypic reactivation of IgM synthesis, such as occurs during human cytomegalovirus (CMV) and EBV infections, may complicate interpretation of the results. With regard to the risk of false-positive results due to the presence of rheumatoid factor, a first reaction step designed for the capture of IgM is known to improve assay specificity.[73] Virus-specific IgG subclass responses have also been investigated with the goal of finding a pattern specific for either latency or reactivation, as illustrated by the report of different isotype (IgG_1 to IgG_4) responses at different stages of HHV-6 infection.[74]

Distinguishing between primary and recurrent HHV-6 infection can be addressed with an antibody avidity test.[75,76] The principle is that low-avidity antibody present during the primary infection can be eluted easily by a mild denaturing agent such as urea, whereas high-avidity antibody synthesized during past or recurrent infection cannot. When the antibody titer is significantly reduced in the presence of urea as compared to the control in the absence of a denaturing agent, sera are considered to contain low-avidity antibody, suggestive of a recent primary infection.

A recurrent question is that of cross-reactivity between the four human betaherpesviruses, CMV, HHV-6A, HHV-6B, and HHV-7, which involves at least two distinct phenomena. The first one is the production of antibodies reacting with common epitopes shared by distinct viruses. The second one is the heterotypic production of antibodies to a given virus when infection by another one occurs, as illustrated by the rise of HHV-6 antibodies in patients with primary CMV infection. In this situation, a cross-reactivity between the glycoproteins gB of CMV and gp116 of HHV-6 has been demonstrated, with the reactivity to gp116 being removed by absorption of sera with gB.[3] Symmetrically, CMV antibody rise has been reported during active HHV-6 infection, and the reactivity to HHV-6 could be removed following absorption with CMV-infected cells.[72] Some cases of seroconversion to HHV-7 have been associated with a simultaneous rise in HHV-6 antibody titer and, experimentally, some polyvalent sera raised to HHV-7 showed a cross-reactivity to HHV-6.[53] Other studies have shown that the IgM response to HHV-7 primary infection was also directed to HHV-6, and cross-reactive antibodies to HHV-6 and HHV-7 were detected in transplant recipients.[77,78] The removal of cross-reacting activity by absorption with heterologous viruses has been proposed as a general strategy to clarify ambiguous serologic results, but the efficiency of this procedure is variable, depending on individual serum specimens.

INDICATIONS AND FINDINGS

The objective of this section is to present the diagnosis process based on the different possible types of HHV-6 and HHV-7 infections; however, except for primary infection, the focus is primarily on HHV-6, for which causal links with clinical manifestations have largely been demonstrated. The procedure for HHV-7 may be extrapolated from these data.

Primary Infection

Primary infections with HHV-6 or HHV-7 may be identified by the presence of specific IgM, a seroconversion of the specific IgG, the isolation of the virus from blood, and the presence of virus DNA in blood, saliva, and stool.[79–81] In practice, with HHV-6 primary infection occurring mainly during the first year of life, the presence of maternal IgGs may interfere with diagnosis in neonates. Thus, in the presence of symptoms compatible with primary HHV-6 infection during the first months, such as seizures accompanied by cutaneous rash, quantitative PCR testing of the blood may be necessary to establish the diagnosis. During early childhood and generally, the biological samples analyzed to establish the diagnosis of an HHV-6 active infection are chosen according to the nature of the symptoms (Table 2.2). In the case of encephalitis, for example, a sample of cerebrospinal fluid is studied. Its analysis, in parallel with a whole blood sample, can trace a gradient from blood through the nervous system that is highly indicative of a causal link.

The primary infection of the embryo or fetus during pregnancy (congenital infection) corresponds to the transmission of a replicative virus from mother to child through the placenta. For HHV-6, virus particles present in the blood of the woman generally correspond to a common active infection, but Hall et al.[82] demonstrated that they may also come from reactivation of a ciHHV-6.[82] The diagnosis of congenital infection is based on the detection of viral DNA, or on the presence of specific IgM at birth to eliminate perinatal contamination.

Given the high seroprevalence in the population, primary HHV-6 and HHV-7 infections in adults are very rare. In such cases, a similar evolution of the antibodies will be observed. By contrast, a reinfection with the same HHV-6 virus or the other viruses may occur without a serological contribution.

Chronic Infection and Reactivations

After primary infection, HHV-6 and HHV-7 remain latent in the body, mainly associated with mononuclear cell fractions and in salivary glands. From time to time, they reactivate in an asymptomatic manner in immunocompetent subjects, but HHV-6 can lead to pathological manifestations in the case of immunosuppression. These reactivations are difficult to diagnose using serological assays, and direct methods are preferred.[83,84] PCR methods offer ease of use and can be applied to all types of matrices, and quantitative analysis allows determining the level of infection. Viral DNA detection, however, is not synonymous with an active infection. For clinical use, blood is largely used to measure the HHV-6 load for diagnosis and follow-up of HHV-6 infections

TABLE 2.2 Overview of Diagnosis Methods: Advantages, Disadvantages, and Indications

Method	Technology	Advantages	Disadvantages	Indications
Indirect methods				
IgM detection	IFA, ELISA, IBA	Demonstrates recent primary infection	Interpretation difficult in immunocompromised patients, link with pathology difficult to establish. No discrimination between HHV-A and B, cross-reactivities with other herpesviruses	Epidemiology
IgG detection and quantitation	IFA, ELISA, ACIF, NTA, IBA	Indicates infection by the virus		
IgG avidity	IFA	Orientates towards recent primary/old infection		
Direct methods				
Virus isolation in cell cultures	Cell cultures	Proves the presence of infectious viral particles	Laborious and long method, low sensitive, expensive	Research activity, evaluation of antiviral activity
Antigen detection and quantitation in blood	Immunocytochemistry	Diagnosis and follow-up of active infection	Laborious, moderate sensitivity, cross-reactivities with other herpesviruses, difficult to use	Diagnosis and follow-up of active infection
Antigen detection in a tissue biopsy	Immunohistochemistry	Localization of the virus into a tissue	Laborious, moderate sensitivity, cross-reactivities with other herpesviruses, difficult development	Localization of the virus in a tissue biopsy and a cell type

Method	Technology	Advantages	Disadvantages	Indications
Viral transcripts detection	PCR	According to the viral transcripts analyzed, characterization of active or latent infection	Low sensitivity, instability of viral transcripts	Evaluation of the type of viral infection
Viral genome detection by in situ hybridization	In situ hybridization	Localization of the viral genome into a tissue	Laborious, low sensitivity, difficult development	Localization of the viral genome into a tissue biopsy and a cell type, identification of ciHHV-6
Viral genome detection by PCR	PCR	High sensitivity, specificity, species identification	High cost	Detection of viral infection, identification of ciHHV-6 by analysis of hair follicle or nail
Viral genome quantitation	Quantitative PCR, real-time PCR	High sensitivity, specificity, species quantitation/ differenciation, evaluation of the level of infection	High cost	Diagnosis and follow-up of active infection, analysis of the causal link with a pathology

IFA: immunofluorescence antibody assay; ELISA: enzyme linked immunosorbent assay; IBA: immunoblot assay; ACIF: anticomplement immunofluorescence assay; NTA: neutralization assay; PCR: polymerase chain reaction.

in patients. According to some laboratories, whole blood, plasma or even serum can be tested.[85–87] Some authors consider that the presence of HHV-6 in plasma or serum may represent an active infection;[88,89] however, we have shown that HHV-6 DNA in plasma reflects the presence of infected blood cells rather than cell-free circulating viral particles, and such a finding is inadequate for estimating the amount of virus produced by active infection of distant lymphoid tissue and organs.[90,91] In that sense, blood is not the best matrix for detecting an active infection in all body compartments, and more adequate specimens should be analyzed (Table 2.3), such as CSF for encephalitis or meningoencephalitis,[92–95] bone marrow for myelosuppression or hemophagocytic syndrome, liver biopsy for hepatitis,[96,97] and myocardial biopsy for myocarditis,[98] among others.

To date, only active HHV-6 infections have been associated with pathologies, and the measure of viral load constitutes a useful tool. However, due to the absence of international standards and the high variability of biological samples tested and methods used, it is very difficult to define a level of viral load at which antiviral treatment should be started. Some studies have analyzed the usual viral loads in healthy blood donors, which are very low, in the range of 3.2 to 3.5 $\log_{10}$ HHV-6 DNA copies per mL,[99] 1.09 to 3.17 $\log_{10}$ HHV-6 DNA copies per 10^6 cells,[100] and 0.73 to 3.70 $\log_{10}$ HHV-7 DNA copies per 10^6 cells.[100] In contrast, in stem cell transplanted patients, the occurrence of HHV-6-related pathological manifestations is associated with a viral load of $>3 \log_{10}$ HHV-6 DNA copies per 10^6 PBMCs or 4 $\log_{10}$ HHV-6 DNA copies per 10^6 cells.[85,101] An active infection is evidenced by a high viral load;[87,102–105] however, the ciHHV-6 phenomenon complicates interpretation of a high viral load, and quantitative analysis of blood in parallel with other biological compartments may be required to identify the presence of ciHHV-6 or to show that the infection is at a higher level in a tissue or an organ than in blood. In solid organ transplant recipients, it is important to search for viral antigens in organ biopsy specimens in cases of graft rejection to obtain more precise information about the localization of the virus into the tissue.[84]

Chromosomal Integration

The phenomenon of ciHHV-6 greatly complicates the diagnosis of active HHV-6 infection. To date, the link between clinical manifestations and the presence of ciHHV-6 remains unclear, and only rare cases of viral reactivation have been reported.[82,106] Before beginning any antiviral treatment, it is necessary to determine whether or not ciHHV-6 is present in the patient. This is particularly the case for stem cell transplant recipients and very fragile patients for whom the administration of antiviral drugs with high toxicity such as ganciclovir or foscarnet should be avoided if their use is not absolutely necessary.[107] Because viral DNA is present at a minimum of one copy in each cell, the viral load will be very high in cellular clinical specimens,

TABLE 2.3 Proposition of Algorithms of Interpretation of HHV-6 Viral Load in Different Biological Matrices

Matrix	Viral Load Expressed as Number of HHV-6 Genome per Million Cells	Clinical Data	Interpretation	Comment
Whole blood	Viral load < threshold	Absence of symptoms	Absence of HHV-6 infection or latent infection without reactivation	
		Presence of symptoms possibly due to HHV-6	Interpretation should include Quantify HHV-6 load in the biological matrix corresponding to the tissue or the organ concerned, or a biopsy of the tissue or the organ concerned	In case of encephalitis, VL in blood may be low. In case of organ infection, a blood sample and an organ biopsy may be tested in parallel in order to correctly interpret the results.
	Threshold $< VL < 4 \log_{10}$		Low active infection	In case of encephalitis, VL in blood may be low. In case suspicion of an organ infection, a blood sample and an organ biopsy may be tested in parallel in order to correctly interpret the results.
	$4 \log_{10} \leq VL < 6 \log_{10}$		Active infection	Monitor once a week and discuss an antiviral treatment if symptoms possibly due to HHV-6 appear.
	$VL \geq 6 \log_{10}$	Presence of symptoms	Highly active infection	Monitor VL once a week to follow the decrease of the viral load under antiviral treatment.
	$6 \log_{10} \leq VL \leq 7 \log_{10}$	No or few symptoms	Possible ciHHV-6	Stability of VL on several sequential samples, no decrease of VL under antiviral treatment, HHV-6 positivity of hair follicle or nail.

(*Continued*)

TABLE 2.3 (Continued)

Matrix	Viral Load Expressed as Number of HHV-6 Genome per Million Cells	Clinical Data	Interpretation	Comment
CSF	$VL < 2 \log_{10}$	Absence of neurological symptoms	Absence of active infection	Control if symptoms appear.
		Presence of neurological symptoms	Difficult to interpret	In case of encephalitis, VL in CSF may be low. Comparison with VL in whole blood helpful. Antiviral treatment may be proposed.
	$2 \log_{10} \leq VL \leq 6 \log_{10}$	Neurological symptoms, imaging data evocative of HHV-6 encephalitis	Possible HHV-6 causal link	Antiviral treatment may be proposed. Comparison with VL in whole blood helpful.
	$VL \geq 6 \log_{10}$	Neurological symptoms, imaging data evocative of HHV-6 encephalitis	HHV-6 active infection highly probable	In case of encephalitis, there is an increased gradient from blood to CSF. Antiviral treatment may be proposed. If no increased gradient exists, a therapeutic test may be proposed, and stopped if no improvement is observed.
	$6 \log_{10} \leq VL \leq 7 \log_{10}$	No symptoms	Possible ciHHV-6	Similar VL in whole blood is evocative of ciHHV-6. Complementary exploration for ciHHV-6 diagnosis on follicles or hairs.
Organ biopsy	$2 \log_{10} \leq VL \leq 4 \log_{10}$	Presence of symptoms in relation with the organ concerned	Difficult to interpret: active infection, or inflammatory process with presence of lymphocytes/macrophages bearing HHV-6 DNA?	Analysis of HHV-6 antigens by histochemistry is greatly useful to localize the virus within the tissue.
	$VL \geq 4 \log_{10}$	Presence of symptoms in relation with the organ concerned	HHV-6 active infection highly probable	Analysis of HHV-6 antigens by histochemistry is greatly useful to localize the virus within the tissue.
	$6 \log_{10} \leq VL \leq 7 \log_{10}$	No symptoms	Possible ciHHV-6	Similar VL in whole blood is evocative of ciHHV-6. Complementary exploration for ciHHV-6 diagnosis on follicles or hairs.

VL: viral load; CSF: cerebrospinal fluid; ciHHV-6: chromosomally inherited HHV-6.

whole blood, leukocytes, and tissue biopsies, which are analyzed to detect an active infection by quantitative PCR, but also in plasma due to the presence of DNA liberated following cell lysis.[90,108] Fluorescence *in situ* hybridization on EBV-transformed B lymphocytes was originally used to identify ciHHV-6;[21] however, this assay is complex and expensive. Alternatively, the presence of HHV-6 DNA in hair follicles or nails is a signature of ciHHV-6.[109,110] These types of samples can be difficult to obtain, but other approaches are more convenient, such as the determination of viral load in whole blood: HHV-6 viral load $>5.5\log_{10}$ DNA copies per mL of whole blood or viral load $>6\log_{10}$ DNA copies per million cells of whole blood.[99,107,111,112] The fact that high viral loads persist over the time is an additional argument for the diagnosis of ciHHV-6.[107,110,113,114] Analyzing the HHV-6 plasma viral load may represent an alternative when the result is expressed per cells or per microgram of cellular DNA.[100,102] Virus differentiation is not necessary.[108]

Epidemiology

After HHV-6 and HHV-7 were isolated on cell cultures, their epidemiology in healthy subjects was studied by the use of serological methods, indicating that the primary infection occurred during early childhood and that there was a very high seroprevalence in adults.[5,6,81,115–117] PCR methods used to characterize their molecular epidemiology have indicated, in association with isolation methods, that infectious HHV-6 particles are present in the saliva and are involved in interhuman transmission.[115]

Therapeutic Follow-Up

To date, there is no consensus on prophylactic or preemptive treatment of HHV-6 infection in immunocompromised patients, but direct virological methods have been evaluated in some studies. In solid organ transplant recipients, the effects of HCMV antiviral prophylaxis with ganciclovir, valganciclovir, or valacyclovir have been studied using DNAemia or antigenemia monitoring, with conflicting results.[118–121] Plasma HHV-6 viral loads have been analyzed to guide preemptive therapy against HHV-6 encephalopathy after allogeneic stem cell transplantation.[87]

To date, there is no consensus or agreed-upon guidelines on antiviral therapy of pathologies related to HHV-6, nor have controlled studies been performed. As a consequence, several virological markers (viral load in whole blood, plasma, or serum; antigenemia; viral DNA in CSF) are used, but no thresholds have been defined. The efficacy of a particular therapy may be evaluated based on a variety of criteria, including a decrease in symptoms, normalization of biological markers (biochemical, hematological), and a reduction in virological markers.[103] In practice, in our hospital, only patients with an active HHV-6 infection and associated clinical symptoms are treated. Except in cases of encephalitis, the evolution of the viral load is followed by PCR in whole blood and expressed as viral genome copies per million cells. It generally decreases by a factor of 10 per week, ultimately becoming undetectable (personal data). Antiviral treatment is stopped after 2 weeks of an undetectable viral load.

CONCLUSIONS: PRACTICAL MANAGEMENT OF DIAGNOSIS

The choice of a virological method to diagnose an HHV-6 infection depends on several factors: the aim of the diagnosis, the age of the patient, the presence or absence of immunosuppression, the nature of the symptoms, and the biological samples available. Indirect diagnosis methods are used mainly for epidemiologic studies, whereas direct methods are preferred when searching for an active infection. Because severe infections due to HHV-6 occur mainly in immunocompromised individuals, it is essential to correctly diagnose an active infection in order to give a specific antiviral treatment if necessary. The phenomenon of ciHHV-6, recently described, raises the problem of differentiating it from an active infection. Earlier, Table 2.3 presented interpretation algorithms for HHV-6 viral load measured in different biological matrices, including whole blood, cerebrospinal fluid, and organ biopsies, and they are expressed as the number of genomic copies per million cells. Of note, these algorithms are not usable for diagnosing HHV-6 infections in the context of grafts, when the donor is bearing ciHHV-6. These cases require specific approaches. Clinical studies are necessary to identify virological markers able to differentiate between an active infection and ciHHV-6, to define consensual protocols to treat an HHV-6 infection, and to clearly confirm a link between HHV-6 and pathologies that are unclear to date.

REFERENCES

1. Flamand L, Komaroff AL, Arbuckle JH, Medveczky PG, Ablashi DV. Review. Part 1. Human herpesvirus-6: basic biology, diagnostic testing, and antiviral efficacy. *J Med Virol* 2010;**82**(9):1560–8.
2. Ward K, Couto Parada X, Passas J, Thiruchelvam A. Evaluation of the specificity and sensitivity of indirect immunofluorescence tests for IgG to human herpesviruses-6 and -7. *J Virol Methods* 2002;**106**(1):107.
3. Adler SP, McVoy M, Chou S, Hempfling S, Yamanishi K, Britt W. Antibodies induced by a primary cytomegalovirus infection react with human herpesvirus 6 proteins. *J Infect Dis* 1993;**168**(5):1119–26.
4. Descamps V, Avenel-Audran M, Valeyrie-Allanore L, Bensaid B, Barbaud A, Al Jawhari M, et al. Saliva polymerase chain reaction assay for detection and follow-up of herpesvirus reactivation in patients with drug reaction with eosinophilia and systemic symptoms (DRESS). *JAMA Dermatol* 2013;**20**:1–5.
5. Salahuddin SZ, Ablashi DV, Markham PD, Josephs SF, Sturzenegger S, Kaplan M, et al. Isolation of a new virus, HBLV, in patients with lymphoproliferative disorders. *Science* 1986;**234**:596–601.

6. Frenkel N, Schirmer EC, Wyatt LS, Katsafanas G, Roffman E, Danovich RM, et al. Isolation of a new herpesvirus from human CD4+ T cells. *Proc Natl Acad Sci USA* 1990;**87**:748–52.
7. Carrigan DR, Knox KK. Human herpesvirus 6: diagnosis of active infection. *Am Clin Lab* 2000;**19**(7):12.
8. Robert C, Aubin JT, Visse B, Fillet AM, Huraux JM, Agut H. Difference in permissiveness of human fibroblast cells to variants A and B of human herpesvirus-6. *Res Virol* 1996;**147**(4):219–25.
9. Manichanh C, Grenot P, Gautheret-Dejean A, Debre P, Huraux JM, Agut H. Susceptibility of human herpesvirus 6 to antiviral compounds by flow cytometry analysis. *Cytometry* 2000;**40**(2):135–40.
10. Ablashi DV, Balachandran N, Josephs SF, Hung CL, Krueger GR, Kramarsky B, et al. Genomic polymorphism, growth properties, and immunologic variations in human herpesvirus-6 isolates. *Virology* 1991;**184**(2):545–52.
11. Challoner PB, Smith KT, Parker JD, MacLeod DL, Coulter SN, Rose TM, et al. Plaque-associated expression of human herpesvirus 6 in multiple sclerosis. *Proc Natl Acad Sci USA* 1995;**92**(16):7440–4.
12. Lautenschlager I, Hockerstedt K, Linnavuori K, Taskinen E. Human herpesvirus-6 infection after liver transplantation. *Clin Infect Dis* 1998;**26**(3):702–7.
13. Helantera I, Loginov R, Koskinen P, Lautenschlager I. Demonstration of HHV-6 antigens in biopsies of kidney transplant recipients with cytomegalovirus infection. *Transpl Int* 2008;**21**(10):980–4.
14. Pitalia AK, Liu-Yin JA, Freemont AJ, Morris DJ, Fitzmaurice RJ. Immunohistological detection of human herpesvirus virus 6 in formalin-fixed, paraffin-embedded lung tissues. *J Med Virol* 1993;**41**:103–7.
15. Kempf W, Adams V, Mirandola P, Menotti L, Di Luca D, Wey N, et al. Persistence of human herpesvirus 7 in normal tissues detected by expression of a structural antigen. *J Infect Dis* 1998;**178**(3):841–5.
16. Fox JD, Briggs M, Ward PA, Tedder RS. Human herpesvirus 6 in salivary glands. *Lancet* 1990;**336**(8715):590–3.
17. Yadav M, Nambiar S, Khoo SP, Yaacob HB. Detection of human herpesvirus 7 in salivary glands. *Arch Oral Biol* 1997;**42**(8):559–67.
18. Lempinen M, Halme L, Arola J, Honkanen E, Salmela K, Lautenschlager I. HHV-6B is frequently found in the gastrointestinal tract in kidney transplantation patients. *Transpl Int* 2012;**25**(7):776–82.
19. Loginov R, Karlsson T, Hockerstedt K, Ablashi D, Lautenschlager I. Quantitative HHV-6B antigenemia test for the monitoring of transplant patients. *Eur J Clin Microbiol Infect Dis* 2010;**29**(7):881–6.
20. Ahtiluoto S, Mannonen L, Paetau A, Vaheri A, Koskiniemi M, Rautiainen P, et al. *In situ* hybridization detection of human herpesvirus 6 in brain tissue from fatal encephalitis. *Pediatrics* 2000;**105**(2):431–3.
21. Daibata M, Taguchi T, Nemoto Y, Taguchi H, Miyoshi I. Inheritance of chromosomally integrated human herpesvirus 6 DNA. *Blood* 1999;**94**(5):1545–9.
22. Siddon A, Lozovatsky L, Mohamed A, Hudnall SD. Human herpesvirus 6 positive Reed-Sternberg cells in nodular sclerosis Hodgkin lymphoma. *Br J Haematol* 2012;**158**(5):635–43.
23. Loginov R, Harma M, Halme L, Hockerstedt K, Lautenschlager I. HHV-6 DNA in peripheral blood mononuclear cells after liver transplantation. *J Clin Virol* 2006;**37**(Suppl 1): S76–81.

24. Cone RW, Huang ML, Ashley R, Corey L. Human herpesvirus 6 DNA in peripheral blood cells and saliva from immunocompetent individuals. *J Clin Microbiol* 1993;**31**(5):1262–7.
25. Chiu SS, Cheung CY, Tse CY, Peiris M. Early diagnosis of primary human herpesvirus 6 infection in childhood: serology, polymerase chain reaction, and virus load. *J Infect Dis* 1998;**178**(5):1250–6.
26. Gautheret A, Aubin JT, Fauveau V, Rozenbaum W, Huraux JM, Agut H. Rate of detection of human herpesvirus-6 at different stages of HIV infection. *Eur J Clin Microbiol Infect Dis* 1995;**14**:820–4.
27. Clark DA, Ait-Khaled M, Wheeler AC, Kidd IM, McLaughlin JE, Johnson MA, et al. Quantification of human herpesvirus 6 in immunocompetent persons and post-mortem tissues from AIDS patients by PCR. *J Gen Virol* 1996;**77**:2271–5. Pt 9.
28. Kidd IM, Clark DA, Ait-Khaled M, Griffiths PD, Emery VC. Measurement of human herpesvirus 7 load in peripheral blood and saliva of healthy subjects by quantitative polymerase chain reaction. *J Infect Dis* 1996;**174**(2):396–401.
29. Fujiwara N, Namba H, Ohuchi R, Isomura H, Uno F, Yoshida M, et al. Monitoring of human herpesvirus-6 and -7 genomes in saliva samples of healthy adults by competitive quantitative PCR. *J Med Virol* 2000;**61**(2):208–13.
30. Secchiero P, Zella D, Crowley RW, Gallo RC, Lusso P. Quantitative PCR for human herpesviruses 6 and 7. *J Clin Microbiol* 1995;**33**(8):2124–30.
31. Pradeau K, Couty L, Szelag JC, Turlure P, Rolle F, Ferrat P, et al. Multiplex real-time PCR assay for simultaneous quantitation of human cytomegalovirus and herpesvirus-6 in polymorphonuclear and mononuclear cells of transplant recipients. *J Virol Methods* 2006;**132**(1–2):77–84.
32. Engelmann I, Petzold DR, Kosinska A, Hepkema BG, Schulz TF, Heim A. Rapid quantitative PCR assays for the simultaneous detection of herpes simplex virus, varicella zoster virus, cytomegalovirus, Epstein–Barr virus, and human herpesvirus 6 DNA in blood and other clinical specimens. *J Med Virol* 2008;**80**(3):467–77.
33. Tanaka T, Kogawa K, Sasa H, Nonoyama S, Furuya K, Sato K. Rapid and simultaneous detection of 6 types of human herpes virus (herpes simplex virus, varicella-zoster virus, Epstein–Barr virus, cytomegalovirus, human herpes virus 6A/B, and human herpes virus 7) by multiplex PCR assay. *Biomed Res* 2009;**30**(5):279–85.
34. Ihira M, Enomoto Y, Kawamura Y, Nakai H, Sugata K, Asano Y, et al. Development of quantitative RT-PCR assays for detection of three classes of HHV-6B gene transcripts. *J Med Virol* 2012;**84**(9):1388–95.
35. Mirandola P, Menegazzi P, Merighi S, Ravaioli T, Cassai E, Di Luca D. Temporal mapping of transcripts in herpesvirus 6 variants. *J Virol* 1998;**72**(5):3837–44.
36. Yoshikawa T, Akimoto S, Nishimura N, Ozaki T, Ihira M, Ohashi M, et al. Evaluation of active human herpesvirus 6 infection by reverse transcription-PCR. *J Med Virol* 2003;**70**(2):267–72.
37. Van den Bosch G, Locatelli G, Geerts L, Faga G, Ieven M, Goossens H, et al. Development of reverse transcriptase PCR assays for detection of active human herpesvirus 6 infection. *J Clin Microbiol* 2001;**39**(6):2308–10.
38. Pradeau K, Bordessoule D, Szelag JC, Rolle F, Ferrat P, Le Meur Y, et al. A reverse transcription-nested PCR assay for HHV-6 mRNA early transcript detection after transplantation. *J Virol Methods* 2006;**134**(1–2):41–7.
39. Andre-Garnier E, Robillard N, Costa-Mattioli M, Besse B, Billaudel S, Imbert-Marcille BM. A one-step RT-PCR and a flow cytometry method as two specific tools for direct evaluation of human herpesvirus-6 replication. *J Virol Methods* 2003;**108**(2):213–22.

40. Menegazzi P, Galvan M, Rotola A, Ravaioli T, Gonelli A, Cassai E, et al. Temporal mapping of transcripts in human herpesvirus-7. *J Gen Virol* 1999;**80**(Pt 10):2705–12.
41. Bonnafous P, Boutolleau D, Naesens L, Deback C, Gautheret-Dejean A, Agut H. Characterization of a cidofovir-resistant HHV-6 mutant obtained by *in vitro* selection. *Antiviral Res* 2008;**77**(3):237–40.
42. Aubin JT, Poirel L, Robert C, Huraux JM, Agut H. Identification of human herpesvirus 6 variants A and B by amplimer hybridization with variant-specific oligonucleotides and amplification with variant-specific primers. *J Clin Microbiol* 1994;**32**(10):2434–40.
43. Aubin JT, Collandre H, Candotti D, Ingrand D, Rouzioux C, Burgard M, et al. Several groups among human herpesvirus 6 strains can be distinguished by southern blotting and polymerase chain reaction. *J Clin Microbiol* 1991;**29**(2):367–72.
44. Li H, Meng S, Levine SM, Stratton CW, Tang YW. Sensitive, qualitative detection of human herpesvirus-6 and simultaneous differentiation of variants A and B. *J Clin Virol* 2009;**46**(1):20–3.
45. Agut H, Collandre H, Aubin JT, Guétard D, Favier V, Ingrand D, Montagnier L, Huraux JM. *In vitro* sensitivity of human herpesvirus-6 to antivirals drugs. *Res Virol* 1989;**140**:219–28.
46. Akesson-Johansson A, Harmenberg J, Wahren B, Linde A. Inhibition of human herpesvirus 6 replication by 9-[4-hydroxy-2-(hydroxymethyl)butyl]guanine (2HM-HBG) and other antiviral compounds. *Antimicrob Agents Chemother* 1990;**34**(12):2417–9.
47. Mace M, Manichanh C, Bonnafous P, Precigout S, Boutolleau D, Gautheret-Dejean A, et al. Real-time PCR as a versatile tool for investigating the susceptibility of human herpesvirus 6 to antiviral agents. *Antimicrob Agents Chemother* 2003;**47**(9):3021–4.
48. Reymen D, Naesens L, Balzarini J, Holy A, Dvorakova H, De Clercq E. Antiviral activity of selected acyclic nucleoside analogues against human herpesvirus 6. *Antiviral Res* 1995;**28**(4):343–57.
49. De Bolle L, Manichanh C, Agut H, De Clercq E, Naesens L. Human herpesvirus 6 DNA polymerase: enzymatic parameters, sensitivity to ganciclovir and determination of the role of the A961V mutation in HHV-6 ganciclovir resistance. *Antiviral Res* 2004;**64**(1):17–25.
50. Isegawa Y, Hara J, Amo K, Osugi Y, Takemoto M, Yamanishi K, et al. Human herpesvirus 6 ganciclovir-resistant strain with amino acid substitutions associated with the death of an allogeneic stem cell transplant recipient. *J Clin Virol* 2009;**44**(1):15–19.
51. Okuno T, Takahashi K, Balachandra K, Shiraki K, Yamanishi K, Takahashi M, et al. Seroepidemiology of human herpesvirus 6 infection in normal children and adults. *J Clin Microbiol* 1989;**27**(4):651–3.
52. Robert C, Agut H, Aubin JT, Collandre H, Ingrand D, Devillechabrolle A, et al. Detection of antibodies to human herpesvirus-6 using immunofluorescence assay. *Res Virol* 1990;**141**:545–55.
53. Foa-Tomasi L, Avitabile E, Ke L, Campadelli-Fiume G. Polyvalent and monoclonal antibodies identify major immunogenic proteins specific for human herpesvirus 7-infected cells and have weak cross-reactivity with human herpesvirus 6. *J Gen Virol* 1994;**75**:2719–27.
54. Sanders VJ, Felisan S, Waddell A, Tourtellotte WW. Detection of herpesviridae in postmortem multiple sclerosis brain tissue and controls by polymerase chain reaction. *J Neurovirol* 1996;**2**(4):249–58.
55. Chou SW, Scott KM. Rises in antibody to human herpesvirus 6 detected by enzyme immunoassay in transplant recipients with primary cytomegalovirus infection. *J Clin Microbiol* 1990;**28**(5):851–4.
56. Saxinger C, Polesky H, Eby N, Grufferman S, Murphy R, Tegtmeir G, et al. Antibody reactivity with HBLV (HHV-6) in U.S. populations. *J Virol Methods* 1988;**21**:199–208.

57. Dahl H, Linde A, Sundqvist VA, Wahren B. An enzyme-linked immunosorbent assay for IgG antibodies to human herpes virus 6. *J Virol Methods* 1990;**29**(3):313–23.
58. Sloots TP, Kapeleris JP, Mackay IM, Batham M, Devine PL. Evaluation of a commercial enzyme-linked immunosorbent assay for detection of serum immunoglobulin G response to human herpesvirus 6. *J Clin Microbiol* 1996;**34**(3):675–9.
59. Yao K, Gagnon S, Akhyani N, Williams E, Fotheringham J, Frohman E, et al. Reactivation of human herpesvirus-6 in natalizumab treated multiple sclerosis patients. *PLoS ONE* 2008;**3**(4) e2028.
60. Zerr DM, Meier AS, Selke SS, Frenkel LM, Huang ML, Wald A, et al. A population-based study of primary human herpesvirus 6 infection. *N Engl J Med* 2005;**352**(8):768–76.
61. Black JB, Schwarz TF, Patton JL, Kite-Powell K, Pellett PE, Wiersbitzky S, et al. Evaluation of immunoassays for detection of antibodies to human herpesvirus 7. *Clin Diagn Lab Immunol* 1996;**3**(1):79–83.
62. LaCroix S, Stewart JA, Thouless ME, Black JB. An immunoblot assay for detection of immunoglobulin M antibody to human herpesvirus 6. *Clin Diagn Lab Immunol* 2000;**7**(5):823–7.
63. Asada H, Yalcin S, Balachandra K, Higashi K, Yamanishi K. Establishment of titration system for human herpesvirus 6 and evaluation of neutralizing antibody response to the virus. *J Clin Microbiol* 1989;**27**(10):2204–7.
64. Suga S, Yoshikawa T, Asano Y, Yazaki T, Ozaki T. Neutralizing antibody assay for human herpesvirus-6. *J Med Virol* 1990;**30**:14–19.
65. Tsukazaki T, Yoshida M, Namba H, Yamada M, Shimizu N, Nii S. Development of a dot blot neutralizing assay for HHV-6 and HHV-7 using specific monoclonal antibodies. *J Virol Methods* 1998;**73**(2):141–9.
66. Moutschen M, Triffaux JM, Demonty J, Legros JJ, Lefebvre PJ. Pathogenic tracks in fatigue syndromes. *Acta Clin Belg* 1994;**49**(6):274–89.
67. Balachandran N, Amelse RE, Zhou WW, Chang CK. Identification of proteins specific for human herpesvirus 6-infected human T cells. *J Virol* 1989;**63**(6):2835–40.
68. Neipel F, Ellinger K, Fleckenstein B. Gene for the major antigenic structural protein (p100) of human herpesvirus 6. *J Virol* 1992;**66**:3918–24.
69. Iyengar S, Levine PH, Ablashi D, Neequaye J, Pearson GR. Sero-epidemiological investigations on human herpesvirus 6 (HHV-6) infections using a newly developed early antigen assay. *Int J Cancer* 1991;**49**(4):551–7.
70. Xu Y, Linde A, Fredrikson S, Dahl H, Winberg G. HHV-6A- or B-specific P41 antigens do not reveal virus variant-specific IgG or IgM responses in human serum. *J Med Virol* 2002;**66**(3):394–9.
71. Caselli E, Boni M, Bracci A, Rotola A, Cermelli C, Castellazzi M, et al. Detection of antibodies directed against human herpesvirus 6 U94/REP in sera of patients affected by multiple sclerosis. *J Clin Microbiol* 2002;**40**(11):4131–7.
72. Sutherland S, Christofinis G, O'Grady J, Williams R. A serological investigation of human herpesvirus 6 infections in liver transplant recipients and the detection of cross-reacting antibodies to cytomegalovirus. *J Med Virol* 1991;**33**(3):172–6.
73. Nielsen L, Vestergaard BF. A mu-capture immunoassay for detection of human herpes virus-6 (HHV-6) IgM antibodies in human serum. *J Clin Virol* 2002;**25**(2):145–54.
74. Carricart SE, Bustos D, Biganzoli P, Nates SE, Pavan JV. Isotype immune response of IgG antibodies at the persistence and reactivation stages of human herpes virus 6 infection. *J Clin Virol* 2004;**31**(4):266–9.
75. Ward KN, Gray JJ, Fotheringham MW, Sheldon MJ. IgG antibodies to human herpesvirus-6 in young children: changes in avidity of antibody correlate with time after infection. *J Med Virol* 1993;**39**(2):131–8.

76. Ward KN, Leong HN, Thiruchelvam AD, Atkinson CE, Clark DA. Human herpesvirus 6 DNA levels in cerebrospinal fluid due to primary infection differ from those due to chromosomal viral integration and have implications for diagnosis of encephalitis. *J Clin Microbiol* 2007;**45**(4):1298–304.
77. Yoshida M, Torigoe S, Yamada M. Elucidation of the cross-reactive immunoglobulin M response to human herpesviruses 6 and 7 on the basis of neutralizing antibodies. *Clin Diagn Lab Immunol* 2002;**9**(2):394–402.
78. Yoshikawa T, Black JB, Ihira M, Suzuki K, Suga S, Iida K, et al. Comparison of specific serological assays for diagnosing human herpesvirus 6 infection after liver transplantation. *Clin Diagn Lab Immunol* 2001;**8**(1):170–3.
79. Caserta MT, Hall CB, Schnabel K, Long CE, D'Heron N. Primary human herpesvirus 7 infection: a comparison of human herpesvirus 7 and human herpesvirus 6 infections in children. *J Pediatr* 1998;**133**(3):386–9.
80. Suga S, Yoshikawa T, Kajita Y, Ozaki T, Asano Y. Prospective study of persistence and excretion of human herpesvirus-6 in patients with exanthem subitum and their parents. *Pediatrics* 1998;**102**:900–4.
81. Yamanishi K, Okuno T, Shiraki K, Takahashi M, Kondo T, Asano Y, et al. Identification of human herpesvirus-6 as a causal agent for exanthem subitum. *Lancet* 1988;**1**(8594):1065–7.
82. Hall CB, Caserta MT, Schnabel KC, Shelley LM, Carnahan JA, Marino AS, et al. Transplacental congenital human herpesvirus 6 infection caused by maternal chromosomally integrated virus. *J Infect Dis* 2010 15;**201**(4):505–7.
83. De Bolle L, Naesens L, De Clercq E. Update on human herpesvirus 6 biology, clinical features, and therapy. *Clin Microbiol Rev* 2005;**18**(1):217–45.
84. Lautenschlager I, Razonable RR. Human herpesvirus-6 infections in kidney, liver, lung, and heart transplantation: review. *Transpl Int* 2012;**25**(5):493–502.
85. Boutolleau D, Fernandez C, Andre E, Imbert-Marcille BM, Milpied N, Agut H, et al. Human herpesvirus (HHV)-6 and HHV-7: two closely related viruses with different infection profiles in stem cell transplantation recipients. *J Infect Dis* 2003;**187**(2):179–86.
86. Ihira M, Akimoto S, Miyake F, Fujita A, Sugata K, Suga S, et al. Direct detection of human herpesvirus 6 DNA in serum by the loop-mediated isothermal amplification method. *J Clin Virol* 2007;**39**(1):22–6.
87. Ogata M, Satou T, Kawano R, Goto K, Ikewaki J, Kohno K, et al. Plasma HHV-6 viral load-guided preemptive therapy against HHV-6 encephalopathy after allogeneic stem cell transplantation: a prospective evaluation. *Bone Marrow Transplant* 2008;**41**(3):279–85.
88. Huang LM, Kuo PF, Lee CY, Chen JY, Liu MY, Yang CS. Detection of human herpesvirus-6 DNA by polymerase chain reaction in serum or plasma. *J Med Virol* 1992;**38**:7–10.
89. Ogata M, Kikuchi H, Satou T, Kawano R, Ikewaki J, Kohno K, et al. Human herpesvirus 6 DNA in plasma after allogeneic stem cell transplantation: incidence and clinical significance. *J Infect Dis* 2006;**193**(1):68–79.
90. Achour A, Boutolleau D, Slim A, Agut H, Gautheret-Dejean A. Human herpesvirus-6 (HHV-6) DNA in plasma reflects the presence of infected blood cells rather than circulating viral particles. *J Clin Virol* 2007;**38**(4):280–5.
91. Fotheringham J, Akhyani N, Vortmeyer A, Donati D, Williams E, Oh U, et al. Detection of active human herpesvirus-6 infection in the brain: correlation with polymerase chain reaction detection in cerebrospinal fluid. *J Infect Dis* 2007;**195**(3):450–4.
92. Hirabayashi K, Nakazawa Y, Katsuyama Y, Yanagisawa T, Saito S, Yoshikawa K, et al. Successful ganciclovir therapy in a patient with human herpesvirus-6 encephalitis after unrelated cord blood transplantation: usefulness of longitudinal measurements of viral load in cerebrospinal fluid. *Infection* 2013;**41**(1):219–23.

93. Muta T, Fukuda T, Harada M. Human herpesvirus-6 encephalitis in hematopoietic SCT recipients in Japan: a retrospective multicenter study. *Bone Marrow Transplant* 2009;**43**(7):583–5.
94. Rieux C, Gautheret-Dejean A, Challine-Lehmann D, Kirch C, Agut H, Vernant JP. Human herpesvirus-6 meningoencephalitis in a recipient of an unrelated allogeneic bone marrow transplantation. *Transplantation* 1998;**65**(10):1408–11.
95. Yao K, Crawford JR, Komaroff AL, Ablashi DV, Jacobson S. Review. Part 2. Human herpesvirus-6 in central nervous system diseases. *J Med Virol* 2010;**82**(10):1669–78.
96. Potenza L, Luppi M, Barozzi P, Rossi G, Cocchi S, Codeluppi M, et al. HHV-6A in syncytial giant-cell hepatitis. *N Engl J Med* 2008;**359**(6):593–602.
97. Pischke S, Gosling J, Engelmann I, Schlue J, Wolk B, Jackel E, et al. High intrahepatic HHV-6 virus loads but neither CMV nor EBV are associated with decreased graft survival after diagnosis of graft hepatitis. *J Hepatol* 2012;**56**(5):1063–9.
98. Leveque N, Boulagnon C, Brasselet C, Lesaffre F, Boutolleau D, Metz D, et al. A fatal case of human herpesvirus 6 chronic myocarditis in an immunocompetent adult. *J Clin Virol* 2011;**52**(2):142–5.
99. Leong HN, Tuke PW, Tedder RS, Khanom AB, Eglin RP, Atkinson CE, et al. The prevalence of chromosomally integrated human herpesvirus 6 genomes in the blood of UK blood donors. *J Med Virol* 2007;**79**(1):45–51.
100. Geraudie B, Charrier M, Bonnafous P, Heurte D, Desmonet M, Bartoletti MA, et al. Quantitation of human herpesvirus-6A, -6B and -7 DNAs in whole blood, mononuclear and polymorphonuclear cell fractions from healthy blood donors. *J Clin Virol* 2012;**53**(2):151–5.
101. Seringe E, Gautheret-Dejean A, Agut H, Vernant JP, Astagneau P Association between active human herpes virus 6 infection and acute cutaneous graft-versus-host disease (GVHD) after hematopoietic stem cell transplantation. Presented at the 50th International Conference on Antimicrobial Agents and Infectious Diseases, Boston, MA, September 12–15, 2010.
102. Caserta MT, Hall CB, Schnabel K, Lofthus G, Marino A, Shelley L, et al. Diagnostic assays for active infection with human herpesvirus 6 (HHV-6). *J Clin Virol* 2010;**48**(1):55–7.
103. de Pagter PJ, Schuurman R, Visscher H, de Vos M, Bierings M, van Loon AM, et al. Human herpes virus 6 plasma DNA positivity after hematopoietic stem cell transplantation in children: an important risk factor for clinical outcome. *Biol Blood Marrow Transplant* 2008;**14**(7):831–9.
104. Gautheret-Dejean A, Henquell C, Mousnier F, Boutolleau D, Bonnafous P, Dhedin N, et al. Different expression of human herpesvirus-6 (HHV-6) load in whole blood may have a significant impact on the diagnosis of active infection. *J Clin Virol* 2009;**46**(1):33–6.
105. Harma M, Loginov R, Piiparinen H, Halme L, Hockerstedt K, Lautenschlager I. HHV-6–DNAemia related to CMV-DNAemia after liver transplantation. *Transplant Proc* 2005;**37**(2):1230–2.
106. Endo A, Imai K, Katano H, Inoue N, Ohye T, Kuruhashi H, et al. Chromosomally integrated human herpesvirus-6 was activated in a patient with X-linked severe combined immunodeficiency. Presented at the 8th International Conference on HHV-6 & 7, April 8–10, 2013, Paris, France.
107. Hubacek P, Maalouf J, Zajickova M, Kouba M, Cinek O, Hyncicova K, et al. Failure of multiple antivirals to affect high HHV-6 DNAaemia resulting from viral chromosomal integration in case of severe aplastic anaemia. *Haematologica* 2007;**92**(10) e98–100.

108. Pellett PE, Ablashi DV, Ambros PF, Agut H, Caserta MT, Descamps V, et al. Chromosomally integrated human herpesvirus 6: questions and answers. *Rev Med Virol* 2011;**22**(3):144–55.
109. Hubacek P, Virgili A, Ward KN, Pohlreich D, Keslova P, Goldova B, et al. HHV-6 DNA throughout the tissues of two stem cell transplant patients with chromosomally integrated HHV-6 and fatal CMV pneumonitis. *Br J Haematol* 2009;**145**(3):394–8.
110. Ward KN, Leong HN, Nacheva EP, Howard J, Atkinson CE, Davies NW, et al. Human herpesvirus 6 chromosomal integration in immunocompetent patients results in high levels of viral DNA in blood, sera, and hair follicles. *J Clin Microbiol* 2006;**44**(4):1571–4.
111. Deback C, Geli J, Ait-Arkoub Z, Angleraud F, Gautheret-Dejean A, Agut H, et al. Use of the Roche LightCycler 480 system in a routine laboratory setting for molecular diagnosis of opportunistic viral infections: evaluation on whole blood specimens and proficiency panels. *J Virol Methods* 2009;**159**(2):291–4.
112. Lee SO, Brown RA, Razonable RR. Clinical significance of pretransplant chromosomally integrated human herpesvirus-6 in liver transplant recipients. *Transplantation* 2011;**92**(2):224–9.
113. Hall CB, Caserta MT, Schnabel K, Shelley LM, Marino AS, Carnahan JA, et al. Chromosomal integration of human herpesvirus 6 is the major mode of congenital human herpesvirus 6 infection. *Pediatrics* 2008;**122**(3):513–20.
114. Gautheret-Dejean A, Hardin R, Nguyen-Quoc S, Bolgert F, Henquell C, Chanzy B, et al. Diagnosing HHV-6 encephalitis: tools and algorithms. Presented at the 8th International Conference on HHV-6 & 7, April 8–10, 2013, Paris, France.
115. Levy JA, Ferro F, Greenspan D, Lennette ET. Frequent isolation of HHV-6 from saliva and high seroprevalence of the virus in the population. *Lancet* 1990;**335**(8697):1047–50.
116. Yoshikawa T, Asano Y, Kobayashi I, Nakashima T, Yazaki T, Suga S, et al. Seroepidemiology of human herpesvirus 7 in healthy children and adults in Japan. *J Med Virol* 1993;**41**(4):319–23.
117. Balachandra K, Ayuthaya PI, Auwanit W, Jayavasu C, Okuno T, Yamanishi K, et al. Prevalence of antibody to human herpesvirus 6 in women and children. *Microbiol Immunol* 1989;**33**(6):515–8.
118. Caïola D, Karras A, Flandre P, Boutolleau D, Scieux C, Agut H, et al. Confirmation of the low clinical effect of human herpesvirus-6 and -7 infections after renal transplantation. *J Med Virol* 2012;**84**(3):450–6.
119. Galarraga MC, Gomez E, de Ona M, Rodriguez A, Laures A, Boga JA, et al. Influence of ganciclovir prophylaxis on cytomegalovirus, human herpesvirus 6, and human herpesvirus 7 viremia in renal transplant recipients. *Transplant Proc* 2005;**37**(5):2124–6.
120. Harma M, Hockerstedt K, Lyytikainen O, Lautenschlager I. HHV-6 and HHV-7 antigenemia related to CMV infection after liver transplantation. *J Med Virol* 2006;**78**(6):800–5.
121. Razonable RR, Brown RA, Humar A, Covington E, Alecock E, Paya CV. Herpesvirus infections in solid organ transplant patients at high risk of primary cytomegalovirus disease. *J Infect Dis* 2005;**192**(8):1331–9.
122. Locatelli G, Santoro F, Veglia F, Gobbi A, Lusso P, Malnati MS. Real-time quantitative PCR for human herpesvirus 6 DNA. *J Clin Microbiol* 2000;**38**(11):4042–8.
123. Gautheret-Dejean A, Manichanh C, Thien-Ah-Koon F, Fillet AM, Mangeney N, Vidaud M, et al. Development of a real-time polymerase chain reaction assay for the diagnosis of human herpesvirus-6 infection and application to bone marrow transplant patients. *J Virol Methods* 2002;**100**(1–2):27–35.

124. Hymas W, Stevenson J, Taggart EW, Hillyard D. Use of lyophilized standards for the calibration of a newly developed real time PCR assay for human herpes type six (HHV6) variants A and B. *J Virol Methods* 2005;**128**(1–2):143–50.
125. Tanaka N, Kimura H, Hoshino Y, Kato K, Yoshikawa T, Asano Y, et al. Monitoring four herpesviruses in unrelated cord blood transplantation. *Bone Marrow Transplant* 2000;**26**(11):1193–7.
126. Ihira M, Yoshikawa T, Enomoto Y, Akimoto S, Ohashi M, Suga S, et al. Rapid diagnosis of human herpesvirus 6 infection by a novel DNA amplification method, loop-mediated isothermal amplification. *J Clin Microbiol* 2004;**42**(1):140–5.
127. Flamand L, Gravel A, Boutolleau D, Alvarez-Lafuente R, Jacobson S, Malnati MS, et al. Multicenter comparison of PCR assays for detection of human herpesvirus 6 DNA in serum. *J Clin Microbiol* 2008;**46**(8):2700–6.
128. Nagate A, Ohyashiki JH, Kasuga I, Minemura K, Abe K, Yamamoto K, et al. Detection and quantification of human herpesvirus 6 genomes using bronchoalveolar lavage fluid in immunocompromised patients with interstitial pneumonia. *Int J Mol Med* 2001;**8**(4):379–83.
129. Gaeta A, Verzaro S, Cristina LM, Mancini C, Nazzari C. Diagnosis of neurological herpesvirus infections: real time PCR in cerebral spinal fluid analysis. *New Microbiol* 2009;**32**(4):333–40.
130. Nitsche A, Muller CW, Radonic A, Landt O, Ellerbrok H, Pauli G, et al. Human herpesvirus 6A DNA is detected frequently in plasma but rarely in peripheral blood leukocytes of patients after bone marrow transplantation. *J Infect Dis* 2001;**183**(1):130–3.
131. Razonable RR, Fanning C, Brown RA, Espy MJ, Rivero A, Wilson J, et al. Selective reactivation of human herpesvirus 6 variant a occurs in critically ill immunocompetent hosts. *J Infect Dis* 2002;**185**(1):110–3.
132. Reddy S, Manna P. Quantitative detection and differentiation of human herpesvirus 6 subtypes in bone marrow transplant patients by using a single real-time polymerase chain reaction assay. *Biol Blood Marrow Transplant* 2005;**11**(7):530–41.
133. Zerr DM, Huang ML, Corey L, Erickson M, Parker HL, Frenkel LM. Sensitive method for detection of human herpesviruses 6 and 7 in saliva collected in field studies. *J Clin Microbiol* 2000;**38**(5):1981–3.
134. Safronetz D, Humar A, Tipples GA. Differentiation and quantitation of human herpesviruses 6A, 6B and 7 by real-time PCR. *J Virol Methods* 2003;**112**(1–2):99–105.
135. Fernandez C, Boutolleau D, Manichanh C, Mangeney N, Agut H, Gautheret-Dejean A. Quantitation of HHV-7 genome by real-time polymerase chain reaction assay using MGB probe technology. *J Virol Methods* 2002;**106**(1):11–16.
136. Broccolo F, Drago F, Careddu AM, Foglieni C, Turbino L, Cocuzza CE, et al. Additional evidence that pityriasis rosea is associated with reactivation of human herpesvirus-6 and -7. *J Invest Dermatol* 2005;**124**(6):1234–40.

Chapter 3

Pathologic Features of HHV-6A, HHV-6B, and HHV-7 Infection (Light and Electron Microscopy)

Gerhard R.F. Krueger[a] and Irmeli Lautenschlager[b]
[a]The University of Texas–Houston Medical School, Houston, Texas, [b]Helsinki University Hospital and University of Helsinki, Helsinki, Finland

INTRODUCTION

Viral Infection and Pathogenesis

This chapter deals with the pathology of roseoloviruses, namely HHV-6 (HHV-6A and HHV-6B) and HHV-7, which belong to the β-herpesvirus subgroup together with human cytomegalovirus (See Figures 3.1 to 3.3). The latter will not be discussed here. Successful infection starts with binding of the viral particles to cellular receptors, which are—at least in terms of being a critical component—CD46 for HHV-6 and CD4 for HHV-7.[1,2] CD46 is expressed

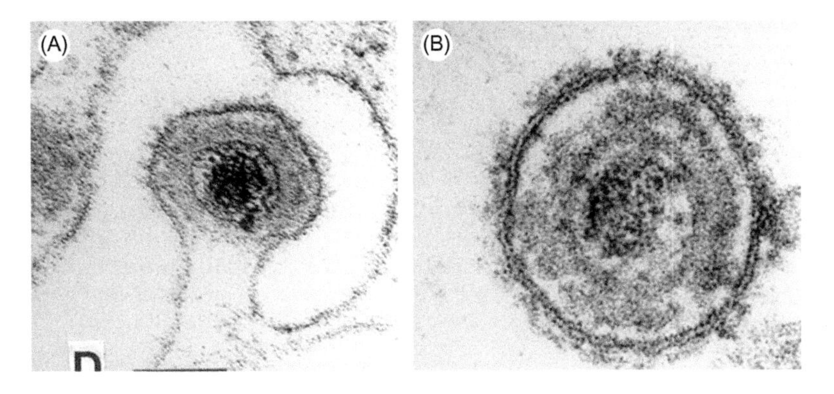

FIGURE 3.1 Electron microscopic details of (A) HHV-6 and (B) HHV-7. Note the same structure of both viruses (from outside to inside): spikes, layered envelope, prominent tegument, nucleocapsid, and nucleocore, but also note the more loosely arranged core of HHV-7 (same preparation technique).

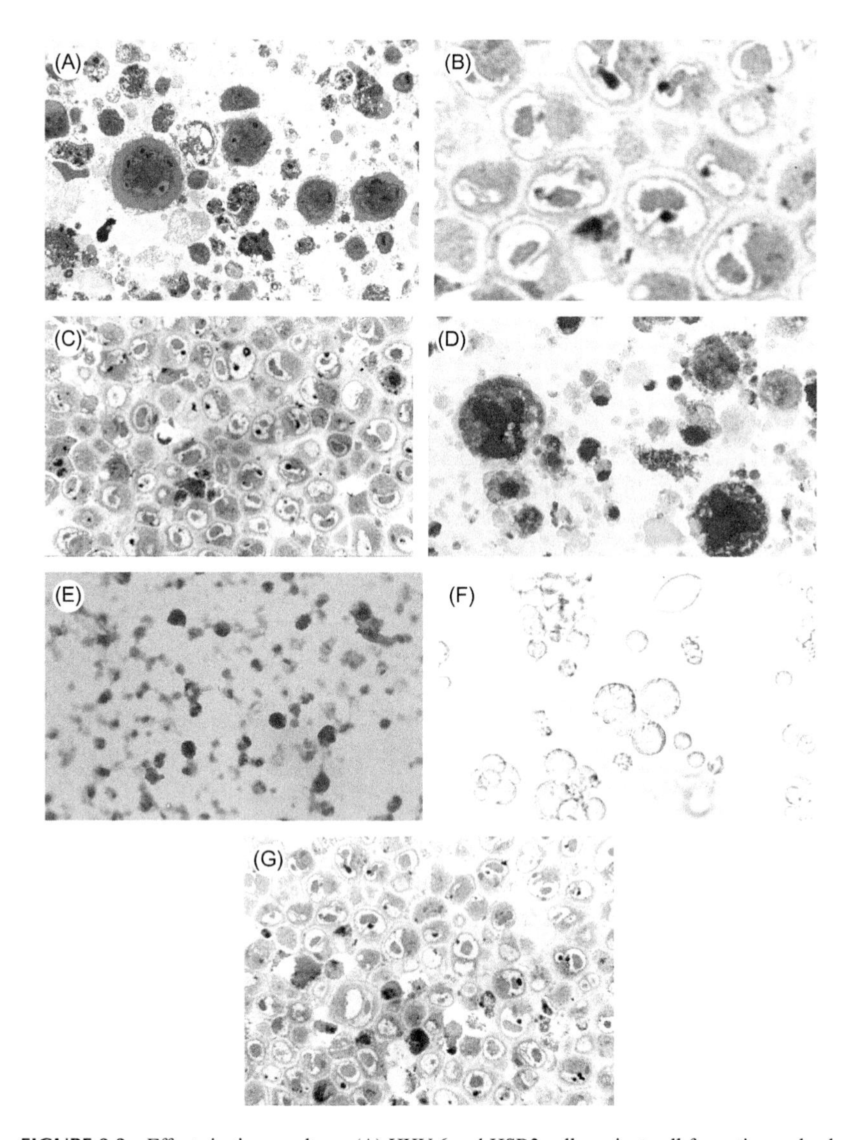

FIGURE 3.2 Effects in tissue culture: (A) HHV-6 and HSB2 cells—giant cell formation and cellular apoptosis. (B) HHV-6 nuclear inclusion bodies. (C) HHV-6 cell proliferation. (D) HHV-6 virus p41 antigen expression by immunocytochemistry. (E) HHV-6 and pZVH14 virus DNA expression by *in situ* hybridization. (F) HHV-7 and MOLT3 cells—blast cell response of infected cells. (G) HHV-7 apoptosis, cell proliferation, and some nuclear inclusions in infected MOLT3 cells.

on nucleated epithelial cells of salivary gland ducts and renal tubules and to some extent on lymphocytes and vascular endothelial cells, yet only weakly on muscle and interstitial cells. It is a regulatory part of the complement system and has cofactor activity for inactivation of complement components C3b and

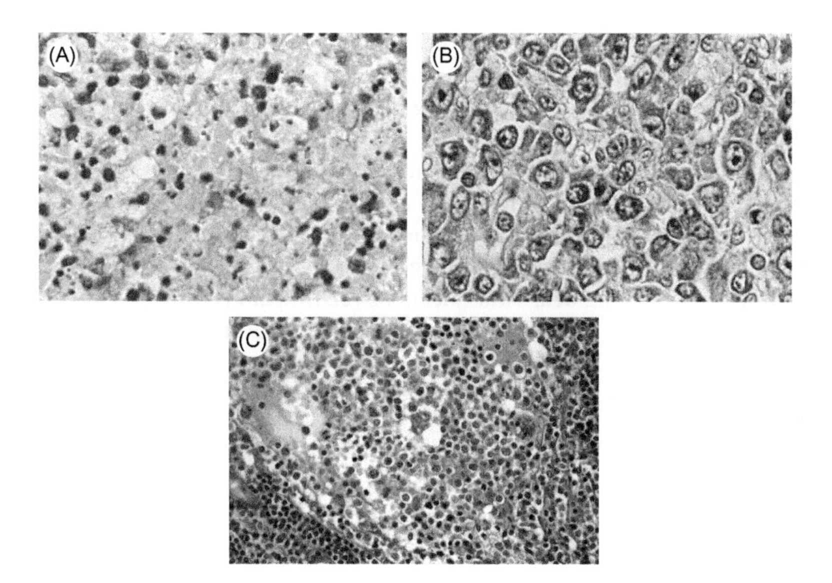

FIGURE 3.3 Effects in lymphoid tissues: (A) HHV-6 prominent cellular apoptosis (note nuclear pyknosis and debris). (B) HHV-6 blastic transformation and hyperplastic response. (C) HHV-7 follicular hyperplasia, blastic response, and cellular degeneration.

C4b by serum factor I. CD46 thus protects the host cell from damage by complement (http://www.ncbi.nlm.nih.gov/gene/4179). Downregulation of CD46 by HHV-6 may interfere with these functions.

The CD4 membrane glycoprotein interacts with major histocompatibility complex class II antigens and is a receptor for the human immunodeficiency virus. Its gene is expressed in T cells, B cells, macrophages, and granulocytes and in specific regions of the brain. It initiates or augments T-cell activation and functions as a mediator of neuronal damage in infectious and immune-mediated diseases of the central nervous system (http://www.ncbi.nlm.nih.gov/gene/920).

After cellular attachment to and replication of HHV-6 and HHV-7 in cells of the salivary gland, viremia causes viral dissemination. Intravenous entrance of the virus produces a short-lived passive viremia of a few hours, with attachment and replication in susceptible cells (e.g., of lymph nodes) producing secondary viremia. A complete replicative cycle of HHV-6 and HHV-7 takes about one week.[3,4]

Viral attachment to the cell membrane and its entrance into the cell are followed by cellular reactions including cell membrane rigidification and alterations of membrane receptor expression, alterations of cytokine/chemokine production, NF-κB and genome activation, and cellular proliferation or apoptosis.[5] Intracellular viruses may undergo replicative cycles, defective replication with expression of only certain antigens, integration into the host cell genome, or latency. All effects have different pathogenic consequences with potential morbidity, thus rendering a simple cause-and-effect relationship difficult to prove.

Pathological changes arise from cytotoxic effects of the virus (necrosis or apoptosis of infected cells), functional alterations of infected cells mediating lesions in other cells and tissues, immune modulation and molecular mimicry causing autoimmune reactions, genomic transactivation, cell membrane receptor expression enabling dual viral infections (e.g., with measles, HIV, EBV, or CMV, among others), and pathogenic cooperation of several viruses.

Morphogenesis of Cell and Tissue Lesions

Cell changes following *in vitro* infection of cells with HHV-6 and HHV-7 have been described before[6] (see also http://www.hhv-6foundation.org/research/pictorial-atlas-of-hhv-6). Susceptible cells undergo blastic transformation with or without giant cell formation, intranuclear inclusions, eventual mitoses, cellular degeneration, and apoptosis.[7,8] Many changes indicate HHV-6A, HHV-6B, or HHV-7 exposure and do not indicate cellular infection with virus replication.[9,10]

Tissue lesions developing after HHV-6 or HHV-7 infection differ in primary infections (PIs) from exogenous or endogenous reinfections (RIs), the latter being reactivation of a latent infection. PI hits a virgin organism that has not developed anti-HHV-6 or HHV-7 immunity. Morphologic lesions are direct toxic effects of the virus (i.e., cell activation, apoptosis, and necrosis) or hyperplasia of the responding immune system (e.g., infectious mononucleosis-like lymphadenitis). RI is met by an antiviral immune response, and lesions are caused by immune reactions against the virus or parts of it. Toxic effects of the virus are potentiated by immune defense reactions (e.g., Kikuchi's lymphadenitis). Immune reactions themselves (including autoimmune reactions) cause tissue lesions even at a distance from the initial viral residence (e.g., collagen-vascular diseases, multiple sclerosis). These lesions differ morphologically and are not specific for HHV-6 or HHV-7; rarely, intranuclear inclusion bodies are noted, as in cytomegalovirus (CMV). Finally, HHV-6 and HHV-7 reactivation may occur in an immune-deficient individual and stay active for a prolonged period of time (i.e., persistent active infection). Under such conditions, extensively necrotic or proliferative lesions develop (e.g., atypical polyclonal lymphoproliferation, malignant lymphomas, possibly myocarditis or Hodgkin's disease).

In essence, proving a cause-and-effect relationship is difficult between HHV-6 or HHV-7 infections and a particular disease. Critical evaluation of all data (clinical, serological, molecular, pathological) may provide conclusive evidence for HHV-6 and HHV-7 disease. Frequently, only successful antiviral therapy will provide the final answer.

ANATOMIC (SYSTEMIC) PATHOLOGY OF TISSUES AND ORGANS

Some representative examples of pathologic lesions in HHV-6 and HHV-7 infections are provided in this section, but more information is available at the HHV-6 Foundation's website (http://www.hhv-6foundation.org/research/pictorial-atlas-of-hhv-6).

Diagnostic Criteria

Pathologic proof of HHV-6 or HHV-7 causing a disease relies upon showing an active viral infection at the site of the tissue lesion.[6] Viral replication or antigen deposits should be considered indicative of even defective replication or cross-reacting tissue antigens (e.g., molecular mimicry). Serologic testing is done in combination with virus isolation, antigen-capture enzyme-linked immunosorbent assay (ELISA) for p41 (follow-up testing in chronic persistent infections). *In situ* hybridization and polymerase chain reaction (PCR) for viral DNA can indicate increased viral load, but not necessarily viral activity in tissues. HHV-6 p41 replication antigen is detected by immunhistochemistry in tissue samples (biopsies or at autopsy) of suspected HHV-6 disease. RNA-PCR further supports such data if adequately preserved tissue is available. Electron microscopy can show herpesvirus particles in diseased tissues, although meticulous fixation and processing of specimens are necessary to identify the herpesvirus particles; use of old paraffin blocks is generally not possible. The type of herpesvirus can be confirmed by serological or molecular techniques.

Pathologic Lesions in Primary Infection

(See Figures 3.4 to 3.11.) As with Epstein–Barr virus, diseases resulting from primary HHV-6 or HHV-7 infection are rare and appear to be the exception rather than the rule; some 80% of HHV-6 and HHV-7 infections remain clinically silent. Primary HHV-6 infection resembles an influenza-like disease with fever, sweats and chills, fatigue, malaise, occasional convulsions, and exanthema, with the latter two occurring preferentially in children (see Chapter 9 for a more in-depth review of primary infection).[11–13] Arthritic symptoms and iridocyclitis may occur.[14]

Few descriptions of pathologic features in primary HHV-6 and HHV-7 infections are available from occasional biopsies and postmortem studies. Exanthem subitum is easily diagnosed by its appearance. Hyperplastic lymph nodes showing blastic transformation of lymphoid cells with intranuclear inclusions and occasional giant cells have been described in acute HHV-6 lymphadenitis by Maric and collaborators.[15] We have not seen giant cells or intranuclear inclusions in either HHV-6 or HHV-7 lymphadenitis. More commonly, HHV-6 lymphadenopathy shows features of Kikuchi–Fujimoto's

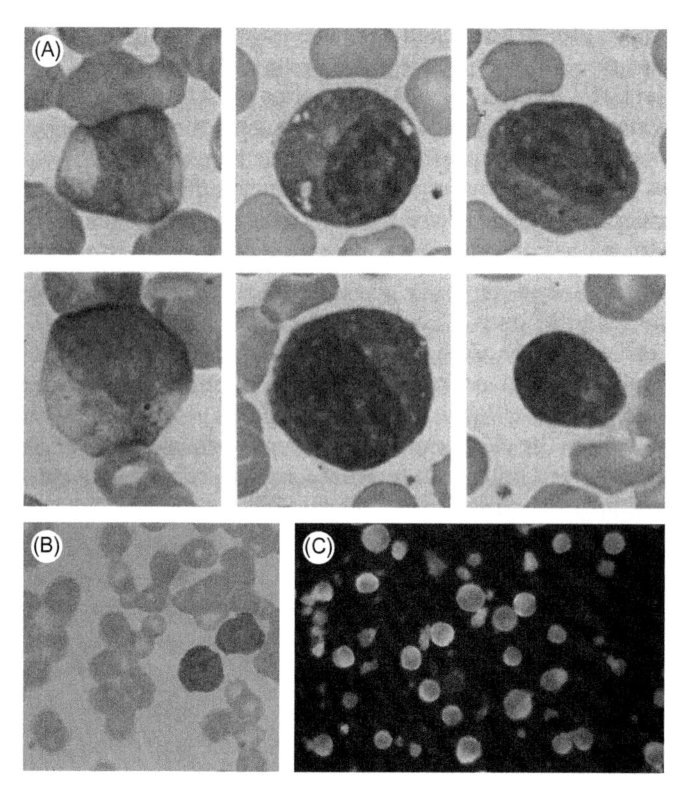

FIGURE 3.4 Changes in peripheral blood: (A, B) Circulating plasmacytoid blast cells of various sizes (A, HHV-6; B, HHV-7). (C) These cells proliferate in culture and contain virus antigen (immunofluorescence with anti-HHV-6).

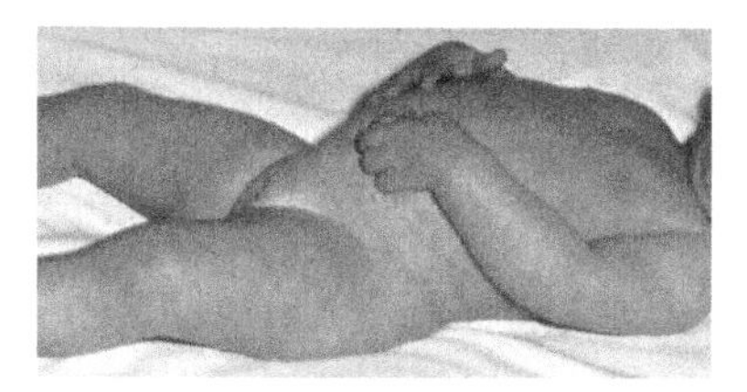

FIGURE 3.5 Typical exanthema subitum in a child with primary HHV-6 infection.

disease (KFD) with apoptosis of infected lymphoid cells and reactive mononuclear cell response.[16,17] HHV-6-induced infectious mononucleosis (IM; EBV- and CMV-negative) shows transformed lymphoid cells in blood, hyperplastic lymph nodes, and tonsils.[18,19] KFD and IM preferentially occur in persons who have already developed immune reactivity against the virus (i.e., not in a clearly virgin primary infection; patients are older than one year). Lytic or

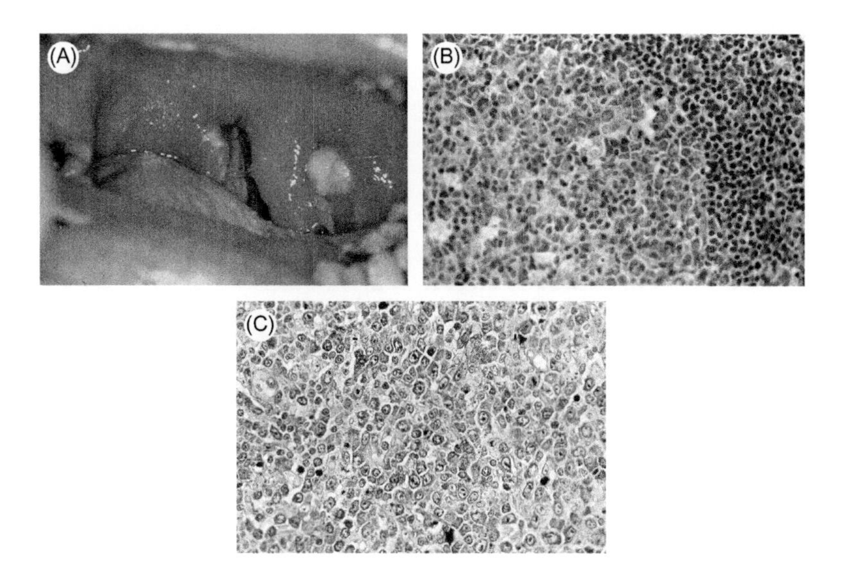

FIGURE 3.6 (A) Acute tonsillitis (angina) with focal pseudomembranes in a child with acute HHV-6 infection. Microscopic changes of lymphoid tissues show (B) prominent follicular hyperplasia and (C) diffuse paracortical hyperplasia.

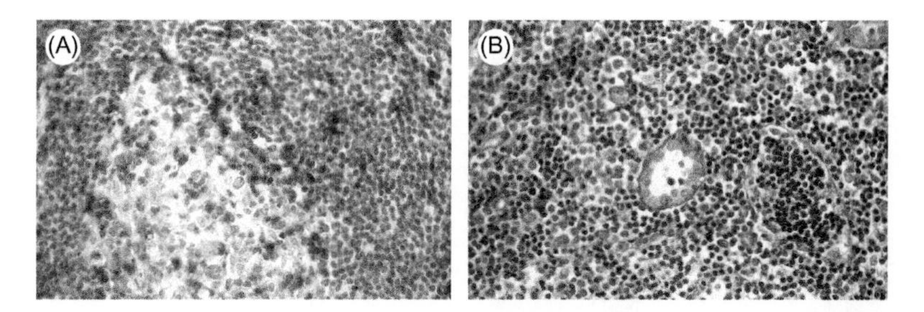

FIGURE 3.7 (A) Lymphadenitis in acute HHV-7 infection shows similar follicular hyperplasia with many cells carrying HHV-7 antigen (red cells in immunostains). (B) Diffuse paracortical hyperplasia is characterized by prominent activation of endothelium in postcapillary venules (center of figure).

apoptotic cell death of infected target cells is seen with or without inflammatory reaction in sporadic cases of primary acute hepatitis, encephalitis, or myocarditis.[20–22] Such cases generally occur during primary HHV-6 infections of immunocompromised individuals.

Viral Persistence and Chromosomal Integration

Primary HHV-6 and HHV-7 infection cause lifelong persistence of the virus with occasional, clinically occult reactivation in some 25% of infected

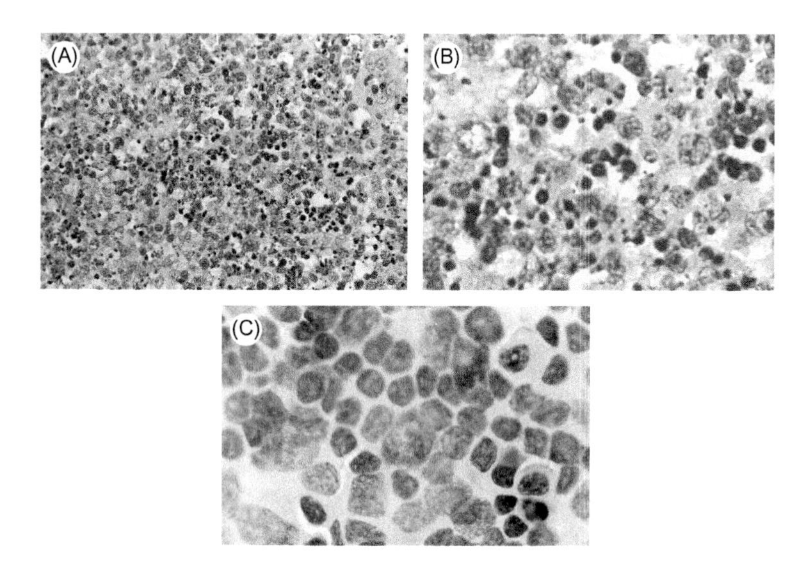

FIGURE 3.8 (A, B) Necrotizing histiocytic lymphadenitis (Kikuchi's syndrome) in a patient with acute HHV-6 infection. Note prominent lymphocyte necrosis and histiocyte reaction. (C) Immunostains on imprints from lymph node cut surfaces show HHV-6 antigen carrying cells (cells stained red).

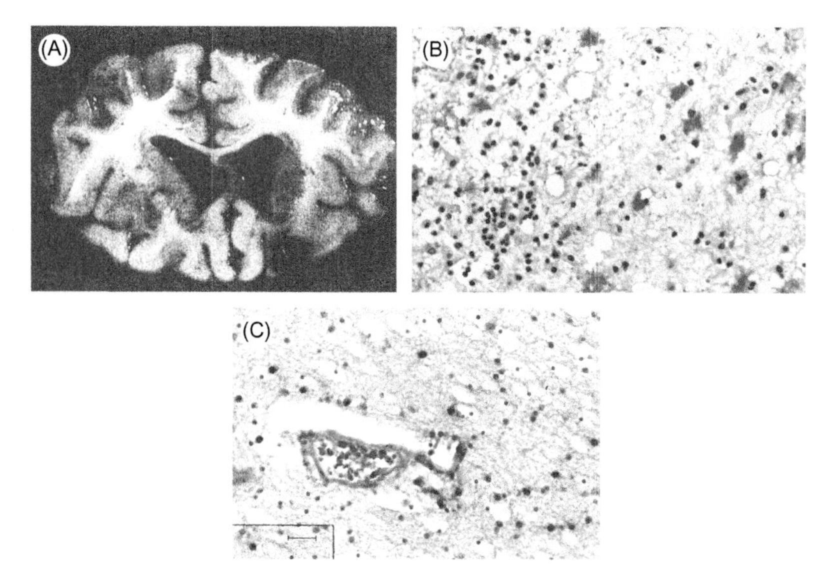

FIGURE 3.9 Lethal necrotizing encephalitis in a child with immune deficiency and acute HHV-6 infection. (A) Cross-section of the brain at autopsy shows multiple dark foci of hemorrhage and necrosis. (B) Immunostains and (C) *in situ* hybridization of brain sections show many HHV-6 antigen (red) and DNA (black) carrying cells in necrotic areas with loose lymphocytic inflammatory response.

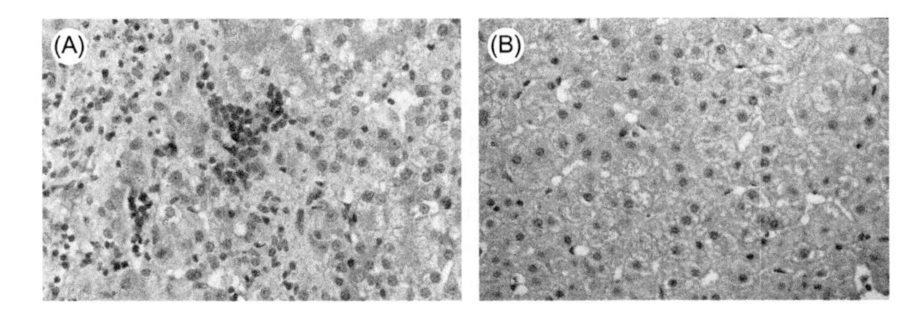

FIGURE 3.10 Acute hepatitis in an immunocompetent adult following late acute HHV-6 infection. The liver biopsy shows (A) mild focal lymphocytic infiltrates and (B) occasional hepatocyte-carrying HHV-6 p41 antigen (immunostain, red cell).

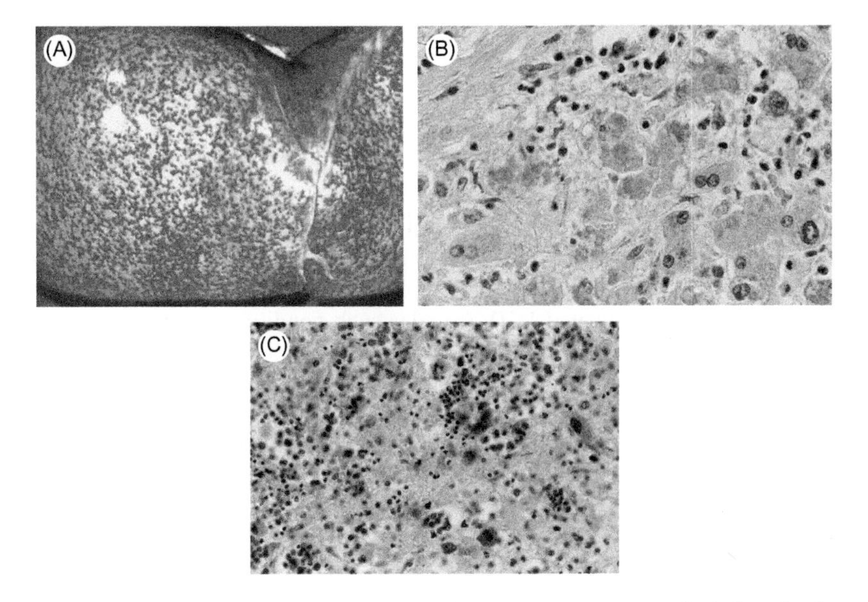

FIGURE 3.11 Lethal necrotizing hepatitis in a baby. (A) Gross appearance of pale liver showing confluent yellowish necrotic areas. (B) Marked degeneration of hepatocytes and scattered inflammatory infiltrate in light microscopy. (C) Immunostains show many cells carrying HHV-6 p41 antigen (paraffin sections; red cells).

individuals at any given time. HHV-6 persists in salivary glands and possibly also in terminal bronchi and neuroglial cells.[23–25] HHV-7 apparently persists in CD4 T lymphocytes and also, to a certain extent, in the epithelia of salivary glands, bronchi, and cells of the skin, liver, and kidney. Neural cells were found to be negative for a tegument antigen when serving as indicators of persistence.[26] Chromosomal integration of HHV-6 (ciHHV-6, both HHV-6A and HHV-6B) was shown in 0.2 to 1.0% of blood donors and in even higher percentages of various patient groups.[27–33] Further details on chromosomal integration can be found in Chapter 15.

The pathophysiological role of viral persistence and chromosomal integration urgently deserves further investigation. Although viral persistence, latency, and chromosomal integration may remain clinically obscure (for endogenous reactivation, see next paragraph), we have repeatedly shown that HHV-6 structural antigens and DNA can be found in various tissues by immunohistochemical and molecular methods, usually at low levels. Also, incomplete viral antigen synthesis and defective viral replication can be observed in such conditions;[34,35] their pathophysiological relevance for the development of diseases (e.g., autoimmune diseases) should be studied.

Pathologic Lesions in Reactivation and Endogenous Reinfection

Reactivation of HHV-6 and HHV-7 may arise due to other infections or exposure to endotoxins, endocrine stimulation (including stress situations), certain cytokines, or plant and food components (e.g., agglutinins, phorbol esters, aflatoxin).[36] Reactivation *per se* does not cause disease unless the virus persists and replicates over longer periods of time secondary to immune deficiency, systemic autoimmune disorders, or tumor growth. In essence, two main conditions support a pathogenic HHV-6 reactivation with reinfection of tissues: (1) continued abnormal stimulation of cells carrying viral genomes and (2) defective host control of virus replication and spread. Pathologic persistent activity of reactivated HHV-6 under such conditions coincides with a number of diseases in various organs.

Morphology of Organ Systems in Reactivated Infections

Cardiovascular System

(See Figures 3.12 and 3.13 and Chapter 16.) HHV-6 infects endothelial cells of various vessels and capillaries.[37–39] It induces endothelial dysfunction, thrombotic microangiopathy, or large vessel arteritis.[38–40] There are reports relating HHV-6 infection to various vascular diseases including leukocytoclastic vasculitis, "coronaritis" in transplant recipients, and Kawasaki's disease.[41–44] Both HHV-6 and HHV-8 have been detected in coronary atherosclerosis.[45] Endothelial dysfunction with subsequent hypoxic cardiac damage is apparently also the cause of myocarditis and dilated cardiomyopathy.[46–48] HHV-6 genomic material was found in coronary arteries of heart allografts, suggesting virus reactivation at this site,[49–51] and fulminant myocarditis has been described in both immunocompetent and incompetent patients.[22,52] HHV-6 antibodies showed a fourfold increase in one patient, and HHV-6 DNA was demonstrated by PCR in the liver and heart tissue. Chromosomally integrated HHV-6 (ciHHV-6) was demonstrated in patients with cardiomyopathy and heart failure syndromes.[53] Strenger and colleagues[35] reported the presence of the HHV-6B genome in cardiac tissues in cases with ciHHV-6.

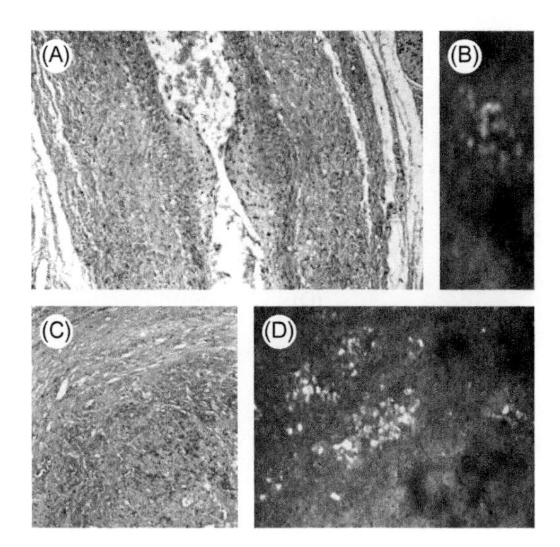

FIGURE 3.12 Vascular changes in HHV-6 infection: (A) Initial endothelial proliferative cushions in cardiac allograft containing (B) HHV-6 DNA (*in situ* PCR; green speckles). Microscopy of artery in Kawasaki's disease showing (C) severe necrotizing and proliferative arteritis and (D) upon *in situ* PCR larger amounts of HHV-6 DNA.

Respiratory System

(See Figure 3.14.) An excellent review of HHV-6 and the respiratory system was published by Schmidt and colleagues[54] in the last edition of this book. Upper respiratory problems are typical for acute (primary) infection, including rhinitis, sinusitis, and pharyngitis accompanied by lymphoid hyperplasia. Diseases of the lower respiratory tract (rare in primary infections but more common in reactivated and persistent infections) include interstitial pneumonia, such as nonspecific interstitial pneumonitis (NSIP) and lymphocytic interstitial pneumonitis (LIP),[55–58] as well as pneumonitis related to organ transplantation[59–60] or to autoimmune disorders (e.g., collagen-vascular diseases, Sjögren's syndrome, systemic lupus erythematosus). A rare case of lethal acute respiratory distress syndrome (ARDS) was described in a young immunocompetent woman with HHV-6 interstitial pneumonitis.[61] In addition, active HHV-6 may add to the pulmonary pathology of *Legionella* and *Pneumocystis* infections[62,63] and probably to other infections such as by EBV, CMV, and measles virus.

Gastrointestinal System

(See Figure 3.15.) Intestinal discomfort and diarrhea are fairly frequent symptoms in HHV-6 infection;[13,64] however, because complaints are transient and biopsies are not taken, pathology is generally not available for individual cases. Bloody diarrhea with HHV-6 DNA in nuclei of large intestinal goblet

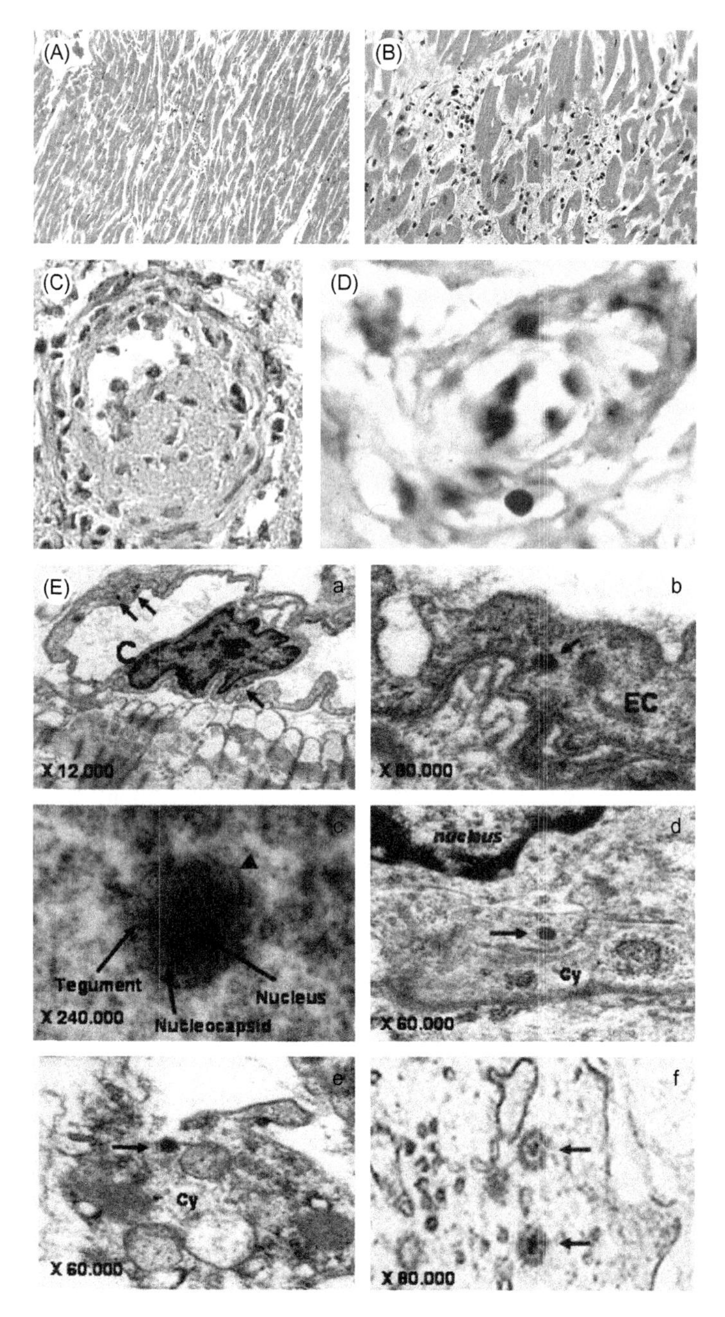

FIGURE 3.13 Myocarditis and dilative cardiomyopathy in HHV-6 infection: (A, B) Scattered interstitial inflammatory infiltrate. (C) Small vessels contain occasional platelet clots, and (D) their endothelial cells may contain HHV-6 DNA (*in situ* hybridization). (E) Electron microscopy may show occasional herpesvirus particles in endothelial cells and in macrophages.

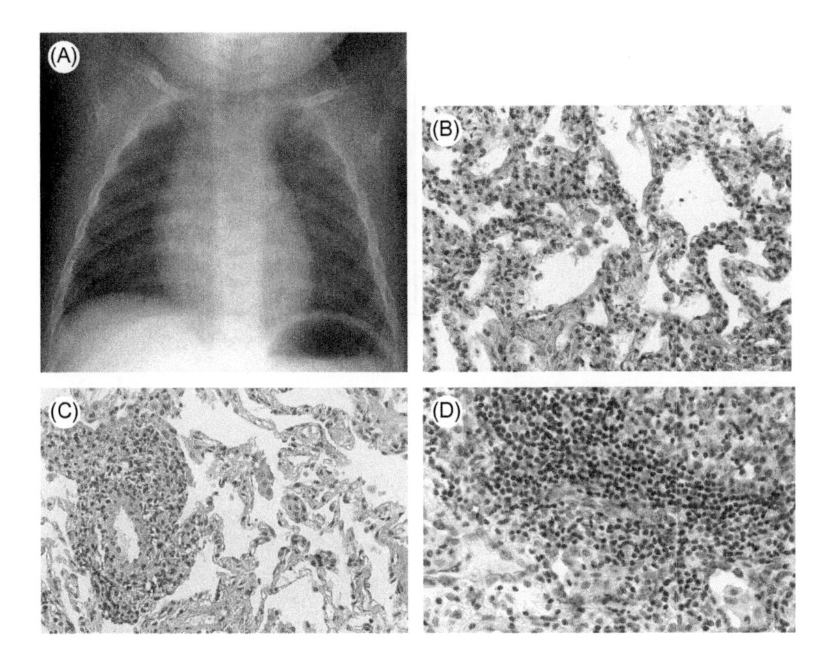

FIGURE 3.14 Interstitial pneumonitis in a child with HHV-6 infection: (A) Thorax X-ray shows the typical symmetrical soft reticular pattern. (B, C) Microscopy reveals interstitial and perivascular lymphocytic infiltrates with or without (D) prominent follicular lymphoid hyperplasia. The latter photograph also shows endothelial cells and alveolar macrophages carrying HHV-6 p41 antigen (red cells).

cells has been reported repeatedly in transplant recipients (e.g., stem cell transplants).[65–67] Similarly, cases of gastroenteritis have been reported in solid organ recipients with virus reactivation.[68–70] Ulcerative colitis is frequently associated with reactivated HHV-6B and CMV in the colonic mucosa, and virus reactivation appears to correlate well with the severity of the disease.[71] HHV-6 is also found in patients with Crohn's disease.[72] HHV-6-induced mesenteric lymphadenopathy has caused intussusception of the small intestine in some pediatric patients.[73,74]

Liver, Biliary System, and Pancreas

(See Figure 3.16.) HHV-6-induced heterophile-negative infectious mononucleosis is commonly accompanied by elevation of the liver enzymes serum glutamic oxaloacetic transaminase (SGOT), gamma-glutamyltransferase (gamma-GT), and lactate dehydrogenase (LDH),[13] indicating some liver involvement in this acute infection. HHV-6A p41 antigen has been shown in sporadic hepatocytes. Similarly, Asano et al.[73] found evidence of hepatic injury in 4% of children with primary HHV-6 infection and exanthema subitum. In addition, HHV-6-related liver damage was reported to accompany

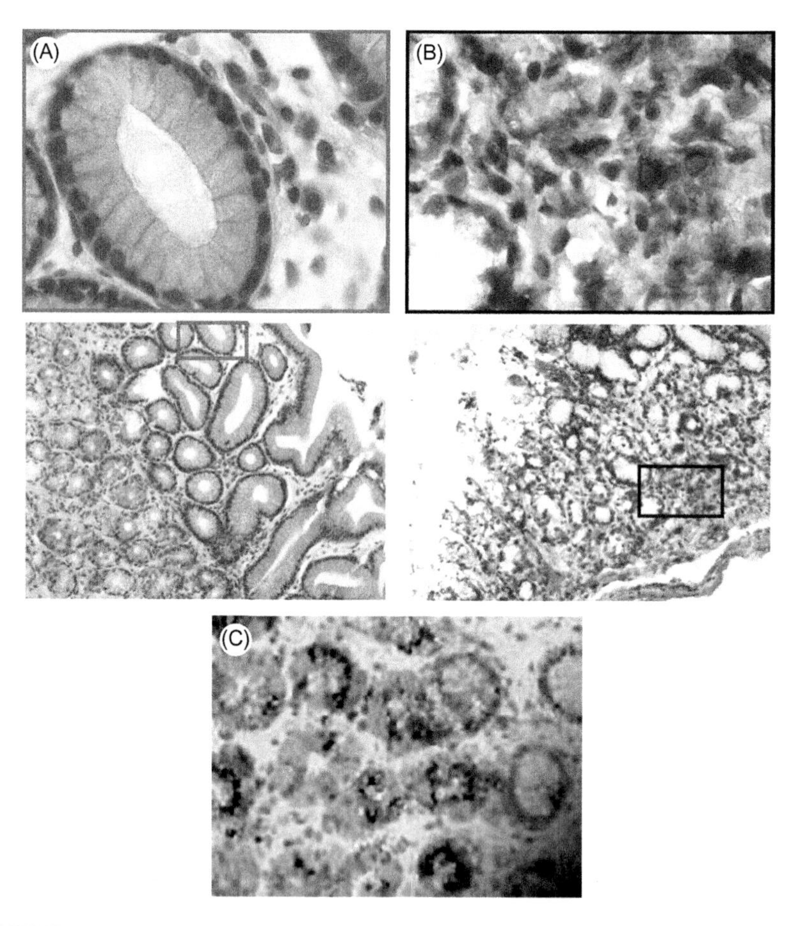

FIGURE 3.15 Intense HHV-6 infection of gastroduodenal mucosa in a 36-year-old patient with primary sclerosing cholangitis. (A, B) Red cells show HHV-6 antigens by immunostaining. (Reproduced with permission from Halme L et al. Human herpesvirus 6 infection of the gastroduodenal mucosa. *Clin Infect Dis* 2008; **46**: 434–439.) (C) *In situ* hybridization in a case of chronic gastritis "of unknown origin" shows deposits of HHV-6 DNA in epithelial cells.

various other disorders such as virus-associated hemophagocytic syndrome (VAHS), drug-induced hypersensitivity syndrome (DHS), and Gianotti–Crosti syndrome.[64,75–77] Some patients were overtly immunodeficient.[78]

The mechanism of liver injury in all such diseases appears to be complex. Inflammatory cytokines (e.g., IL-8, IL-6, TNF) stimulated by virus infection can induce liver damage as well as direct virus toxicity to hepatocytes.[64,79] HepG2 cells in culture are permissive for HHV-6 infection,[80] and HHV-6 DNA and antigens have been detected in hepatocytes and in vascular endothelium.[20,81,82] Consequently, primary viral hepatitis (acute and chronic) as well as autoimmune hepatitis (AIH) can be expected to develop in HHV-6 infections.

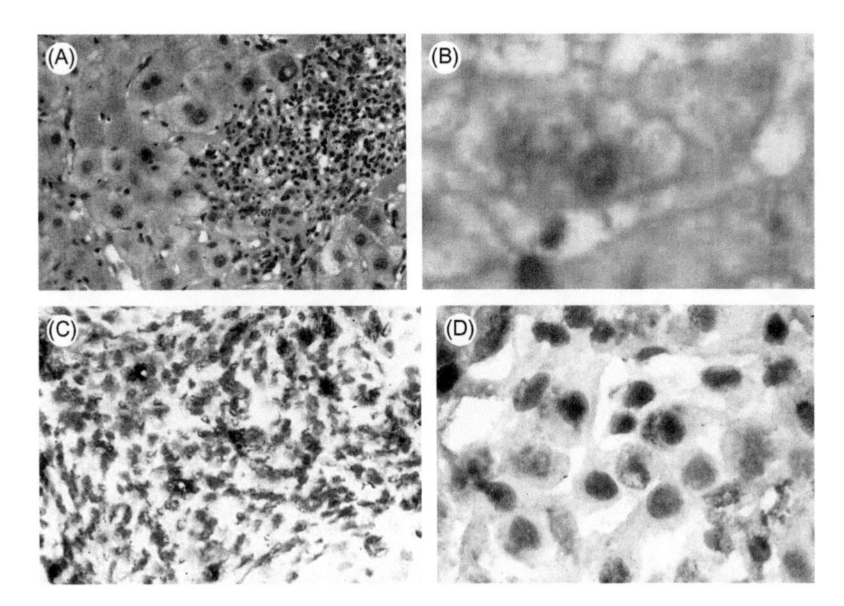

FIGURE 3.16 Severe autoimmune hepatitis shows (A) deposits of HHV-6 p41 antigen by (B) immunostaining (red stain in hepatocyte). A case of acute liver failure (ALF) in an adult shows (C) extensive hepatocyte HHV-6 antigen deposits and (D) numerous HHV-6 antigen-positive mononuclear cells in the portal inflammatory infiltrate (frozen sections, immunostains). *(Reproduced with permission from Härmä M et al. Transplantation 2003;* **76***: 536–539).*

Acute fulminant hepatitis caused by HHV-6 has been repeatedly observed in both children and adults.[64,82] HHV-6 has been associated with acute liver failure, and viral antigens or DNA have been demonstrated in hepatic tissue.[83,84] A case where HHV-6 DNA was shown in vascular endothelium instead of hepatocytes mimicked Reye's syndrome both clinically and pathologically.[20] In addition, there are many reports of liver failure and chronic hepatitis in HHV-6 and occasionally in HHV-7 infection, most frequently in transplants.[85–93] Grima and colleagues[94] described an 18-year-old woman suffering from HHV-6 infection and hepatitis with autoimmune features. Potenza and coworkers[95] reported a case of post-infantile giant cell hepatitis in HHV-6 infection. Both reports support the notion of an autoimmune pathogenesis of HHV-6-induced liver disease. In addition, we have shown[96] that HHV-6 seropositivity with HHV-6 antigen in the liver apparently supports the development of autoantibodies—antinuclear antibodies (ANAs), antimitochondrial antibodies (AMAs), and smooth muscle antibodies (SMAs)—and thus may complicate the course of hepatitis B virus (HBV) and alcoholic liver disease (hepatitis C virus was not overtly influenced).

High intrahepatic HHV-6 viral loads, but not the presence of CMV or EBV, are associated with liver allograft hepatitis and decreased graft survival.[97] In the biliary system, sclerosing cholangitis can occur in allograft recipients and

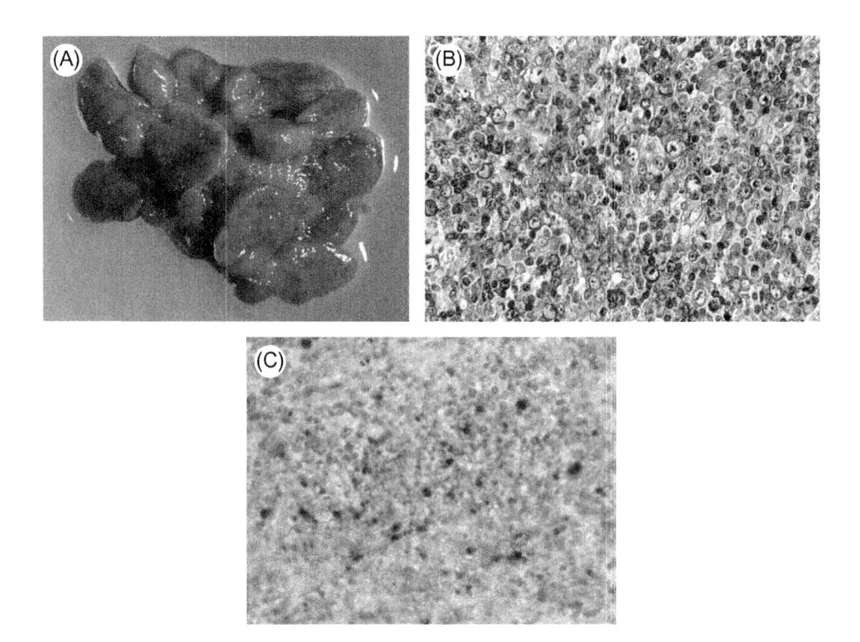

FIGURE 3.17 Atypical polyclonal lymphoproliferation in a patient with persistent active HHV-6 infection and commonly certain immune deficiency. The disease may proceed to overt lymphoma (e.g., in allograft recipients or in patients with HIV infection). It is frequently accompanied by coincident Epstein–Barr virus reactivation. (A) Gross appearance of resected axillary lymph nodes; lymph nodes are enlarged while their structure is widely preserved. (B) Microscopy shows the severe diffuse blast cell proliferation typical of persistent infectious mononucleosis. (C) *In situ* hybridization reveals scattered cells containing HHV-6 DNA.

in HIV-positive individuals secondary to CMV and HHV-6 reactivation.[98] Whether HHV-6 is directly involved or aggravates the effects of CMV still requires further clarification. The same pertains to acute pancreatitis in these patients.

Hematopoietic and Lymphatic System

(See Figures 3.17 to 3.19 and Chapter 17.) In chronic persistent HHV-6 infection, viral DNA load and cellular changes show certain cyclic changes, suggesting some fluctuation in viral replication.[99] In about 6% of cases, heterophile-negative infectious mononucleosis is caused by HHV-6A or B infection.[18,19,100] More frequent is the reactivation of latent HHV-6 in patients with classical EBV-induced infectious mononucleosis resulting in a more protracted course of the disease with elevated liver enzymes.[101] Occasionally caused by HHV-6 (preferentially HHV-6B) are angioimmunoblastic lymphadenopathy,[102,103] hemophagocytic syndromes,[104,105] and Langerhans cell histiocytosis.[106]

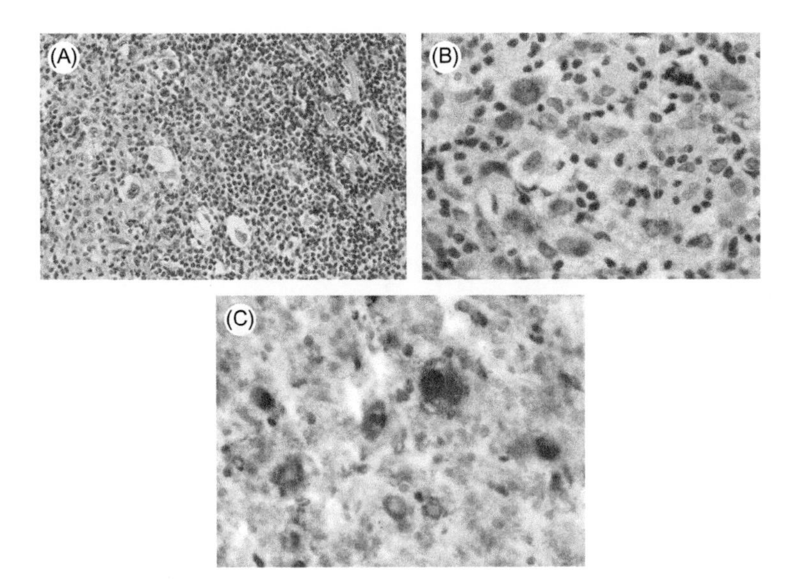

FIGURE 3.18 (A) Nodular sclerosing Hodgkin's disease shows (B) HHV-6 antigen (red cells) and (C) Epstein–Barr virus antigen in immunostaining.

There is an ever-growing number of (somewhat controversial) publications on HHV-6 in malignant lymphomas, including subtypes of Hodgkin's disease,[107–121] as well as in prelymphomatous atypical polyclonal lymphoproliferations, including Canale–Smith syndrome and Rosai–Dorfman syndrome.[122,123] Araujo and collaborators[124] reported an association between HHV-6 infection and chronic granulomatous disease in children. Part of the controversy is of a schematic nature, as etiologic and co-pathogenic activities of HHV-6 in lymphomagenesis have not been clearly distinguished.[125,126] This aspect will require more serious attention in future studies. Because oncogenic effects of the HHV-6 genome[127,128] have not been unequivocally proven, other mechanisms may account for atypical lymphoproliferation such as persistent antigenic stimulation or immune dysfunction.[125,129]

Both subtypes of HHV-6 and HHV-7 infect hematopoietic stem cells and suppress their engraftment in transplant recipients.[130–136] Frequent HHV-6 infections with high virus load have been reported in bone marrow and stem cell transplant recipients.[137–140] In addition to failure of engraftment, lymphocytopenia, suppression of myelopoiesis, or erythrocytopenia may result from HHV-6 reactivation, and the virus can even be transmitted by the graft itself.[135,141] HHV-6 encephalitis was observed to follow hematopoietic stem cell transplantation.[142,143] Also, HHV-6 antigen has been found in the bone marrow cells of patients with myelodysplasia and certain chronic myeloproliferative disorders.[144,145] Because this finding pertains equally to EBV and

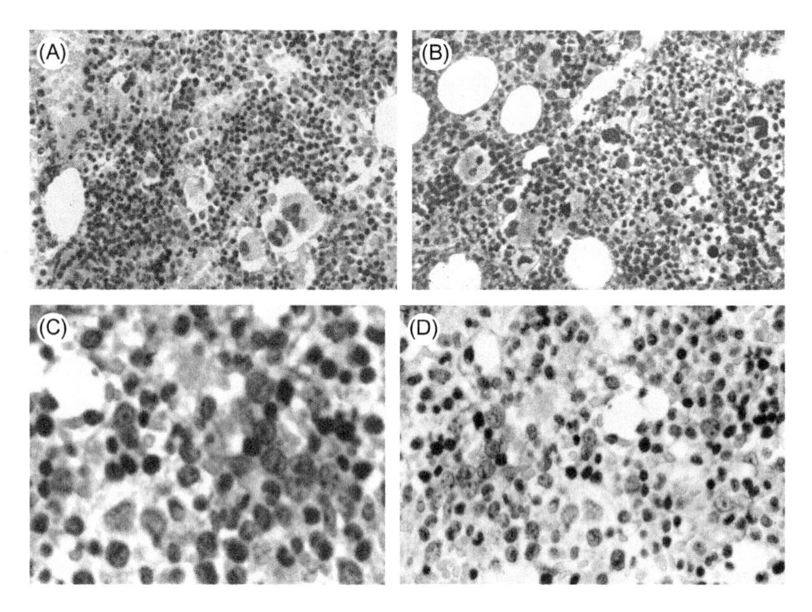

FIGURE 3.19 (A) Myelodysplastic syndrome and (B) chronic myeloproliferative diseases may contain groups of scattered hematopoietic cells containing (C, D) HHV-6 p41 antigen by immunostaining (red intracytoplasmic deposits).

HHV-7, a deficient clearance of virus may be responsible for this phenomenon rather than it indicating some etiologic relationships.

Endocrine System

Both HHV-6 and HHV-7 have been demonstrated in thyroid tissue of patients with autoimmune thyroiditis (e.g., Hashimoto's disease), yet their pathogenetic involvement in these diseases still requires elucidation.[146,147] Autoimmune thyroiditis may also occur in conjunction with other diseases, such as drug-associated hypersensitivity diseases.[148]

Urogenital System

(See Figure 3.20.) Although tubular epithelial cells of the kidney express the receptor CD46 for HHV-6 and HHV-6 antigen has been shown in these cells,[6,149] primary renal diseases being caused by HHV-6 or HHV-7 has not yet been documented. There is accumulating evidence, however, that HHV-6 reactivation frequently occurs in renal transplantation and may add to CMV pathology.[150–153] Schroeder et al.[154] suspected that, occasionally, HHV-6 may also cause lesions in renal allograft recipients independent of CMV. Viruses including HHV-6 may invade the amniotic cavity in pregnancy and cause abortion.[155,156]

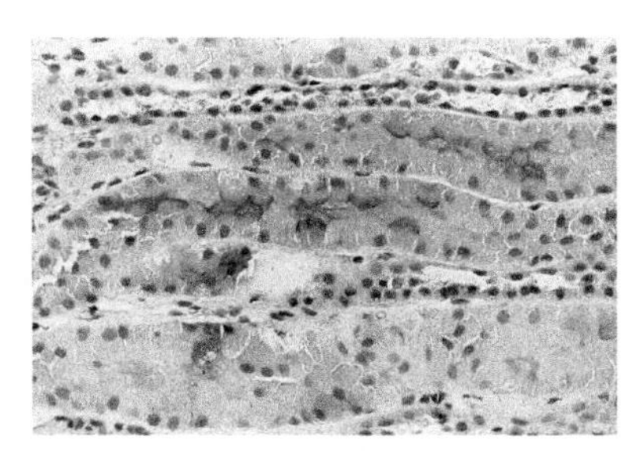

FIGURE 3.20 HHV-6 p41 antigen is occasionally shown in renal tubular cells, yet no clear-cut association to renal diseases has been established so far. (In renal allografts, HHV-6 appears to enhance pathologic effects by human cytomegalovirus reactivation).

Skin and Soft Tissues

(See Figure 3.21.) The characteristic skin reaction exanthema subitum (ES) in primary HHV-6 and HHV-7 infections was mentioned above; HHV-6B is isolated most frequently in ES, and occasionally also HHV-6A or HHV-7.[157,158] HHV-6 reactivation, such as in transplant recipients, in AIDS or chronic fatigue syndrome (CFS) patients, or in other immune-deficient or autoimmune patients (e.g., systemic lupus erythematosus) can cause nonspecific skin reactions that are infectious mononucleosis like or mimic a graft versus host reaction (GVHR).[157] Other skin afflictions related to HHV-6 are leukocytoclastic vasculitis[41] and Langerhans' cell histiocytosis.[106] We have observed a 20-year-old sportsman who developed symptoms of CFS and suffered from severe panniculitis nodularis non suppurativa (Pfeiffer–Weber–Christian's disease). HHV-6 IgA antibody titers were 1:320 and HHV-6A p41 antigen was shown in sweat gland cells and in some interstitial mononuclear cells. The patient recovered completely after 8 months on steroid therapy. Finally, there are accumulating reports of drug reaction with eosinophilia and systemic symptoms (DRESS) occurring preferentially in patients with HHV-6 reactivation.[159–166]

Soft-tissue diseases in active HHV-6 infections are discussed in more detail in the following section on autoimmune disorders. Rhabdomyolysis was observed in a patient with primary HHV-6 infection.[167] In addition, the effect of HHV-6 on skeletal muscles and connective tissue is indicated by symptoms found in chronic fatigue syndrome and in the ill-defined fibromyalgia syndrome. We have identified single fiber degeneration in skeletal muscle biopsies from patients with a very mild inflammatory response.

Central Nervous System

(See Figure 3.22 and Chapters 5 and 7.) HHV-6 replicates with low efficiency in neuroglial cells.[168] Viral DNA and antigen have been successfully demonstrated in human brain tissue, both in healthy organs and in diseased tissues, with HHV-6A being about three times more frequent than HHV-6B.[168–170]

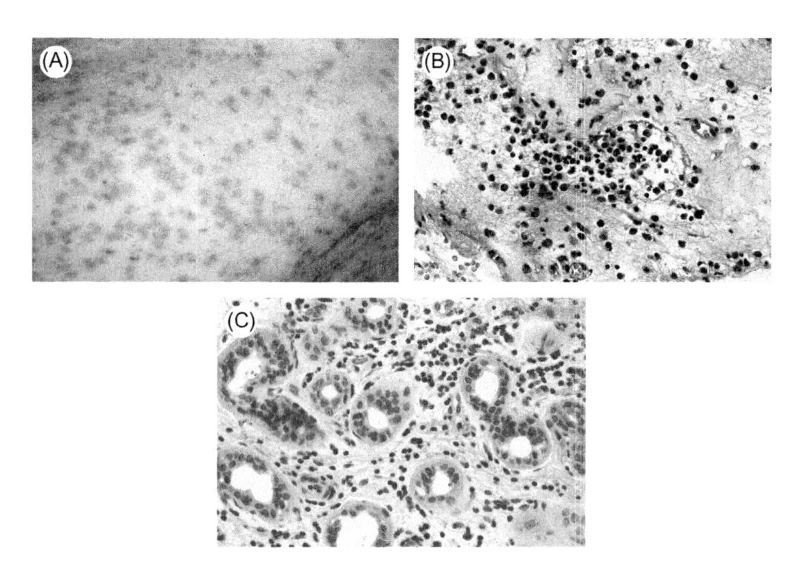

FIGURE 3.21 Drug reaction with eosinophilia and systemic syndromes (DRESS) frequently coincides with HHV-6 reactivation. (A) Allergic exanthema in such a patient. (B) Severe eosinophilia in nasal mucosa with coincident allergic rhinitis. (C) In a case with severe systemic panniculitis, HHV-6 p41 antigen was shown in sweat gland epithelia.

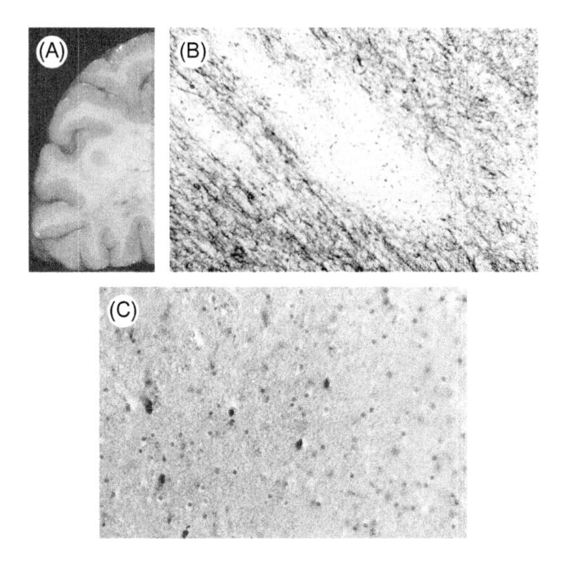

FIGURE 3.22 Acute exacerbations of multiple sclerosis (MS) are often accompanied by reactivation of HHV-6. (A) Gross section at autopsy of the brain of an MS patient; the arrow points to a pink focus of fresh demyelinization. (B) Kluwer–Barrera staining for myelin sheaths (blue-black) outlines the extent of demyelinization. (C) *In situ* hybridization at autopsy may identify HHV-6 DNA.

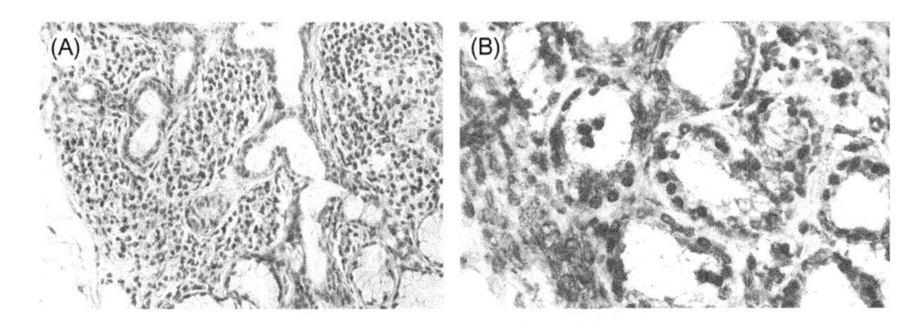

FIGURE 3.23 Sjögren's syndrome (SS) is a systemic autoimmune disorder in which HHV-6 reactivation may often occur. (A) Early diagnosis is assisted by biopsy of lip salivary glands, which show the typical inflammatory and degenerative pattern. (B) Immunostaining for HHV-6 p41 antigen reveals deposits of the antigen in salivary epithelia and in some interstitial cells.

HHV-6 may enter the brain through the olfactory pathway.[171] There are increasing reports of HHV-6-associated meningitis, encephalitis in children with febrile seizures,[172–179] acute necrotizing or hemorrhagic encephalitis, limbic encephalitis, epilepsy, and demyelinating brain diseases in immune-deficient patients and in persons suffering from multiple sclerosis.[21,180–190] CNS infections with HHV-6 appear more frequently in patients with T-cell immune deficiency.[191] A child with Dravet syndrome was observed to also be suffering from HHV-6 associated encephalopathy.[192]

AUTOIMMUNE DISORDERS

(See Figures 3.23 and 3.24.) Persistent HHV-6 may cause functional disturbances of the immune system as indicated by elevated antibody titers against HHV-6 in allergies and drug-induced hypersensitivity reactions, as well as in systemic lupus erythematosus, Sjögren's syndrome, and progressive systemic sclerosis.[13,157,159,160,193–196] Idiopathic arthritis, autoimmune hemolytic anemia, and neutropenia are other disorders reported.[141,197] Autoimmune thyroiditis was mentioned above under endocrine diseases. Virus persistence may constitute a risk factor for additional immune dysregulation and for increasing the severity of adverse reactions.[198]

VIRUS INTERACTIONS

It is important to note that HHV-6 can apparently activate other viral infections, such as those induced by EBV, CMV, HIV-1, and measles, and also apparently papillomavirus and parvovirus. HHV-6 may thus contribute to the pathologic effects of these viruses.[199–209] Dual active infections appear especially frequently with other herpesviruses (CMV, EBV, HHV-7) as well as with HIV-1. We have observed such EBV and HHV-6 coinfection and suggestive cooperation in Hodgkin's disease.[116]

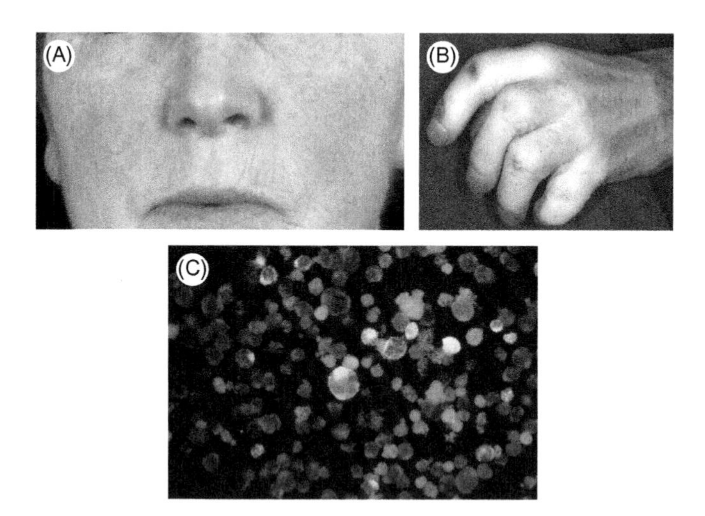

FIGURE 3.24 Other systemic autoimmune disorders in which HHV-6 reactivations are reported are (A) systemic lupus erythematosus (SLE) and (B) progressive systemic sclerosis (PSS, or scleroderma). When peripheral blood lymphocytes from both patient groups are co-cultured with susceptible (e.g., HSB2) cells, HHV-6 virus can be isolated in a fair percentage. (C) Indirect immunofluorescence for HHV-6 antigen of cultured lymphocytes from such patients.

REFERENCES

1. Santoro F, Kennedy PE, Locatelli G, et al. CD46 is a cellular receptor for human herpesvirus 6. *Cell* 1999;**99**:817–27.
2. Lusso P, Secchiero P, Crowley RW, et al. CD4 is a critical component of the receptor for human herpesvirus 7: interference with human immunodeficiency virus. *PNAS* 1994;**91**:3872–7.
3. Zhou ZH, Stoops JK, Krueger GRF. Ultrastructure and assembly of human herpesvirus-6 (HHV-6). In: Krueger G, Ablashi DV, editors. *Human Herpesvirus-6: General Virology, Epidemiology and Clinical Pathology*. Amsterdam: Elsevier; 2006. p. 11–21.
4. Klussmann JP, Krueger E, Sloots T, et al. Ultrastructural study of human herpesvirus-7 in tissue culture. *Virchows Arch* 1997;**430**:417–26.
5. Krueger GRF, Ablashi DV. Human herpesvirus-6: a short review of its biological behavior. *Intervirology* 2003;**46**:257–69.
6. Krueger GRF, Schneider B. Pathologic features of HHV-6 disease. In: Krueger G, Ablashi DV, editors. *Human Herpesvirus-6: General Virology, Epidemiology and Clinical Pathology*. Amsterdam: Elsevier; 2006. p. 133–48.
7. Biberfeld P, Kramarsky B, Salahuddin SZ, et al. Ultrastructural characterization of a new B-lymphotropic DNA virus (human herpesvirus 6) isolated from patients with lymphoproliferative disease. *J Nat Cancer Inst* 1987;**79**:933–41.
8. Kramarsky B, Sander C. Electron microscopy of human hyerpesvirus-6 (HHV-6). In: Ablashi DV, Krueger GRF, Salahuddin SZ, editors. *Human Herpesvirus 6*. Amsterdam: Elsevier; 1992. p. 59–68.

9. Boehmer S. Biological Effect of Human Herpesvirus-6 on Cell Populations of Different Phenotypes, doctoral thesis at University of Cologne Medical Faculty, Hundt Druck, Cologne, Germany, 1987, pp. 1–219.
10. Kirn E, Krueger E, Boehmer S, Klussmann JP, Krueger GRF. In vitro cytobiological effects of human herpesvirus 6 and 7: immunohistological monitoring of apoptosis, differentiation and cell proliferation. *Anticancer Res* 1997;**17**:4623–32.
11. Asano Y, Yoshikawa T, Suga S, et al. Severity of human herpesvirus-6 viremia and clinical findings in infants with exanthema subitum. *J Pediat* 1991;**118**:891–5.
12. Hall CB, Long CE, Schnabel KC, et al. Human herpesvirus 6 infection in children. A prospective study of complications and reactivation. *N Engl J Med* 1994;**331**:432–8.
13. Krueger GRF, Klueppelberg U, Hoffmann A, et al. Clinical correlates of infection with human herpesvirus-6. *In Vivo* 1994;**8**:457–86.
14. Wiersbitzky S, Ratzmann GW, Bruns R, et al. Reactivation in children of juvenile chronic arthritis and chronic iridocyclitis associated with human herpesvirus 6. *Pädiatr Grenzgeb* 1993;**31**:203–5.
15. Maric I, Bryant R, Abu-Asab M, et al. Human herpesvirus 6 associated acute lymphadenitis in immunocompetent adults. *Mod Pathol* 2004;**17**:1427–33.
16. Kikuchi M, Sumiyoshe Y, Minamishima Y. Kikuchi's disease (histiocytic necrotizing lymphadenitis). In: Ablashi DV, Krueger GRF, Salahuddin SZ, editors. *Human Herpesvirus-6: Epidemiology, Molecular Biology, and Clinical Pathology*. Amsterdam: Elsevier; 1992. p. 175–83.
17. Sumiyoshi Y, Kikuchi M, Ohshima K, et al. Human herpesvirus-6 genomes in histiocytic necrotizing lymphadenitis (Kikuchi's disease). *Am J Clin Pathol* 1993;**99**:609–14.
18. Steeper TA, Horwitz CA, Ablashi DV, et al. The spectrum of clinical and laboratory findings resulting from human herpesvirus-6 (HHV-6) in patients with mononucleosis-like illnesses not resulting from Epsein–Barr virus or cytomegalovirus. *Am J Clin Pathol* 1990;**93**:776–83.
19. Horwitz CA, Krueger GRF, Steeper TA, et al. HHV-6 induced mononucleosis-like illnesses. In: Ablashi DV, Krueger GRF, Salahuddin SZ, editors. *Human Herpesvirus-6: Epidemiology, Molecular Biology, and Clinical Pathology*. Amsterdam: Elsevier; 1992. p. 159–74.
20. Aita K, Jin Y, Irie H, et al. Are there histopathological characteristics particular to fulminant hepatic failure caused by human herpesvirus 6 infection? A case report and discussion. *Hum Pathol* 2001;**32**:887–9.
21. Wagner M, Müller-Berghaus J, Schroeder R, et al. Human herpesvirus-6 (HHV-6) associated necrotizing encephalitis in Griscelli's syndrome. *J Med Virol* 1997;**53**:306–12.
22. Fukae S, Ashizawa N, Morikawa S, et al. A fatal case of fulminant myocarditis with human herpesvirus 6 infection. *Intern Med* 2000;**39**:632–6.
23. Fox JD, Briggs M, Ward PA, et al. Human herpesvirus 6 in salivary glands. *Lancet* 1990;**336**:590–3.
24. Krueger GR, Wassermann K, De Clerck LS, et al. Latent herpesvirus-6 in salivary and bronchial glands. *Lancet* 1990;**336**:1255–6.
25. Donati D, Martinelli E, Cassiani-Ingoni R, et al. Variant specific tropism of HHV-6 in human astrocytes. *J Virol* 2005;**79**:9439–48.
26. Kempf W, Adams V, Mirandola P, et al. Persistence of human herpesvirus 7 in normal tissues detected by expression of a structural antigen. *J Infect Dis* 1998;**178**:841–5.
27. Pellett PE, Ablashi DV, Ambros PF, et al. Chromosomally integrated human herpesvirus 6: questions and answers. *Rev Med Virol* 2012;**22**:144–55.

28. Arbuckle JH, Medveczky PG. The molecular biology of human herpesvirus-6 latency and telomere integration. *Microbes Infect* 2011;**13**:731–41.
29. Giroux M., Gautheret-Dejean A., Dewilde A., et al.Human herpesvirus-6 chromosomal integration and Bing–Neel syndrome revealing a Waldenstroem's disease: incidental association? Presented at the 7th International Conference on HHV-6 & 7, Reston, VA, February 27–March 2, 2011.
30. Malnati M., Broccolo F., Di Marco E., et al.A simple and fast method to diagnose chromosomal integration of human herpesvirus 6 in circulating mononuclear cells. Presented at the 7th International Conference on HHV-6 & 7, Reston, VA, February 27–March 2, 2011.
31. Gravel A, Sinnett D, Flamand LFrequency of chromosomally integrated human herpesvirus 6 in children with acute lymphoblastic leukemia. Presented at the 7th International Conference on HHV-6 & 7, Reston, VA, February 27–March 2, 2011.
32. Kaufer BB, Egerer A, Jarosinski KW, et al.Herpesvirus telomeric repeats facilitate genomic integration into host telomeres and mobilization of viral DNA during reactivation. Presented at the 7th International Conference on HHV-6 & 7, Reston, VA, February 27–March 2, 2011.
33. Arbuckle JH, Medveczky MM, Pantry S,et al.The characterization and mapping of the integrated human herpesvirus-6A and 6B genome. Presented at the 7th International Conference on HHV-6 & 7, Reston, VA, February 27–March 2, 2011.
34. Strenger V, Lautenschalegr I, Richter S, et al.Detection of HHV-6 antigen and herpesvirus particles in individuals with chromosomally integrated HHV-6 (ciHHV-6). Presented at the 7th International Conference on HHV-6 & 7, Reston, VA, February 27–March 2, 2011.
35. Strenger V, Aberle SW, Wendelin G, et al. Chromosomal integration of the HHV-6 genome as a possible cause of HHV-6 detection in cardiac tissues. *J Clin Pathol* 2010;**63**:1129–30.
36. Krueger GRF, Ablashi DV, Whitman JI, et al. Clinical pathology and reactivation of human lymphotropic herpesviruses. *Rev Med Hosp Gen Mexico* 1998;**61**:226–40.
37. Rotola A, DiLuca D, Cassai E, et al. Human herpesvirus 6 infects and replicates in aortic endothelium. *J Clin Microbiol* 2000;**38**:3135–6.
38. Wu CA, Shanley JD. Chronic infection of human umbilical vein endothelial cells by human herpesvirus-6. *J Gen Virol* 1998;**79**:1247–56.
39. Takatsuka H, Wakae T, Mori A, et al. Endothelial damage caused by cytomegalovirus and human herpesvirus 6. *Bone Marrow Transplant* 2003;**31**:475–9.
40. Toyabe S, Harada W, Suzuki H, et al. Large vessel arteritis associated with human herpesvirus 6 infections. *Clin Rheumatol* 2002;**21**:528–32.
41. Drago F, Rampini P, Brusati C, et al. Leukocytoclastic vasculitis in a patient with HHV 6 infection. *Acta Derm Venerol (Stockh)* 1999;**80**:68.
42. Hagiwara K, Yoshida T, Komura H, et al. Isolation of human herpesvirus 6 from an infant with Kawasaki's disease. *Eur J Pediatr* 1993;**152**:176.
43. Luka J, Gubin J, Auerbach C, et al. Detection of human herpesvirus-6 (HHV-6) genome by in situ PCR in tissues from Kawasaki disease and in coronary arteries of transplant hearts (abstract). Presented at the Third International Conference on Modern Methods in Analytical Morphology, Atlantic City, NJ, June 5–9, 1995.
44. Yoshikawa T, Akimoto S, Nishimura N, et al. Evaluation of active human herpesvirus 6 infection by reverse transcription-PCR. *J Med Virol* 2003;**70**:267–72.
45. Magnoni M, Malnati M, Cristell N, et al. Molecular study of human herpesvirus 6 and 8 involvement in coronary atherosclerosis and coronary instability. *J Med Virol* 2012;**84**:1961–6.

46. Comar M, D'Agaro P, Campello C, Poli A, Breinholt 3rd JP, Towbin JA, et al. Human herpes virus 6 in archival cardiac tissues from children with idiopathic dilated cardiomyopathy or congenital heart disease. *J Clin Pathol* 2009;**62**:80–3.
47. Krueger GRF, Rojo J, Buja LM, et al. Human herpesvirus-6 (HHV-6) is a possible cardiac pathogen: an immunopathological and ultrastructural study. *Rev Med Hosp Gen Mexico* 2008;**71**:187–91.
48. Leveque N, Boulagnon C, Brasselet C, et al. A fatal case of human herpesvirus 6 chronic myocarditis in an immunocompetent adult. *J Clin Virol* 2011;**52**:142–5.
49. Luka J, Auerbach C, Carson SD, et al. Detection of human herpesvirus-6 (HHV-6) genomes by *in situ* PCR and *in situ* hybridization in tissues from Kawasaki's disease and in coronary arteries of transplanted hearts. *J Clin Virol* 2006;**37**(Suppl. 1):S111.
50. Vallbracht KB, Schwimmbeck PL, Kuehl U, et al. Endothelium-dependent flow-mediated vasodilatation of systemic arteries is impaired in patients with myocardial virus persistence. *Circulation* 2004;**110**:2938–45.
51. Vallbracht KB, Schwimmbeck PL, Kuehl U, et al. Differential aspects of endothelial function of the coronary microcirculation considering myocardial virus persistence, endothelial activation and leukocyte infiltrates. *Circulation* 2005;**111**:1784–91.
52. Chang YL, Parker ME, Nuovo G, et al. Human herpesvirus 6-related fulminant myocarditis and hepatitis in an immunocompetent adult with fatal outcome. *Human Pathol* 2009;**40**:740–5.
53. Lassner D, Rhode M, Kuehl U, et al.Chromosomally integrated human herpesvirus 6 in patients with acquired cardiomyopathies and heart failure symptoms. Presented at the 7th International Conference on HHV-6 & 7, Reston, VA, February 27–March 2, 2011.
54. Schmidt SM, Wiersbitzky H, Wiersbitzky SKW. HHV-6 and the respiratory system. In: Krueger G, Ablashi DV, editors. *Human Herpesvirus-6: General Virology, Epidemiology, and Clinical Pathology*. Amsterdam: Elsevier; 2006. p. 173–83.
55. Cone RW, Hackman RC, Meei-Li WH, et al. Human herpesvirus 6 in lung tissue from patients with pneumonitis after bone marrow transplantation. *N Engl J Med* 1993;**329**:156–61.
56. Know KK, Pietryga D, Harrington DJ, et al. Progressive immunodeficiency and fetal pneumonitis associated with human herpesvirus 6 infection in an infant. *Clin Infect Dis* 1995;**20**:406–13.
57. Hammerling JA, Lambrecht RS, Kehl KS, et al. Prevalence of human herpesvirus 6 in lung tissue from children with pneumonitis. *J Clin Pathol* 1996;**49**:802–4.
58. Yamamoto K, Yoshikawa T, Okamoto S, et al. HHV-6 and 7 DNA loads in lung tissues collected from patients with interstitial pneumonia. *J Med Virol* 2005;**75**:70–5.
59. Buchbinder S, Elmaagacli AH, Schaefer UW, et al. Human herpesvirus 6 is an important pathogen in infectious lung disease after allogeneic bone marrow transplantation. *Bone Marrow Transplant* 2000;**26**:639–44.
60. Costa C, Curtoni A, Bergallo M, et al. Quantitative detection of HHV-6 and HHV-7 in transbronchial biopsies from lung transplant recipients. *New Microbiol* 2011;**34**:275–80.
61. Merk J, Schmid FX, Fleck M, et al. Fatal pulmonary failure attributable to viral pneumonia with human herpes virus 6 (HHV6) in a young immunocompetent woman. *J Intensive Care Med* 2005;**20**:302–6.
62. Russler SK, Tapper ME, Knox KK, et al. Pneumonitis associated with coinfection by human herpesvirus 6 and *Legionella* in an immunocompetent adult. *Am J Pathol* 1991;**138**:1405–11.

63. Vuorinen T, Kotilainen P, Lautenschlager I, et al. Interstitial pneumonitis and coinfection of human herpesvirus 6 and *Pneumocystis carinii* in a patient with hypogammaglobulinemia. *J Clin Microbiol* 2004;**42**:5415–8.
64. Yoshikawa T. HHV-6, the liver and the gastrointestinal tract. In: Krueger G, Ablashi DV, editors. *Human Herpesvirus-6: General Virology, Epidemiology, and Clinical Pathology*. Amsterdam: Elsevier; 2006. p. 243–9.
65. Amo K, Tanaka-Taya K, Inagi R, et al. Human herpesvirus 6B infection of the large intestine of patients with diarrhea. *Clin Infect Dis* 2003;**36**:120–3.
66. Yoshikawa T. Human herpesvirus 6 and 7 infection in transplantation. *Pediatr Transplant* 2003;**7**:11–17.
67. Mousset S, Martin H, Berger A, et al. Human herpesvirus 6 in biopsies from patients with gastrointestinal symptoms after allogeneic stem cell transplantation. *Ann Hematol* 2012;**91**:737–42.
68. Halme L, Arola J, Höckerstedt K, et al. Human herpesvirus 6 infection of the gastroduodenal mucosa. *Clin Infect Dis* 2008;**46**:434–9.
69. Lamoth F, Jayet PY, Aubert JD, et al. Case report: human herpesvirus 6 reactivation associated with colitis in a lung transplant recipient. *J Med Virol* 2008;**80**:1804–7.
70. Lempinen M, Halme L, Arola J, et al. HHV-6B is frequently found in the gastrointestinal tract in kidney transplantation patients. *Transplant Int* 2012;**25**:776–82.
71. Sipponen T, Turunen U, Lautenschlager I, et al. Human herpesvirus 6 and cytomegalovirus in ileocolonic mucosa in inflammatory bowel disease. *Scand J Gastroenterol* 2011;**46**:1324–33.
72. Sura R, Gavrilov B, Flamand L, et al. Human herpesvirus-6 in patients with Crohn's disease. *APMIS* 2010;**118**:394–400.
73. Asano Y, Yoshikawa T, Suga S, et al. Simultaneous occurrence of human herpesvirus-6 infection and intussusception in three infants. *Pediat Infect Dis J* 1991;**10**:335–7.
74. Komura E, Hashida T, Otsuka T, et al. Herpesvirus-6 and intussusception. *Pediat Infect Dis J* 1993;**12**:788–9.
75. Huang LM, Lee CY, Kin KH, et al. Human herpesvirus-6 associated with fatal hemophagocytic syndrome. *Lancet* 1990;**336**:60–1.
76. Syruckowa Z, Stary J, Sedlacek P, et al. Infection-associated hemophagocytic syndrome complicated by infectious lymphoproliferation: a case report. *Pediatr Hematol Oncol* 1996;**13**:143–50.
77. Yasumoto S, Tsujita J, Imayama S, et al. Case report: Gianotti–Crosti syndrome associated with human herpesvirus-6 infection. *J Dermatol* 1996;**23**:499–501.
78. Karras A, Thervet E, Legendre C. for the Groupe comparatif de transplantation de'Ile de France. Hemophagocytic syndrome in renal transplant recipients: report of 17 cases and review of literature. *Transplantation* 2004;**77**:238–43.
79. Krueger GRF, Ramon AMC, editors. *Immunopathology of Liver Diseases: An Overview*. The Netherlands: Bayer Diagnostics; 1999.
80. Cermelli C, Concari M, Carubbi F, et al. Growth of human herpesvirus 6 in HEPG2 cells. *Virus Res* 1996;**45**:75–85.
81. Mason A, Sallie R, Perrillo R, et al. Prevalence of Herpesviridae and hepatitis B virus DNA in the liver of patients with non-A, non-B fulminant hepatic failure. *Hepatology* 1996;**24**:1361–5.
82. Ishikawa K, Hasegawa K, Naritomi T, et al. Prevalence of Herpesviridae and hepatitis virus sequences in the livers of patients with fulminant hepatitis of unknown etiology in Japan. *J Gastroenterol* 2002;**37**:523–30.

83. Härmä M, Höckerstedt K, Lautenschlager I. Human herpesvirus-6 and acute liver failure. *Transplantation* 2003;**76**:536–9.
84. Chevret L, Boutolleau D, Halimi-Idri N, et al. Human herpesvirus-6 infection: a prospective study evaluating HHV-6 DNA levels in liver from children with acute liver failure. *J Med Virol* 2008;**80**:1051–7.
85. Tajiri H, Tanaka-Taja K, Ozaki Y, et al. Chronic hepatitis in an infant in association with human herpesvirus-6 infection. *J Pediat* 1997;**131**:473–5.
86. Lautenschlager I, Höckerstedt K, Linnavuori K, et al. Human herpesvirus-6 infection after liver transplantation. *Clin Infect Dis* 1998;**26**:702–7.
87. Griffiths PD, Ait-Khaled M, Bearcroft CP, et al. Human herpesviruses 6 and 7 as potential pathogens after liver transplant: prospective comparison with the effect of cytomegalovirus. *J Med Virol* 1999;**59**:496–501.
88. Massih RCA, Razonable RR. Human herpesvirus 6 infections after liver transplantation. *World J Gastroenterol* 2009;**15**:2561–9.
89. Feldstein AE, Razonable RR, Boyce TG, et al. Prevalence and clinical significance of human herpesviruses 6 and 7 active infection in pediatric liver transplant patients. *Pediatr Transplant* 2003;**7**:125–9.
90. Razonable RR, Lautenschlager I. Impact of human herpes virus 6 in liver transplantation. *World J Hepatol* 2010;**2**:345–53.
91. Costa FA, Soki MN, Andrade PD, et al. Simultaneous monitoring of CMV and human herpesvirus 6 infections and diseases in liver transplant patients: one-year follow-up. *Clinics (Sao Paulo)* 2011;**66**:949–53.
92. Sampaio AM, Thomasini RL, Guardia AC, et al. Cytomegalovirus, human herpesvirus-6, and human herpesvirus-7 in adult liver transplant recipients: diagnosis based on antigenemia. *Transplant Proc* 2011;**43**:1357–9.
93. Lee SO, Brown RA, Razonable RR. Chromosomally integrated human herpesvirus-6 in transplant recipients. *Transplant Infect Dis* 2012;**14**:346–54.
94. Grima P, Chiavaroli R, Calabrese P, et al. Severe hepatitis with autoimmune features following a HHV-6: a case report. *Cases J* 2008;**1**:110.
95. Potenza L, Luppi M, Barozzi P, et al. HHV-6A in syncytial giant-cell hepatitis. *N Engl J Med* 2008;**359**:593–602.
96. Rojo J, Simoes P, Krueger GRF, et al. Human herpesvirus 6 has no apparent influence on course of HCV hepatitis, but may complicate HBV hepatitis and alcoholic liver disease. A pilot study. *In Vivo* 2003;**17**:29–34.
97. Pischke S, Goessling J, Engelmann I, et al. High intrahepatic HHV-6 virus loads but neither CMV nor EBV are associated with decreased graft survival after diagnosis of graft hepatitis. *J Hepatol* 2012;**56**:1063–9.
98. Voigtländer T, Negm AA, Schneider AS, et al. Secondary sclerosing cholangitis in critically ill patients: model of end-stage liver disease score and renal function predict outcome. *Endoscopy* 2012;**44**:1055–8.
99. Krueger GRF, Koch B, Hoffmann A, et al. Dynamics of chronic active herpesvirus 6 infection in patients with chronic fatigue syndrome: data acquisition for computer modeling. *In Vivo* 2001;**15**:461–6.
100. Akashi K, Eizuru Y, Sumiyoshi Y, et al. Brief report: severe infectious mononucleosis-like syndrome and primary human herpesvirus 6 infection in an adults. *N Engl J Med* 1993;**329**:168–71.
101. Bertram G, Dreiner N, Krueger GR, et al. Frequent double infection with Epstein–Barr virus and human herpesvirus-6 in patients with acute infectious mononucleosis. *In Vivo* 1991;**5**:271–9.

102. Luppi M, Marasca R, Barozzi P, et al. Frequent detection of human herpesvirus-6 sequences by polymerase chain reaction in paraffin-embedded lymph nodes from patients with angioimmunoblastic lymphadenopathy and angioimmunoblastic-like lymphoma. *Leukemia Res* 1993;**17**:1003–11.
103. Daibata M, Ido E, Murakami K, et al. Angioimmunoblastic lymphadenopathy with disseminated human herpesvirus 6 infection in a patient with acute myeloblastic leukemia. *Leukemia* 1997;**11**:882–5.
104. Sugita K, Kurumada H, Eguchi M, et al. Human herpesvirus 6 infections associated with hemophagocytic syndrome. *Acta Haematol* 1995;**93**:108–9.
105. Tanaka H, Nishimura T, Hakui M, et al. Human herpesvirus 6-associated hemophagocytic syndrome in a healthy adult. *Emerg Infect Dis* 2000;**8**:87–8.
106. Leahy MA, Krejci SM, Friednash M, et al. Human herpesvirus 6 is present in lesions of Langerhans cell histiocytosis. *J Invest Dermatol* 1993;**101**:642–5.
107. Krueger GR, Manak M, Bourgeois N, et al. Persistent active herpesvirus infection associated with atypical polyclonal lymphoproliferation (APL) and malignant lymphoma. *Anticancer Res* 1989;**9**:1457–76.
108. Torelli G, Marasca R, Luppi M, et al. Human herpesvirus 6 in human lymphomas: identification of specific sequences in Hodgkin's lymphomas by polymerase chain reaction. *Blood* 1991;**77**:2251–8.
109. Di Luca D, Dolcetti R, Mirandola P, et al. Human herpesvirus 6: a survey of presence and variant distribution in normal peripheral lymphocytes and lymphoproliferative disorders. *J Infect Dis* 1994;**170**:211–5.
110. Razzaque A, Francillon Y, Jilly PN, et al. Detection of human herpesvirus 6 sequences in lymphoma tissues by immunohistochemistry and polymerase chain reactions. *Cancer Lett* 1996;**106**:221–6.
111. Bandobashi K, Daibata M, Kamioka M, et al. Human herpesvirus 6 (HHV-6)-positive Burkitt's lymphoma: establishment of a novel cell line infected with HHV-6. *Blood* 1997;**90**:1200–7.
112. Braun DK, Pellett PE, Hanson CA. Presence and expression of human herpesvirus 6 in peripheral blood mononuclear cells of S100-positive T-cell chronic lymphoproliferative disease. *J Infect Dis* 1995;**171**:1351–5.
113. Lin WC, Moore JO, Mann KP, et al. Posttransplant CD8 gamma-delta T cell lymphoma associated with human herpesvirus 6 infection. *Leukemia Lymphoma* 1999;**33**:377–84.
114. Hallas C, Neipel F, Huettner C, et al. Presence of human herpesvirus type 6 in sporadic lymphoproliferative disorders. *Diagn Mol Pathol* 1996;**5**:166–72.
115. Luppi M, Barozzi P, Garber R, et al. Expression of human herpesvirus 6 antigen in benign and malignant lymphoproliferative syndromes. *Am J Pathol* 1998;**153**:815–23.
116. Krueger GRF, Huetter ML, Rojo J, et al. Human herpesvirus-4 (EBV) and HHV-6 in Hodgkin's and Kikuchi's diseases and their relation to proliferation and apoptosis. *Anticancer Res* 2001;**21**:2155–62.
117. Hermouet S, Sutton CA, Rose TM, et al. Qualitative and quantitative analysis of human herpesviruses in chronic and acute B cell lymphocytic leukemia and in multiple myeloma. *Leukemia* 2003;**17**:185–95.
118. Nakayama-Ichiyama S, Yokote T, Oka S, et al. Diffuse large B-cell lymphoma, not otherwise specified, associated with coinfection of human herpesvirus 6 and 8. *J Clin Oncol* 2011;**29**:e636–7.
119. Nakayama-Ichiyama S, Yokote T, Kobayashi K, et al. Primary effusion lymphoma of T-cell origin with t(7;8)(q32;q13) in an HIV-negative patient with HCV-related liver cirrhosis and hepatocellular carcinoma positive for HHV6 and HHV8. *Ann Hematol* 2011;**90**:1229–31.

120. Nakayama-Ichiyama S, Yokote T, Iwaki K, et al. Co-infection of human herpesvirus-6 and human herpesvirus-8 in primary cutaneous diffuse large B-cell lymphoma, leg type. *Br J Haematol* 2011;**155**:514–6.
121. Siddon A, Lozovatsky L, Mohamed A, et al. Human herpesvirus 6 positive Reed–Sternberg cells in nodular sclerosis Hodgkin lymphoma. *Br J Haematol* 2012;**158**:635–43.
122. Krueger GRF, Brandt ME, Wang G, et al. A computational analysis of Canale–Smith syndrome: chronic lymphadenopathy simulating malignant lymphoma. *Anticancer Res* 2002;**22**:2365–72.
123. Arakaki N, Gallo G, Majluf R, et al. Rosai–Dorfman disease: renal involvement with HHV6 detection. Description of a pediatric case report. *Pediatr Dev Pathol* 2012;**15**:324–8.
124. Araujo A, Pagnier A, Frange P, et al. Lymphohistiocytic activation syndrome and *Bukholderia cepacia* complex infection in a child revealing chronic granulomatous disease and chromosomal integration of the HHV-6 genome. *Arch Pediatr* 2011;**18**:416–9.
125. Krueger GR, Ferrer Argote V. A unifying concept of viral immunopathogenesis of proliferative and aproliferative diseases (working hypothesis). *In Vivo* 1994;**8**:493–9.
126. Yoshikawa T, Kato Y, Ihira M, et al. Kinetics of cytokine and chemokine responses in patients with primary human herpesvirus 6 infection. *J Clin Virol* 2011;**50**:65–8.
127. Razzaque A. Oncogenic potential of human herpesvirus-6 DNA. *Oncogene* 1990;**5**:1365–70.
128. Thompson J, Choudhury S, Kashanchi F, et al. A transforming fragment with the direct repeat region of human herpesvirus type 6 that transactivates HIV-1. *Oncogene* 1994;**9**:1167–75.
129. Kakimoto M, Hasegawa A, Fujita S, et al. Phenotypic and functional alterations of dendritic cells induced by human herpesvirus 6 infection. *J Virol* 2002;**76**:10338–45.
130. Carrigan DR, Knox KK. Human herpesvirus 6 (HHV-6) isolation from bone marrow: HHV-6-associated bone marrow suppression in bone marrow transplant patients. *Blood* 1994;**84**:3307–10.
131. Singh N, Carrigan DR. Human herpesvirus-6 in transplantation: an emerging pathogen. *Ann Intern Med* 1996;**124**:1065–71.
132. Isomura H, Yamada M, Yoshida M, et al. Suppressive effects of human herpesvirus 6 on *in-vitro* colony formation of hematopoietic progenitor cells. *J Med Virol* 1997;**52**:406–12.
133. Penchansky L, Jordan JA. Transient erythroblastopenia of childhood associated with human herpesvirus type 6, variant B. *Am J Clin Pathol* 1997;**108**:127–32.
134. Wang FZ, Linde A, Dahl H, et al. Human herpesvirus 6 infection inhibits specific lymphocyte proliferation responses and is related to lymphocytopenia after allogeneic stem cell transplantation. *Bone Marrow Transplant* 1999;**24**:1201–6.
135. Lau YL, Peiris M, Chan GC, et al. Primary human herpesvirus 6 infection transmitted from donor to recipient through bone marrow infusion. *Bone Marrow Transplant* 1998;**21**:1063–6.
136. Zerr DM, Corey L, Kim HW, et al. Clinical outcomes of human herpesvirus 6 reactivation after hematopoietic stem cell transplantation. *Clin Infect Dis* 2005;**40**:932–40.
137. Clark DA. Human herpesvirus 6. *Rev Med Virol* 2000;**10**:155–73.
138. Sashihara J, Tanaka-Taya K, Tanaka S, et al. High incidence of human herpesvirus 6 infection with a high viral load in cord blood stem cell transplant recipients. *Blood* 2002;**100**:2005–11.
139. Ihara M, Yoshikawa T, Suzuki K, et al. Monitoring of active human herpesvirus 6 infections in bone marrow transplant recipients by real time PCR. *Microbiol Immunol* 2002;**46**:701–5.
140. Boutolleau D, Fernandez C, André E, et al. Human herpesvirus (HHV)-6 and HHV-7: two closely related viruses with different infection profiles in stem cell transplantation recipients. *J Infect Dis* 2003;**187**:179–86.

141. Yagasaki H, Kato M, Shimizu N, et al. Autoimmune hemolytic anemia and autoimmune neutropenia in a child with erythroblastopenia of childhood (TEC) caused by human herpesvirus-6 (HHV-6). *Ann Hematol* 2011;**90**:851–2.
142. Zerr DM. Human herpesvirus 6 and central nervous system disease in hematopoietic cell transplantation. *J Clin Virol* 2006;**37**(Suppl 1):S52–6.
143. Ishiyama K, Katagiri T, Hoshino T, et al. Preemptive therapy of human herpesvirus-6 encephalitis with foscarnet sodium for high-risk patients after hematopoietic SCT. *Bone Marrow Transplant* 2011;**46**:863–9.
144. Krueger GRF, Kudlimay D, Ramon A, et al. Demonstration of active and latent Epstein–Barr virus and human herpesvirus-6 infections in bone marrow cells of patients with myelodysplasia and chronic myeloproliferative disease. *In Vivo* 1994;**8**:533–42.
145. Rojo J, Cruz-Ortiz H. Krueger GRF. Los virus herpes linfotropicos humanos: Epstein–Barr (VEB), 6 y 7 (HHV-6 y HHV-7) en la patogenia de enfermadades mielodisplasicas (SMDn) y mieloproliferativas (SMP). *Rev Med Hosp Gen Mexico* 2000;**63**:18–24.
146. Caselli E, Zatelli MC, Rizzo R, et al. Virologic and immunologic evidence supporting an association between HHV-6 and Hashimoto's thyroiditis. *PLOS Pathogens* 2012;**8**:e1002951.
147. Nora-Krukle Z., Sultanova A., Gravelsina S., et al. Presence and activity of human herpesvirus 6 and 7 in patients with autoimmune thyroid disease. Presented at the 7th International Conference on HHV–6 & 7, Reston, VA, February 27–March 2, 2011.
148. Funck-Brentano E, Duong T, Family D, et al. Auto-immune thyroiditis and drug reaction with eosinophilia and systemic symptoms (DRESS) associated with HHV-6 viral reactivation. *Ann Dermatol Venereol* 2011;**138**:580–5.
149. Chen T, Hudnall SD. Anatomical mapping of human herpesvirus reservoirs of infection. *Mod Pathol* 2006;**19**:726–37.
150. Kaden J, May G, Wagner M, et al. Nachweis von Antikoerpern gegen humanes herpesvirus-6 (HHV-6) bei Nierentransplantatempfaengern mit Cytomegalievirus (CMV)-Infektion. *Transplantationsmedizin* 1997;**9**:28–32.
151. Helanterä I, Loginov R, Koskinen P, et al. Demonstration of HHV-6 antigens in biopsies of kidney transplant recipients with cytomegalovirus infection. *Transpl Int* 2008;**21**:980–4.
152. Csoma E, Mészáros B, Gáll T, et al. Dominance of variant A in human herpesvirus 6 viraemia after renal transplantation. *Virol J* 2011;**8**:403–8.
153. Caïola D, Karras A, Flandre P, et al. Confirmation of the low clinical effect of human herpesvirus-6 and -7 infections after renal transplantation. *J Med Virol* 2012;**84**:450–6.
154. Schroeder RB, Michelon TF, Garbin G, et al. Early HHV-6 replication is associated with morbidity non-related to CMV infection after kidney transplantation. *Braz J Infect Dis* 2012;**16**:146–52.
155. Gervasi MT, Romero R, Bracalente G, et al. Viral invasion of the amniotic cavity (VIAC) in the midtrimester of pregnancy. *J Matern Fetal Neonatal Med* 2012;**25**:2002–13.
156. Al-Buhtori M, Moore L, Benbow EW, et al. Viral detection in hydrops fetalis, spontaneous abortion, and unexplained fetal death in utero. *J Med Virol* 2011;**83**:679–84.
157. Lasch JA, Klussmann JP, Krueger GRF. Die humanen Herpesviren 6 und 7. Grundlagen und moegliche Bedeutung fuer die Dermatologie [Human herpesviruses 6 and 7: a brief review and possible implications for dermatology]. *Hautarzt* 1996;**47**:341–50.
158. Magalhães Ide M, Martins RV, Vianna RO, et al. Detection of human herpesvirus 7 infection in young children presenting with exanthema subitum. *Mem Inst Oswaldo Cruz* 2011;**106**:371–3.

159. Descamps V, Bouscarat F, Laglenne S, et al. Human herpesvirus 6 infection associated with anticonvulsant hypersensitivity syndrome and reactive hemophagocytic syndrome. *Br J Dermatol* 1997;**137**:605–8.
160. Descamps V, Valance A, Edlinger C, et al. Association of human herpesvirus 6 infection and drug reaction with eosinophilia and systemic symptoms. *Arch Dermatol* 2001;**137**:301–4.
161. Descamps V. Drug reaction with eosinophilia and systemic symptoms: a multiorgan antiviral T-cell response. Presented at the 7th International Conference on HHV-6 & 7, Reston, VA, February 27–March 2, 2011.
162. Wolz MM, Sciallis GF, Pittelkow MR. Human herpesviruses 6, 7, and 8 from a dermatologic perspective. *Mayo Clin Proc* 2012;**87**:1004–14.
163. Ushigome Y, Kano Y, Hirahara K, et al. Human herpesvirus 6 reactivation in drug-induced hypersensitivity syndrome and DRESS validation score. *Am J Med* 2012;**125**:e9–10.
164. Morito H, Kitamura K, Fukumoto T, et al. Drug eruption with eosinophilia and systemic syndrome associated with reactivation of human herpesvirus 7, not human herpesvirus 6. *J Dermatol* 2012;**39**:669–70.
165. Riyaz N, Sarita S, Arunkumar G, et al. Drug-induced hypersensitivity syndrome with human herpesvirus-6 reactivation. *Indian J Dermatol Venerol Leprol* 2012;**78**:175–7.
166. Cacoub P, Musette P, Descamps V, et al. The DRESS syndrome: a literature review. *Am J Med* 2011;**124**:588–97.
167. Fujino M, Ohashi M, Tanaka K, et al. Rhabdomyolysis in an infant with primary human herpesvirus-6 infection. *Pediatr Infect Dis J* 2012;**31**:1202–3.
168. Luppi M, Barozzi P, Maiorana A, et al. Human herpesvirus-6: a survey of presence and distribution of genomic sequences in normal brain and neuroglial tumors. *J Med Virol* 1995;**47**:105–11.
169. Cuomo L, Trivedi P, Cardillo MR, et al. Human herpesvirus 6 infection in neoplastic and normal brain tissue. *J Med Virol* 2001;**63**:45–51.
170. Hall CB, Caserta MT, Schnabel KC, et al. Persistence of human herpesvirus 6 according to site and variant: possible greater neutropism of variant A. *Clin Infect Dis* 1998;**26**:132–7.
171. Harberts E, Yao K, Wohler JE, et al. Human herpesvirus-6 entry into the central nervous system through the olfactory pathway. *Proc Natl Acad Sci USA* 2011;**108**:13734–9.
172. Caserta MT, Hall CB, Schnabel K, et al. Neuroinvasion and persistence of human herpesvirus 6 in children. *J Infect Dis* 1994;**170**:1586–9.
173. Wilborn F, Schmidt CA, Brinkmann V, et al. A potential role of human herpesvirus 6 in nervous system disease. *J Neuroimmunol* 1994;**49**:213–4.
174. Knox KK, Carrigan DR. Active human herpesvirus 6 (HHV-6) infection of the central nervous system in patients with AIDS. *J Acquir Immune Defic Syndr Hum Retrovirol* 1995;**9**:66–73.
175. Bonthius DJ, Karacay B. Meningitis and encephalitis in children. An update. *Neurol Clin* 2002;**20**:1013–38.
176. Eeg-Olofsson O. Virological and immunological aspects of seizure disorders. *Brain Dev* 2003;**25**:9–13.
177. Nahdi I, Boukoum H, Nabil Ben Salem A, et al. Detection of herpes simplex virus (1 and 2), varicella-zoster virus, cytomegalovirus, human herpesvirus 6 and enterovirus in immunocompetent Tunisian patients with acute neuromeningeal disorder. *J Med Virol* 2012;**84**:282–9.
178. Hosseininasab A, Alborzi A, Ziyaeyan M, et al. Viral etiology of aseptic meningitis among children in southern Iran. *J Med Virol* 2010;**83**:884–8.

179. Kawamura Y, Sugata K, Ihira M, et al. Different characteristics of human herpesvirus 6 encephalitis between primary infection and viral reactivation. *J Clin Virol* 2011;**51**:12–19.
180. Challoner PB, Smith KT, Parker JD, et al. Plaque-associated expression of human herpesvirus 6 in multiple sclerosis. *Proc Natl Acad Sci USA* 1995;**92**:7440–4.
181. Soldan SS, Leist TP, Juhng KN, et al. Increased lymphoproliferative response to the herpesvirus type 6A variant in multiple sclerosis patients. *Ann Neurol* 2000;**47**:306–13.
182. An SF, Groves M, Martinian L, et al. Detection of infectious agents in brain of patients with acute hemorrhagic leukoencephalitis. *J Neurovirol* 2002;**8**:439–46.
183. Cirone M, Cuomo L, Zompetta C, et al. Human herpesvirus 6 and multiple sclerosis: a study of T-cell cross-reactivity to viral and myelin basic proteins. *J Med Virol* 2002;**68**:268–72.
184. Chapenko S, Millers A, Nora Z, et al. Correlation between HHV-6 reactivation and multiple sclerosis disease activity. *J Med Virol* 2003;**69**:111–7.
185. Tejada-Simon MV, Zang YC, Hong J, et al. Cross-reactivity with myelin basic protein and human herpsevirus-6 in multiple sclerosis. *Ann Neurol* 2003;**53**:189–97.
186. Hubele F, Bilger K, Kremer S, et al. Sequential FDG PET and MRI findings in a case of human herpes virus 6 limbic encephalitis. *Clin Nucl Med* 2012;**37**:716–7.
187. Howell KB, Tiedemann K, Haeusler G, et al. Symptomatic generalized epilepsy after HHV6 posttransplant acute limbic encephalitis in children. *Epilepsia* 2012;**3**:1528–67.
188. Nora-Krukle Z, Chapenko S, Logina I, et al. Human herpesvirus 6 and 7 reactivation and disease activity in multiple sclerosis. *Medicina (Kaunas)* 2011;**47**:527–31.
189. Behzad-Behbahani A, Mikaeili MH, Entezam M, et al. Human herpesvirus-6 viral load and antibody titer in serum samples of patients with multiple sclerosis. *J Microbiol Immunol Infect* 2011;**44**:247–51.
190. Hill JA, Koo S, Guzman Suarez BB, et al. Cord-blood hematopoietic stem-cell transplant confers an increased risk for human herpesvirus-6-associated acute limbic encephalitis: a cohort analysis. *Biol Blood Marrow Transplant* 2012;**18**:1638–48.
191. Pruitt AA. Nervous system infections in patients with cancer. *Neurol Clin* 2003;**21**:193–219.
192. Hiraiwa-Sofue A, Ito Y, Ohta R, Kimura H, Okumura A. Human herpesvirus 6-associated encephalopathy in a child with Dravet syndrome. *Neuropediatrics* 2013;**44**:155–8.
193. Tohyama M, Yahata Y, Yasukawa M, et al. Severe hypersensitivity syndrome due to sulfasalazine associated with reactivation of human herpesvirus 6. *Arch Dermatol* 1998;**134**:1113–7.
194. Conilleau V, Dompmartin A, Verneuil L, et al. Hypersensitivity syndrome due to 2 anticonvulsant drugs. *Contact Dermatitis* 1999;**41**:141–4.
195. Klueppelberg U. Der systemische Lupus erythematodes (SLE): Uebersicht und eigene Arbeiten zur Herpesvirus-6 Infektion, doctoral thesis University of Cologne, Cologne, Germany, 1994, pp. 1–95.
196. De Clerck LSAU: Kindly check the given name, Bourgeois N, Krueger GRF, et al. Human herpesvirus-6 in Sjoegren's syndrome. In: Ablashi DV, Krueger GRF, Salahuddin SZ, editors. *Human Herpesvirus-6*. Amsterdam: Elsevier; 1992. p. 303–15.
197. Kawada J, Iwata N, Kitagawa Y, et al. Prospective monitoring of epstein–Barr virus and other herpesviruses in patients with juvenile idiopathic arthritis treated with methotrexate and tocilizumab. *Mod Rheumatol* 2012;**22**:565–70.
198. Suzuki Y, Inagi R, Aono T, et al. Human herpesvirus 6 infection as a risk factor for the development of severe drug-induced hypersensitivity syndrome. *Arch Dermatol* 1998;**134**:1108–12.

199. Krueger GRF, Schonnebeck M, Braun M. Changes in cell membrane fluidity and in receptor expression following infection with HHV-6 may influence superinfection with other viruses. *AIDS Res Human Retrovir* 1990;**6**:148–9.
200. Schonnebeck M, Krueger GRF, Braun M, et al. Human herpesvirus-6 infection may predispose cells for superinfection by other viruses. *In Vivo* 1991;**5**:255–64.
201. Lusso P, Ablashi DV, Luka J. Interaction between HHV-6 and other viruses. In: Ablashi DV, Krueger GRF, Salahuddin SZ, editors. *Human Herpesvirus-6*. Amsterdam: Elsevier; 1992. p. 121–33. pp.
202. Flamand L, Stefanescu L, Ablashi DV, et al. Activation of the Epsein–Barr virus replicative cycle by human herpesvirus 6. *J Virol* 1993;**67**:6768–77.
203. Al-Kaldi N, Watson AR, Harris A, et al. Dual infection with human herpesvirus type 6 and parvovirus B19 in renal transplant recipient. *Pediatr Nephrol* 1994;**8**:349–50.
204. Chen M, Wang H, Woodworth CD, et al. Detection of human herpesvirus 6 and papillomavirus 16 in cervical carcinoma. *Am J Pathol* 1994;**145**:1509–16.
205. Lusso P, Garzino-Demo A, Crowley RW, et al. Infection of gamma/delta T lymphocytes by human herpesvirus-6: transcriptional induction of CD4 and susceptibility to HIV infection. *J Exp Med* 1995;**181**:1303–10.
206. Ablashi DV, Bernbaum J, DiPaolo JA. Human herpesvirus 6 as copathogen. *Trends Immunobiol* 1995;**3**:324–6.
207. Como L, Trivedi P, deGarzia U, et al. Upregulation of Epstein–Barr virus-encoded latent membrane protein by human herpesvirus 6 superinfection of EBV carrying Burkitt lymphoma cells. *J Med Virol* 1998;**55**:219–26.
208. Singh VK, Lin SX, Yang VC. Serological association of measles virus and human herpesvirus-6 with brain antibodies and autism. *Clin Immunol Immunopathol* 1998;**89**:105–8.
209. Lautenschlager I, Lappalainen M, Linnavuori K, et al. CMV infection is usually associated with concurrent HHV-6 and HHV-7 antigenemia in liver transplant patients. *J Clin Virol* 2002;**25**(Suppl. 2):S57–61.

Chapter 4

HHV-6A, HHV-6B, and HHV-7 in Febrile Seizures and Status Epilepticus

Leon G. Epstein and John J. Millichap
Ann & Robert Lurie H. Children's Hospital of Chicago, Northwestern University Feinberg School of Medicine, Chicago, Illinois

INTRODUCTION

Febrile seizure is defined as a seizure associated with a febrile illness without a history of afebrile seizures or evidence of an acute central nervous system infection or other neurological insult.[1,2] Most febrile seizures are considered benign and are referred to as "simple." They are characterized by duration of less than 15 minutes and an absence of focal features. Some febrile seizures are prolonged (>15 minutes) or have focal features and are referred to as "complex"; 5 to 8% of febrile seizures meet the criteria for febrile status epilepticus, with a duration greater than 30 minutes.[3] Febrile seizure is the single most common seizure type, occurring in 2 to 5% of children under age 5, with a peak incidence in the second year of life.[4] Febrile status epilepticus accounts for only 5% of febrile seizures but 25% of all childhood status epilepticus and >70% of status epilepticus in the second year of life.[5] There is often acute hippocampal injury following febrile status epilepticus, which may increase the risk of developing temporal lobe epilepsy.[4,6–10]

Human herpesvirus 6A (HHV-6A), HHV-6B, and HHV-7 are now classified as distinct viruses within the *Roseolovirus* genus of the Betaherpesvirinae subfamily.[11] HHV-6 was discovered in 1986 by virologists at the National Institutes of Health,[12] and was later identified as the cause of roseola infantum in 1988.[13] HHV-7, discovered in 1990, is also associated with febrile seizures, but to a lesser degree.[14] Although three distinct viruses exist, most primary infections in children are caused by HHV-6B; thus, this virus is the main focus of this chapter. Clinical manifestations of primary infection with HHV-6B include febrile seizures, roseola (exanthem subitum, sixth disease), febrile illness without rash, limbic encephalitis, and mononucleosis-like syndromes. Independent reports of

Human Herpesviruses HHV-6A, HHV-6B & HHV-7.

HHV-6 isolation in the United States, Japan, United Kingdom, Australia, and Africa demonstrate that the virus is widespread.[16] The fever is characteristically high (>39.5°C; >103.0°F), and seizures occur during the acute febrile period in approximately 10 to 15% of primary infections. The virus persists and sometimes reactivates. As the concentration of maternal antibody decreases during the first year of life, the risk of infection increases, peaking between 6 and 24 months of age.

Reviewing studies that associate HHV-6A, HHV-6B, or HHV-7 with seizures is often difficult due to variability in methods used by investigators. It is important to distinguish between HHV-6A and HHV-6B, although this would not have been possible in older studies. Most seizures associated with HHV-6B and HHV-7 occur in children, many of whom are below the age of 3 years. In this population, it is important to distinguish between primary infection and reactivated infection. In older children, primary infection is rare, as infection with these viruses is nearly universal in early childhood. In adult patients, active infection with HHV-6A, HHV-6B, or HHV-7 would be a reactivation. The definitive diagnosis of primary HHV-6 infection necessitates isolation of the virus from peripheral blood. Polymerase chain reaction (PCR)-based assays have largely replaced cultures because of their widespread availability and quick results, often within days. PCR-based assays can employ specific primers or probes to distinguish HHV-6A, HHV-6B, and HHV-7.[17,18] The presence of HHV-6A, HHV-6B, or HHV-7 DNA and RNA in the blood at baseline indicates active viral replication, referred to as *viremia*.[18–21] Primary infection with HHV-6 or HHV-7 can be defined as the absence of HHV-6- or HHV-7-specific antibody in the presence of viremia. Characteristic of a reactivated HHV-6 or HHV-7 infection are persistent viral specific antibodies after 6 months of age in a patient with viremia. Maternal HHV-6B antibodies may still be present up to 6 months of age.[22] Prior infection is characterized by the presence of HHV-6 or HHV-7 antibodies in the absence of viremia. The absence of viral specific antibodies or viremia indicates no previous infection.[23]

A prospective study is currently underway to investigate the consequences of prolonged febrile seizures (FEBSTAT) that will also determine the relationship between HHV-6A, HHV-6B, and HHV-7 infections and febrile status epilepticus, hippocampal injury, and subsequent temporal lobe epilepsy.[23]

HISTORY OF HHV-6, ROSEOLA INFANTUM, AND FEBRILE SEIZURES

Studies published before the identification of HHV-6 reported the incidence of roseola infantum as a cause of febrile convulsions and the incidence of convulsions in patients with roseola infantum.[24,25] In a review of the literature from 1924 to 1964,[26] roseola infantum (exanthem subitum) was recognized as the cause of fever most commonly associated with febrile convulsions. There were 3168 febrile seizure patients reported in 13 publications, and roseola infantum

TABLE 4.1 Incidence of HHV-6 Infection and/or Exanthem Subitum (Roseola Infantum) in Children with Febrile Seizures

Authors	Year	Country	Febrile Seizure Patients	HHV-6 Infection	
			n	*n*	%
Millichap[26]	1967	Worldwide	3168	128	4
Barone et al.[27]	1995	United States	42	11	26
Bertolani et al.[28]	1996	Italy	65	23	35
Chua et al.[29]	1997	Malaysia	31	5	16
Hukin et al.[30]	1998	Canada	35	15	43
Suga et al.[33]	2000	Japan	105	21	20
Pancharoen et al.[32]	2000	Thailand	82	7	9
Murakami et al.[31]	2004	Japan	56	19	34
Epstein et al.[23]	2012	United States	169	54	32
Total patients	1995–2012	—	585	155	26

Adapted with permission from Millichap JG, Millichap JJ. Pediatr Neurol 2006; **35***(3): 165–72.*

was cited as the cause of the fever and convulsion in a mean of 4% of these patients (range, 0.6 to 7.6%). A subsequent review found that 585 patients less than 3 years of age with febrile seizures were reported in the literature between 1995 and 2012; 155 (26%) of these had a primary HHV-6 infection (Table 4.1).[23,26–33] Of 902 children under 3 years of age with primary HHV-6 infection and fever, 149 (~17%) had a seizure (Table 4.2).[22,26,27,30,34–40]

Although early investigators considered roseola-associated encephalitis as the cause of the convulsion, only 3% of cerebrospinal fluid (CSF) examinations showed a mild pleocytosis. Of 581 patients with roseola reported in 11 publications, the average incidence of convulsions was 22%. In these early reports, 10% of patients with roseola-associated febrile convulsions were less than 6 months of age, an unusual age period for febrile seizures to occur. The remarkably high body temperature that accompanies roseola infantum was considered a more likely explanation for the accompanying seizure than roseola encephalitis.[34]

HHV-6B AND FEBRILE SEIZURES

Several studies have linked primary HHV-6B infection with febrile seizures in young children. A large prospective study performed at the University of Rochester in 1994 followed 1653 infants and young children with acute febrile

TABLE 4.2 Incidence of Febrile Seizures in Children with HHV-6 Infection of Exanthem Subitum (Roseola Infantum)

Authors	Year	Country	HHV-6 Patients	Seizures	
			n	*n*	%
Millichap[26]	1967	Worldwide	581	128	22
Hall et al.[22]	1994	United States	160	21	13
Asano et al.[36]	1994	Japan	176	14	8
Ward et al.[39]	1994	United Kingdom	25	5	20
Barone et al.[27]	1995	United States	11	11	100
Hukin et al.[30]	1998	Canada	68	35	51
Jee et al.[37]	1998	United States	316	45	14
Ansari et al.[35]	2004	United States	54	7	13
Zerr et al.[40]	2005	United States	81	0	0
Ward et al.[38]	2005	United Kingdom	11	11	100
Total patients	1994–2005	—	902	149	17

Adapted with permission from Millichap JC, Millichap JJ. Pediatr Neurol 2006; **35***(3): 165–72.*

illnesses. Primary HHV-6 infection was determined by the presence of viremia (viral culture) and seroconversion in 160 (9.7%), and 21 (13%) of these hospitalized patients had seizures. Furthermore, HHV-6B infections accounted for one third of all febrile seizures in children up to the age of 2 years.[22] In another study, evidence of HHV-6B DNA in the CSF of children during and after primary infection was reported, and it was characterized by persistence of HHV-6B in the absence of primary HHV-6B infection. HHV-6B DNA persisted in the CNS of 28.9% children.[41] In 1995, the association of HHV-6B infection with first febrile seizures was demonstrated by viral culture in 8 (19%) of 42 children. Three additional cases were identified by acute and convalescent-phase serologic studies.[27] Interestingly, the frequency of HHV-6 infection with a febrile seizure was similar to the frequency of convulsion with roseola infantum recorded in the literature in 1968.[26] In a study of the incidence of HHV-6 infection in children with febrile convulsion, HHV-6 was confirmed virologically and/or serologically as the cause of the fever in only 5 (16.1%) of 31 children admitted to the hospital.[29] Clinical characteristics of first febrile convulsions during primary HHV-6B infection were also studied in Japan. Of the 105 patients with febrile convulsions, 21 had HHV-6B (17 with exanthem subitum). First febrile convulsions associated with primary HHV-6B infection occurred at a significantly lower age than among those without infection. In addition, those patients with primary HHV-6B infection had increased frequency of clustering seizures, prolonged and partial seizures, and postictal paralysis. A subsequent study in Japan found HHV-6 or -7 was the cause of 34% of initial febrile seizures in a cohort of 35 boys and 21 girls who presented under the age of 3 years old.[31]

In contrast to these studies with positive correlations, several investigators have concluded that HHV-6 and exanthem subitum are not a major cause of febrile seizures. In a population-based, prospective study of HHV-6B infection among children under 2 years old, 130 of 277 children acquired the infection at a peak age of 9 to 21 months, none had seizures, and rash was rare.[40] A case-control study in Vancouver found that the incidence of primary HHV-6 infection was similar in 35 patients with febrile seizures and in 33 age-matched controls with fever and without seizures. Evidence of past HHV-6 infection was demonstrated in 13 febrile seizure patients and in 8 controls. HHV-6 did not correlate with first and second febrile seizures in this study.[30] These studies show that HHV-6B primary infection commonly occurs without fever, rash, or seizures; however, they do not address the question of how often febrile seizures or febrile status epilepticus occur in association with HHV-6B infection in symptomatic patients.

Italian children with a first febrile seizure were compared with control patients with fever but without seizures. HHV-6B was identified in 35% patients with seizures and in 50% of those without seizures. HHV-7 was identified in 2.3% with seizures and in 7.7% of those without seizures.[28] Other viral infections involved in a first febrile seizure in this Italian population included adenovirus (13.8%), respiratory syncytial virus (10.7%), herpes simplex virus (9.2%), and cytomegalovirus (3%). Barone et al.,[27] in the United States, using both virological and serologic methods, identified primary HHV-6B infection in 11 (26%) febrile seizure patients. A geographic variation in incidence and lesser association of HHV-6 infection with febrile seizures is found in Thailand[32] and in Malaysia[29] compared with the United States and Canada (Table 4.1).

The mean maximum fever in infants with primary HHV-6 infection is generally high, at or above 39.5°C (103.0°F). The seizure threshold varies with age, neurological maturation, and inherited susceptibility.[42,43] Although fever is common with HHV-6B infection, the detection of virus in the CSF and the occurrence of seizures are variable, suggesting that direct HHV-6B neuroinvasion or encephalitis is not the sole etiologic factor for febrile seizures.

HHV-6B AND RECURRENCE OF FEBRILE SEIZURES

In a study from researchers at the University of Rochester, children with the first febrile seizure were evaluated in a tertiary care emergency department and followed for at least 12 months (average, 36 months). The study included 36 HHV-6B culture-positive children and 86 HHV-6B culture-negative controls. Twenty percent of the HHV-6B culture-positive children compared to 40% of the HHV-6B culture-negative children ($p<0.038$) experienced a recurrent seizure within 1 year of their first febrile seizure. Mean time to recurrence within

12 months was 8.6 months for HHV-6B-positive and 3.8 months for HHV-6B-negative children ($p<0.001$). The results of this study suggest that the risk of recurrent seizure after a primary HHV-6B-induced febrile seizure is not greater than that in children whose first febrile seizure is due to other etiologies.[37]

HHV-6A, HHV-6B, AND HHV-7 IN CHILDREN WITH FEBRILE STATUS EPILEPTICUS

Ward and colleagues[38] performed a 3-year prospective study of 205 children ranging in age from 2 to 35 months who were hospitalized with suspected encephalitis and/or severe illness with convulsions reported via the British Paediatric Surveillance Unit Network. These authors found that 17% (26/156) of these children had primary infection with either HHV-6 (11) or HHV-7 (13), or both (2). All 26 had fever, 25/26 had convulsions, and 18/26 had status epilepticus. This report was the first report to highlight the importance of HHV-6 and HHV-7 infection as a cause of febrile status epilepticus in children under 2 years of age.

Subsequent results from a prospective multicenter study (FEBSTAT) of the consequences of febrile status epilepticus in children provided additional evidence that HHV-6B in particular and HHV-7 are important causes of febrile status epilepticus. This study was designed to determine whether febrile status epilepticus is a cause of hippocampal injury and intractable temporal lobe epilepsy and whether electroencephalography (EEG), brain magnetic resonance imaging (MRI), or viral infection with HHV-6 or HHV-7 is associated with these outcomes.[44] The FEBSTAT study team determined the frequency of HHV-6A, HHV-6B, and HHV-7 infection as a cause of febrile status epilepticus in children ages 1 month to 5 years. Of 199 children evaluated, HHV-6B or HHV-7 was assayed in 169 (84.9%), using quantitative polymerase chain reaction and antibody titers. HHV-6A was not found in any subjects with febrile status epilepticus. HHV-6B viremia occurred in 54 children (32%), including 38 with primary infection and 16 with reactivated infection. HHV-7 viremia was observed in 12 children (7.1%). HHV-6B and HHV-7 infection combined accounted for one-third of cases of febrile status epilepticus in young children, establishing a possible link between HHV-6B and HHV-7 and subsequent temporal lobe epilepsy.[23]

In addition to viral studies, acute EEG was examined as part of the FEBSTAT study. The 199 children had EEGs performed within 72 hours of febrile status epilepticus, and 90 (45.2%) were abnormal.[45] Focal slowing (23.6%) and focal attenuation (12.6%) were the most common findings, and both were associated with focal MRI abnormalities (Figure 4.1). No association was observed between the EEG findings and the presence of HHV-6B or HHV-7 infection. The FEBSTAT study results suggest that the EEG may be a sensitive noninvasive biomarker for acute injury following febrile status

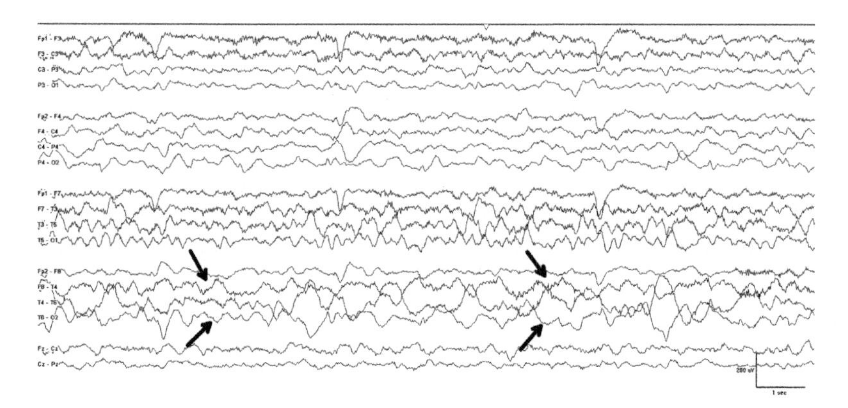

FIGURE 4.1 Electroencephalograph (longitudinal bipolar montage). EEG recorded during wake in a 1-year-old following febrile status epilepticus shows continuous right temporal slowing (arrows).

epilepticus, but the predictive value for subsequent temporal lobe epilepsy remains to be determined.

HHV-6 AND HHV-7 AND TEMPORAL LOBE EPILEPSY

Human herpesvirus 6 DNA is present in surgical mesial temporal lobe sections and is localized to astrocytes.[46] An investigation of temporal lobectomy specimens (a study headed by the Clinical Epilepsy Section of the NINDS) provides evidence of active HHV-6B but not HHV-6A replication in hippocampal astrocytes in two-thirds of patients with mesial temporal sclerosis, but not other causes of epilepsy. HHV-6B may cause "excitotoxicity" by interfering with astrocyte excitatory amino acid transport. In further work published in 2007, a team of investigators identified HHV-6B infection in astrocytes and resected brain material from 11 of 16 (69%) mesial temporal lobe epilepsy patients but none of 14 patients with other epilepsy syndromes.[47] These investigators have also demonstrated HHV-6B DNA in the olfactory bulb/tract regions of autopsy specimens and in nasal mucous samples. The nasal cavity is a reservoir for HHV-6B, and the olfactory pathway is a route of CNS entry for the virus.[48] Researchers from Germany found HHV-6 DNA in surgically resected temporal lobe tissue in 55.6% of patients with a history of encephalitis, but not in those patients with idiopathic mesial temporal sclerosis or lesion-associated temporal lobe epilepsy and a history of complex febrile seizures.[49] In a study from West China published in 2011, researchers found HHV-6B DNA in 28.1% of hippocampal samples from patients with mesial temporal lobe epilepsy. There was a positive history of febrile seizure in 78% of HHV-6B-positive patients. Additional studies will be necessary to determine the relationships among viral infections, complex or prolonged febrile seizures, and the development of intractable temporal lobe epilepsy.[46,50]

HHV-6 ENCEPHALITIS IN IMMUNOCOMPETENT PATIENTS

Encephalopathy may be defined as the presence of seizures or alteration of clinical neurologic function. Abnormal cytokine responses or altered seizure susceptibility may be related to the response to acute HHV-6 infection in otherwise immunocompetent individuals. For example, in a study from Japan, serum and CSF levels of cytokines in infants with HHV-6 encephalopathy or neurologic sequelae were significantly higher than those without sequelae or in infants with febrile seizures. Elevated levels of serum and CSF IL-6 may predict neurological sequelae of HHV-6 encephalopathy.[51] The case of a 13-month-old girl with Dravet syndrome who had neurological sequelae as a result of HHV-6-associated encephalopathy was reported by Hiraiwa Hospital, Nagoya, Japan.[52] Seizure susceptibility due to SCN1A mutation may have contributed to the acute encephalopathy.

HHV-6 AND LIMBIC ENCEPHALITIS IN IMMUNOSUPPRESSED PATIENTS

Prior to 2001, HHV-6B-associated limbic encephalitis was not well characterized, but that changed with publication of a report from Duke University Medical Center regarding five young patients who demonstrated a clinically identifiable syndrome following stem cell transplantation.[53] The patients exhibited short-term memory loss, limbic seizures, confusion, and sleep disturbance. MRI of the brain was significant for increased T2 signal intensity in one or both hippocampal regions (Figure 4.2). CSF was available from three of the five patients, and studies revealed HHV-6B. In one of these patients, expression of HHV-6 P41/P101 was demonstrated in the hippocampus using immunohistochemistry.[53]

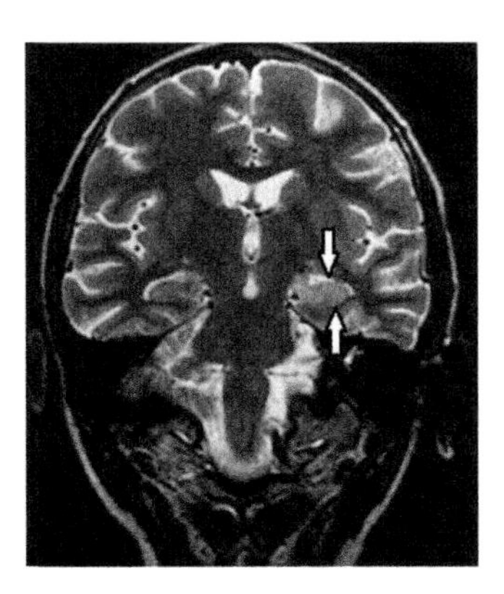

FIGURE 4.2 MRI of the brain. Fast spin echo T2-weighted oblique coronal image shows increased T2 signal and swelling of left hippocampus (arrows) in a patient with HHV-6B limbic encephalitis after stem cell transplant. *Adapted with permission from Wainwright MS et al. Ann Neurol 2001;* ***50****(5): 612–9.*

The results from this study suggested that reactivation of latent HHV-6B in immunocompromised patients resulted in limbic encephalitis and hippocampal injury.[53] Subsequently, other studies supported this initial observation. Researchers from Japan examined the characteristics of HHV-6 encephalitis in children at the time of primary infection compared to those at the time of reactivation in transplant recipients. Low copy number of CSF HHV-6 DNA was detected in 7 of 22 patients with HHV-6 encephalitis as primary infection, whereas high viral DNA copy numbers were found in all 7 CSF samples from posttransplant HHV-6 encephalitis patients ($p<0.001$).[54] In a case series of three Australian children with HHV-6-associated posttransplant acute limbic encephalitis (PALE), the children had associated epilepsy, and serial MRI revealed atrophy in the medial temporal lobe.[55] Researchers in Finland have advocated a trial of high-dose ganciclovir in the treatment of HHV-6 CNS infection in immunocompetent as well as immunocompetent children. They reported that the initial recovery of a 15-month-old child with severe HHV-6B encephalitis was followed by neurologic sequelae including epilepsy and ataxia.[56] HHV-6 DNA detectable in the CSF and peripheral blood may provide a screening test in high-risk patients, and treatment for viral encephalitis may be beneficial for prevention of sequelae.[53,57]

Human herpesvirus 6B encephalitis in immunosuppressed patients may be distinguished by a combination of characteristic clinical features and imaging abnormalities in the hippocampus, amygdala, and limbic structures beyond the medial temporal lobe. Clinical and MRI findings indicative of HHV-6B encephalitis were studied in Duke University Medical Center. CSF was positive for HHV-6B DNA by PCR in 9 of 16 immunosuppressed patients tested. Clinical characteristics such as insomnia, agitation, and hallucinations correlated with the presence of HHV-6B. However, neuroimaging abnormalities were present in the hippocampus of both HHV-6B-positive and -negative patients. Extrahippocampal involvement was more common in HHV-6B-positive patients, and HHV-6 in the CSF had a positive correlation with clinical signs of insomnia, hallucinations, or seizures.[58] One explanation for the findings reported in this study is that HHV-6B reactivation in limbic structures of the brain associated with immune suppression may not always result in viral DNA being shed in the CSF.

CONCLUSIONS

Roseola infantum was historically associated with febrile seizures. More recent studies have found that both HHV-6B and HHV-7 are frequently associated with febrile seizures. The question remains whether the seizures are due to the high fever related to the host response to the systemic viral infection or are due to transient HHV-6B encephalitis. Studies have shown HHV-6B infection is likely the cause of one-third of cases of febrile status epilepticus.[23,26,34] Febrile status epilepticus is suspected to lead to subsequent temporal lobe epilepsy in 30 to 40% of cases within 8 to 11 years.[2,4,59,60] That HHV-6B may have a role in intractable epilepsy is further supported by the presence of this virus in temporal lobe resections from epilepsy patients and the finding of HHV-6B reactivation in the temporal lobe of an immunosuppressed patient with limbic encephalitis. If the suspected association between HHV-6B infection and temporal lobe epilepsy is proven, then antiviral or anti-inflammatory therapies could potentially prevent epilepsy in these patients.

REFERENCES

1. National Institutes of Health Consensus development conference on febrile seizures, May 19–21, 1980. *Epilepsia* 1981;**22**(3):377–81.
2. Commission on Epidemiology and Prognosis, International League Against Epilepsy Guidelines for epidemiologic studies on epilepsy. *Epilepsia* 1993;**34**(4):592–6.
3. Hesdorffer DC, Benn EK, Bagiella E, et al. Distribution of febrile seizure duration and associations with development. *Ann Neurol* 2011;**70**(1):93–100.
4. Shinnar S. Febrile seizures and mesial temporal sclerosis. *Epilepsy Currents* 2003;**3**:115–8.
5. Shinnar S, Pellock JM, Moshe SL, et al. In whom does status epilepticus occur: age-related differences in children. *Epilepsia* 1997;**38**(8):907–14.
6. VanLandingham KE, Heinz ER, Cavazos JE, Lewis DV. Magnetic resonance imaging evidence of hippocampal injury after prolonged focal febrile convulsions. *Ann Neurol* 1998;**43**(4):413–26.
7. Lewis DV, Barboriak DP, MacFall JR, Provenzale JM, Mitchell TV, VanLandingham KE. Do prolonged febrile seizures produce medial temporal sclerosis? Hypotheses, MRI evidence and unanswered questions. *Prog Brain Res* 2002;**135**:263–78.
8. Scott RC, King MD, Gadian DG, Neville BG, Connelly A. Hippocampal abnormalities after prolonged febrile convulsion: a longitudinal MRI study. *Brain* 2003;**126**:2551–7. Pt 11.
9. Provenzale JM, Barboriak DP, VanLandingham K, MacFall J, Delong D, Lewis DV, et al. Signal hyperintensity after febrile status epilepticus is predictive of subsequent mesial temporal sclerosis. *AJR Am J Roentgenol* 2008;**190**(4):976–83.
10. Scott RC, Gadian DG, King MD, et al. Magnetic resonance imaging findings within 5 days of status epilepticus in childhood. *Brain* 2002;**125**:1951–9. Pt 9.
11. King AMQ, Adams MJ, Carstens EB, Lefkowitz EJ, editors. *Virus Taxonomy: Ninth Report of the International Committee on Taxonomy of Viruses*. London: Academic Press; 2012.
12. Salahuddin SZ, Ablashi DV, Markham PD, et al. Isolation of a new virus, HBLV, in patients with lymphoproliferative disorders. *Science* 1986;**234**(4776):596–601.
13. Yamanishi K, Okuno T, Shiraki K, et al. Identification of human herpesvirus-6 as a causal agent for exanthem subitum. *Lancet* 1988;**1**(8594):1065–7.
14. Pickering LK. *Red Book: 2009 Report of the Committee on Infectious Diseases*, 28th ed. Elk Grove Village, IL: American Academy of Pediatrics; 2009. xxxvi.
15. Dewhurst S, McIntyre K, Schnabel K, Hall CB. Human herpesvirus 6 (HHV-6) variant B accounts for the majority of symptomatic primary HHV-6 infections in a population of U.S. infants. *J Clin Microbiol* 1993;**31**(2):416–8.
16. Wahren B, Linde A. Virological and clinical characteristics of human herpesvirus 6. *Scand J Infect Dis Suppl* 1991;**80**:105–9.
17. Zerr DM, Huang ML, Corey L, Erickson M, Parker HL, Frenkel LM. Sensitive method for detection of human herpesviruses 6 and 7 in saliva collected in field studies. *J Clin Microbiol* 2000;**38**(5):1981–3.
18. Boutolleau D, Duros C, Bonnafous P, et al. Identification of human herpesvirus 6 variants A and B by primer-specific real-time PCR may help to revisit their respective role in pathology. *J Clin Virol* 2006;**35**(3):257–63.

19. Norton RA, Caserta MT, Hall CB, Schnabel K, Hocknell P, Dewhurst S. Detection of human herpesvirus 6 by reverse transcription-PCR. *J Clin Microbiol* 1999;**37**(11):3672–5.
20. Caserta MT, Hall CB, Schnabel K, et al. Diagnostic assays for active infection with human herpesvirus 6 (HHV-6). *J Clin Virol* 2010;**48**(1):55–7.
21. Hall CB, Caserta MT, Schnabel KC, et al. Characteristics and acquisition of human herpesvirus (HHV) 7 infections in relation to infection with HHV-6. *J Infect Dis* 2006;**193**(8):1063–9.
22. Hall CB, Long CE, Schnabel KC, et al. Human herpesvirus-6 infection in children. A prospective study of complications and reactivation. *N Engl J Med* 1994;**331**(7):432–8.
23. Epstein LG, Shinnar S, Hesdorffer DC, et al. Human herpesvirus 6 and 7 in febrile status epilepticus: the FEBSTAT study. *Epilepsia* 2012;**53**(9):1481–8.
24. Berenberg W, Wright S, Janeway CA. Roseola infantum (exanthem subitum). *N Engl J Med* 1949;**241**:253–9.
25. Burnstine RC, Paine RS. Residual encephalopathy following roseola infantum. *AMA J Dis Child* 1959;**98**(2):144–52.
26. Millichap JG. *Febrile Convulsions*. New York: Macmillan; 1967.
27. Barone SR, Kaplan MH, Krilov LR. Human herpesvirus-6 infection in children with first febrile seizures. *J Pediatr* 1995;**127**(1):95–7.
28. Bertolani MF, Portolani M, Marotti F, et al. A study of childhood febrile convulsions with particular reference to HHV-6 infection: pathogenic considerations. *Childs Nerv Syst* 1996;**12**(9):534–9.
29. Chua KB, Lam SK, AbuBakar S, Koh MT, Lee WS. The incidence of human herpesvirus 6 infection in children with febrile convulsion admitted to the University Hospital, Kuala Lumpur. *Med J Malaysia* 1997;**52**(4):335–41.
30. Hukin J, Farrell K, MacWilliam LM, et al. Case-control study of primary human herpesvirus 6 infection in children with febrile seizures. *Pediatrics* 1998;**101**(2):E3.
31. Murakami K. [A study of the relationship between initial febrile seizures and human herpes virus 6, 7 infections]. *No To Hattatsu* 2004;**36**(3):248–52.
32. Pancharoen C, Chansongsakul T, Bhattarakosol P. Causes of fever in children with first febrile seizures: how common are human herpesvirus-6 and dengue virus infections?. *Southeast Asian J Trop Med Publ Health* 2000;**31**(3):521–3.
33. Suga S, Suzuki K, Ihira M, et al. Clinical characteristics of febrile convulsions during primary HHV-6 infection. *Arch Dis Child* 2000;**82**(1):62–6.
34. Millichap JG, Millichap JJ. Role of viral infections in the etiology of febrile seizures. *Pediatr Neurol* 2006;**35**(3):165–72.
35. Ansari A, Li S, Abzug MJ, Weinberg A. Human herpesviruses 6 and 7 and central nervous system infection in children. *Emer Infect Dis* 2004;**10**(8):1450–4.
36. Asano Y, Yoshikawa T, Suga S, et al. Clinical features of infants with primary human herpesvirus 6 infection (exanthem subitum, roseola infantum). *Pediatrics* 1994;**93**(1):104–8.
37. Jee SH, Long CE, Schnabel KC, Sehgal N, Epstein LG, Hall CB. Risk of recurrent seizures after a primary human herpesvirus 6-induced febrile seizure. *Pediatr Infect Dis J* 1998;**17**(1):43–8.
38. Ward KN, Andrews NJ, Verity CM, Miller E, Ross EM. Human herpesviruses-6 and -7 each cause significant neurological morbidity in Britain and Ireland. *Arch Dis Child* 2005;**90**(6):619–23.
39. Ward KN, Gray JJ. Primary human herpesvirus-6 infection is frequently overlooked as a cause of febrile fits in young children. *J Med Virol* 1994;**42**(2):119–23.
40. Zerr DM, Meier AS, Selke SS, et al. A population-based study of primary human herpesvirus 6 infection. *N Engl J Med* 2005;**352**(8):768–76.

41. Caserta MT, Hall CB, Schnabel K, et al. Neuroinvasion and persistence of human herpesvirus 6 in children. *J Infect Dis* 1994;**170**(6):1586–9.
42. Berg AT, Shinnar S, Shapiro ED, Salomon ME, Crain EF, Hauser WA. Risk factors for a first febrile seizure: a matched case-control study. *Epilepsia* 1995;**36**(4):334–41.
43. Millichap JG. Studies in febrile seizures. I. Height of body temperature as a measure of the febrile-seizure threshold. *Pediatrics* 1959;**23**:76–85. 1 Part 1.
44. Shinnar S, Hesdorffer DC, Nordli Jr. DR, et al. Phenomenology of prolonged febrile seizures: results of the FEBSTAT study. *Neurology* 2008;**71**(3):170–6.
45. Nordli Jr. DR, Moshe SL, Shinnar S, et al. Acute EEG findings in children with febrile status epilepticus: results of the FEBSTAT study. *Neurology* 2012;**79**(22):2180–6.
46. Donati D, Akhyani N, Fogdell-Hahn A, et al. Detection of human herpesvirus-6 in mesial temporal lobe epilepsy surgical brain resections. *Neurology* 2003;**61**(10):1405–11.
47. Fotheringham J, Donati D, Akhyani N, et al. Association of human herpesvirus-6B with mesial temporal lobe epilepsy. *PLoS Med* 2007;**4**(5):e180.
48. Harberts E, Yao K, Wohler JE, et al. Human herpesvirus-6 entry into the central nervous system through the olfactory pathway. *Proc Natl Acad Sci USA* 2011;**108**(33):13734–9.
49. Niehusmann P, Mittelstaedt T, Bien CG, et al. Presence of human herpes virus 6 DNA exclusively in temporal lobe epilepsy brain tissue of patients with history of encephalitis. *Epilepsia* 2010;**51**(12):2478–83.
50. Theodore WH, Epstein L, Gaillard WD, Shinnar S, Wainwright MS, Jacobson S. Human herpes virus 6B: a possible role in epilepsy?. *Epilepsia* 2008;**49**(11):1828–37.
51. Ichiyama T, Ito Y, Kubota M, Yamazaki T, Nakamura K, Furukawa S. Serum and cerebrospinal fluid levels of cytokines in acute encephalopathy associated with human herpesvirus-6 infection. *Brain Dev* 2009;**31**(10):731–8.
52. Hiraiwa-Sofue A, Ito Y, Ohta R, Kimura H, Okumura A. Human herpesvirus 6-associated encephalopathy in a child with Dravet syndrome. *Neuropediatrics* 2012.
53. Wainwright MS, Martin PL, Morse RP, et al. Human herpesvirus 6 limbic encephalitis after stem cell transplantation. *Ann Neurol* 2001;**50**(5):612–9.
54. Kawamura Y, Sugata K, Ihira M, et al. Different characteristics of human herpesvirus 6 encephalitis between primary infection and viral reactivation. *J Clin Virol* 2011;**51**(1):12–19.
55. Howell KB, Tiedemann K, Haeusler G, et al. Symptomatic generalized epilepsy after HHV6 posttransplant acute limbic encephalitis in children. *Epilepsia* 2012;**53**(7):e122–6.
56. Olli-Lahdesmaki T, Haataja L, Parkkola R, Waris M, Bleyzac N, Ruuskanen O. High-dose ganciclovir in HHV-6 encephalitis of an immunocompetent child. *Pediatr Neurol* 2010;**43**(1):53–6.
57. Gewurz BE, Marty FM, Baden LR, Katz JT. Human herpesvirus 6 encephalitis. *Curr Infect Dis Rep* 2008;**10**(4):292–9.
58. Provenzale JM, van Landingham K, White LE. Clinical and imaging findings suggesting human herpesvirus 6 encephalitis. *Pediatr Neurol* 2010;**42**(1):32–9.
59. French JA, Williamson PD, Thadani VM, et al. Characteristics of medial temporal lobe epilepsy. I. Results of history and physical examination. *Ann Neurol* 1993;**34**(6):774–80.
60. Mathern GW, Pretorius JK, Babb TL. Influence of the type of initial precipitating injury and at what age it occurs on course and outcome in patients with temporal lobe seizures. *J Neurosurg* 1995;**82**(2):220–7.

Chapter 5

HHV-6A, HHV-6B, and HHV-7 in Encephalitis

Joseph Ongrádi[a], Balázs Stercz[a], Tetsushi Yoshikawa[b], and Dharam V. Ablashi[c]

[a]*Institute of Medical Microbiology, Semmelweis University, Budapest, Hungary,* [b]*Fujita Health University School of Medicine, Toyoake, Aichi, Japan,* [c]*The HHV-6 Foundation, Santa Barbara, California*

INTRODUCTION

Viral encephalitis is a medical emergency. The spectrum of brain involvement and prognosis of the disease manifestation depend on the pathogen, the physical state of the host, and a range of environmental factors.[1,2] Viral encephalitis is an aseptic inflammatory process of the brain parenchyma associated with clinical evidence of brain dysfunction and significant morbidity and mortality.[3] Its outcome directly correlates with a rapid diagnosis and the onset of antiviral and supportive therapy.[4,5] Encephalopathy is mediated via metabolic processes and can be caused by systemic organ dysfunction or systemic infection that spares the brain. Viral encephalitis frequently involves the meninges and the spinal cord.[4,6]

Lately, human herpesvirus 6 viruses A and B (HHV-6A, HHV-6B)[7] and HHV-7[8] have been identified and regarded as prominent neurotropic viruses.[2,9] They could be etiological agents for multiple pathological conditions of the central nervous system (CNS), including different forms of acute, sporadic encephalitis; epilepsy; acute and prolonged febrile convulsions; meningitis; meningoencephalitis; encephalopathy; chronic fatigue syndrome (CFS); and several types of tumors.[10,11] These viruses establish a life-long latency in the immune system and persist in the brain.[12] Their single or simultaneous reactivation upon the effect of risk factors elicits debilitating CNS diseases.[13,14] HHV-6 viruses are able to transactivate other viruses, especially human immunodeficiency virus (HIV)-1,[15] and some viruses of human endogenous retroviruses (HERVs), which are the immediate cause of brain damage.[16,17] There are scarce data that chromosomally integrated HHV-6 (ciHHV-6) viruses[18,19] can induce severe CNS diseases including encephalitis[20] and encephalomyelitis.[21]

Human Herpesviruses HHV-6A, HHV-6B & HHV-7.

Unfortunately, correct virus diagnosis is found late and the initial treatment is usually inadequate.[21,22] The mortality rate for HHV-6 encephalitis is high, and surviving patients often display lingering neurological compromise.[22]

GENERAL ASPECTS AND NEUROTROPISM OF HUMAN HERPESVIRUS 6A INFECTION

Human herpesvirus 6A very rarely infects children in the developed world, but prevalence increases from adolescence on.[23] In sub-Saharan Africa,[24] many children below the age of 18 months carry this virus. Transmission might occur from mother to child.[24] The symptoms of acute infection are unknown, but febrile conditions in children and severe inflammatory or neurological disease in primary adult infections were observed.[14,25–27] Persistent HHV-6A infection in the brain may contribute to dementia associated with acquired immunodeficiency syndrome (AIDS). Primary infection in adults might trigger the onset of multiple sclerosis (MS).[25,28] HHV-6A infection activates HIV-1, human papillomavirus (HPV),[15] and Epstein–Barr virus (EBV),[29] which contributes to brain lymphomas of AIDS patients.[30]

GENERAL ASPECTS OF HUMAN HERPESVIRUS 6B INFECTION

The salivary glands serve as a reservoir for symptomless shedding of HHV-6B. This virus is regarded as a distinct species (see references in Ongrádi et al.[15] and De Bolle et al.[30]). By the age of 2 years, almost all children become seropositive. The majority of infections are symptomless, but children develop exanthem subitum (ES) at different ratios depending on geographical region: 17% in Britain,[26] 30% in the United States, and 60 to 80% in Japan.[30,31] Perinatal transmission is unlikely. HHV-6B frequently reactivates from latency in CD4+ immune cells among immunocompromised conditions followed by HHV-7 and CMV reactivation.[14] HHV-6B might act as a cofactor in the pathogenesis of several chronic debilitating immunological or neurological diseases such as Hodgkin's lymphomas, MS, mesial temporal lobe epilepsy (MTLE), CFS, and drug-induced hypersensitivity syndrome (DIHS).[30,32] Its DNA is commonly detected in the brain of deceased AIDS patients, and HHV-6B proteins are often located in the demyelinated areas, suggesting an active role in their neurological complications.[33]

Importance of Chromosomally Integrated HHV-6 in Neurological Conditions

Both viruses of HHV-6 can integrate into the telomeres of human chromosomes.[19,34,35] ciHHV-6 is inherited by the standard Mendelian inheritance.[19] Cell, blood, and organ transplantation transmits ciHHV-6. Approximately 0.2 to 0.85% of infants experience vertical transmission of HHV-6A or

HHV-6B.[36,37] Among individuals with reactivated ciHHV-6 some suffer from severe neurological symptoms for several years with remittences and relapses.[25] In children with encephalitis, ciHHV-6 prevalence was 3.3%, four times the rate in normal controls.[35] ciHHV-6 can be made to reactivate by chemically stimulating the integrated cells.[34] ciHHV-6 patients with neurological problems did not[18] or did[20,21,35] respond to antivirals. ciHHV-6 might be a risk factor for familial glioma cases.[38]

HUMAN HERPESVIRUS 7

Human herpesvirus 7 was isolated from activated T lymphocytes of a healthy donor.[8] HHV-7 is shed in the saliva of healthy subjects. It establishes latent infection in CD4+ lymphocytes and salivary glands. HHV-7 is ubiquitous worldwide, and approximately 70% of children by the age of 4 years contract infection via droplets or breast milk.[23] Occassionally, primary HHV-7 infection induces febrile convulsions and can invade the brain.[39] In young adults, a primary infection might elicit pityriasis rosea (PR).[40,41] HHV-7 reactivates alone or along with HHV-6B and CMV in immunocompromised patients.[5,13]

CLINICAL PRESENTATION OF HHV-6- OR HHV-7-ASSOCIATED ENCEPHALITIS

An acute encephalitis in an immunocompetent individual might occur after a primary infection or after becoming immunocompromised due to cell or organ transplantation, or another underlying disease.[1,3–5,21]

Encephalitis Following a Primary HHV-6 or HHV-7 Infection in Immunocompetent Individuals

In small children, encephalitis is suspected when an altered level of consciousness, significant change in personality, cognitive dysfunction, or focal neurological symptoms not explained by cranial nerve paralysis persist for more than 24 hours and other cases are excluded. The disease is accompanied by headache, nausea, temperature of ≥38°C, and specific laboratory results.[5] Since the first reported fatal case of encephalitis,[42] occassionally primary HHV-6B infection with accompanying neurological complications, including febrile seizures with or without manifestation of ES, and rarely acute encephalitis, meningitis, or demyelinating disease in children and adults have been described (Figure 5.1).[1,43] The exact fatality rate of HHV-6B encephalitis at the time of primary infection remains unclear. Convulsions are frequently atypical, sometimes resulting in status epilepticus.[42] Very rarely, invalidating sequele (e.g., speech disturbance, visual deficit, persistent hemiplegia) might follow HHV-6 meningoencephalitis in children. The exact fatality rate of HHV-6B encephalitis at the time of primary infection remains unknown.[44]

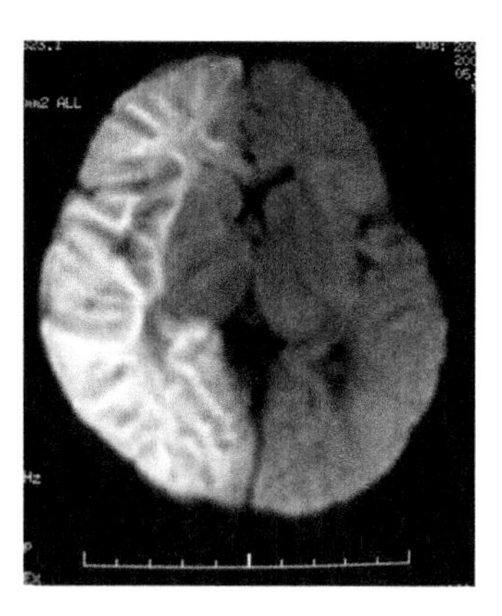

FIGURE 5.1 Brain MRI of HHV-6B encephalitis at the time of primary viral infection. This 1-year-old girl had hemorrhagic shock and encephalopathy syndrome caused by primary HHV-6B infection. She had high fever and left hemiconvulsion at the time of admission to the hospital. Magnetic resonance imaging showed a high-intensity signal in the right cerebral hemisphere.

Acute HHV-7 infection and detection in the CSF of children and adults were associated with encephalopathy, meningitis, viral encephalitis/meningo-encephalitis, facial palsy, vestibular neuritis, febrile or focal seizure, severe headache, fatigue, nausea, vomiting, fever up to 39.8°C, photosensitivity, lethargy, impaired orientation, comatous state, somnolence, difficulties in walking, tiredness, sudden onset of dizzines, and tendency to fall to one side (straddle gait).[45–47] Complications in adults include urinary retention, flaccid paralyis of the limbs progressing to quadriplegia, and respiratory failure. There is no difference between the neurological effects of primary infection with HHV-6 and HHV-7 in children. Unexplained encephalitis associated with flaccid paralysis[46] or meningoencephalitis with paraparesis[6] in adults might be induced by primary HHV-7 infection.

There is a paucity of data on the onset of encephalitis after simultaneous infection with HHV-6 and HHV-7. Simultaneous HHV-7 and CMV infection induced encephaloradiculomyelitis.[48] In 22 cases of HSV encephalitis, three patients had both HSV and HHV-6 in their CSF; two of these patients died. HHV-6 reactivation was regarded as the consequence rather than a cause, of established brain damage.[46]

Encephalitis Following Reactivation of HHV-6 or HHV-7 in Immunocompromised Patients

Although HHV-6 reactivation is common among stem cell transplant (SCT), bone marrow transplant (BMT), and solid organ transplant (SOT) recipients, a minority of patients develop HHV-6-associated life-threatening encephalitis.[49,22,50] Recipients of heart or lung transplant,[14] allogeneic stem cell

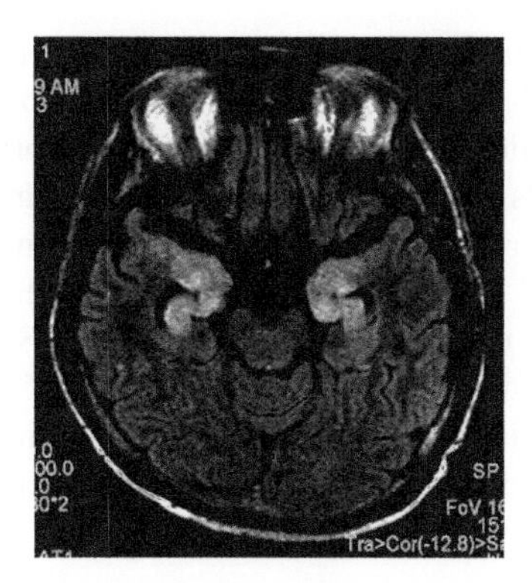

FIGURE 5.2 Brain MRI of posttransplant HHV-6B encephalitis. A 33-year-old male with malignant lymphoma received a hematopoietic stem cell transplant. He had convulsion and stupor 19 days after transplant; subsequently, he also experienced amnesia. As a high copy number of HHV-6B DNA was detected in the cerebrospinal fluid, he was finally diagnosed with posttransplant HHV-6B encephalitis.

transplants, and liver transplants[51–53] seem to be at an increased risk. Viral reactivation in the brain might cause encephalitis,[46,54,55] but viremia is also detected 2 to 4 weeks after transplantation.[53,56] Patients were generally young (median 35 years). Onset of encephalitis began at a median of 24 days posttransplantation. CNS symptoms were characterized by depressed consciousness, confusion, disorientation, insomnia, problems with memory (especially short-term memory), clinical and electrographic seizures, and imaging abnormalities (Figure 5.2).[52,54,56] The most severe manifestation is the syndrome of posttransplantation acute limbic encephalitis (HHV-6-PALE), which has a high death toll,[57] especially after hematopoietic SCT (HSCT).[22] Posterior reversible encephalopathy syndrome (PRES) has been associated with a number of medical conditions, and the first typical cases associated with HHV-6B reactivation have been described recently.[2,58] High viral loads up to 2×10^6/mL are shown throughout the brain.[54] Small amounts of DNA may signify an active infection and can be amplified from the CSF.[59] Other studies have not found an association between HHV-6 and CNS dysfunction.[53] Lack of standardized HHV-6 detection or a small number of patients might explain conflicting results.[56]

Human herpesvirus 7 DNA has been detected in normal brain tissues through nested polymerase chain reaction (PCR), proving that a low copy number persists in the brain of healthy people.[47] HHV-7 has been associated with encephalitis in BMT patients.[49] CNS diseases in immunocompromised patients are reported to occur with HHV-7 reactivation from neurolatency. After receiving HSCT for acute lymphoblastic leukemia (ALL), a small girl developed visual and hearing impairment, bulbar dysfunction, confusion, and cardiorespiratory insuffiency. HHV-7 was detected in the CSF, but in spite of ganciclovir therapy she died. CSF samples from control leukemic children without neurological complications were shown to be negative for HHV-7, suggesting that HHV-7 is likely to be more than a bystander in the CSF.[47] Most SCT recipient children are HHV-6 seropositive prior to transplant. After transplant, higher levels of either HHV-6 or HHV-7 antibodies, antigenemia, and DNA detection were more frequent in the allogeneic than in autologous SCT recipients. HHV-7 reactivation has not been observed at a constant time post-SCT. Both viruses reactivate in 30 to 70% of all SCT recipients. Patients with total body irradiation (TBI) faced reactivation of both viruses more often than those not irradiated.[9] HHV-7-reactivation-associated encephalitis might follow a first course of encephalitis induced by HHV-6 after HSCT. Several weeks of renewed foscarnet therapy have resulted in neurological improvements.[13]

Other reactivated herpesviruses might induce simultaneous infection and encephalitis with HHV-6 or HHV-7. Double infections of HSV/CMV and HSV/HHV-6 were found in 2% of patients. DNA of HSV-1, CMV, and HHV-6 is found exclusively in the CSF of encephalitis cases, and DNA of Varicella zoster virus (VZV) can be detected in both encephalitis and meningitis patients. Simultaneous infection by HSV-2 and HHV-7 was not detected.[5] HHV-7 has been shown to induce indirect effects in the body that may become risk factors for CMV diseases.[49] Simultaneous HHV-7, HHV-6, and CMV antigenemia is frequently found in immunocompromised hosts.[13] The transactivator potential of HHV-7 could be different from the CMV or EBV activating effect of HHV-6A.[15] HHV-7 may increase immunodeficiency through aggravating cytokine dysfunction.[48] Its high IL-10 inducing ability might contribute to CNS disorders.[28] Studies on patients with influenza-associated encephalopathy showed that there is likely no association of primary or reactivated HHV-6 or HHV-7 infection with their clinical symptoms.[60]

PATHOLOGY AND PATHOGENESIS

Human herpesvirus 6 gains access to the CNS by crossing the blood–brain barrier (BBB) through the olfactory pathway.[32] Olfactory ensheathing cells (OECs), specialized glial cells that possess characteristics of astrocytes and Schwann cells, are located in the nasal cavity and have been found to support HHV-6A replication but not HHV-6B infection.[32] HHV-6A triggers production of proinflammatory cytokines, interleukin (IL)-6, and chemokines CCCL-1 and RANTES.[32] As with other viruses utilizing the olfactory route, the presence of HHV-6 in limbic tissues can lead to limbic encephalitis.[61] Active viral replication in different cell types may contribute to different clinical symptoms.[59] Infection of oligodendrocytes may be associated with MS, whereas infection of astrocytes may be more common in other CNS disorders such as MTLE and encephalitis.[62] HHV-6B establishes latent infection in an astrocytoma cell line.[63] HHV-6 displays tropism for hippocampal astrocytes,[22,56] and neuronal loss in these infected areas has been demonstrated after HSCT-induced encephalitis.[33,59,57,64] HHV-6A induces productive infection of glial cells and is harbored in 72% of gliomas, while HHV-6B causes persistent infections.[59] Cytokine pattern alterations in CSF were found to be different in encephalitis cases occuring after primary infections and reactivation in transplant recipients,[65] suggesting the importance of immune-mediated mechanisms.[66] Infants with encephalitis associated with primary HHV-6 infection show higher serum and CSF levels of IL-6. SCT recipients have a tendency to display hypercytokinemia. Before HHV-6 reactivation and progression toward encephalopathy, an elevated level of IL-6 was detected.[22] High CSF levels of IL-8,[11] IL-10, interferon (IFN)-γ, IL-8, and tissue-regenerating matrix metalloproteinase-9 (MMP-9) were found in encephalitis patients but not in healthy individuals.[11,22,66] *In vitro* reactivation of HHV-6B abortive infection in an astrocytoma cell line also resulted in the overproduction of IL-6 and IL-1β proinflammatory cytokines.[63] High CSF levels of IL-1β and IL-8 trigger vasogenic edema in the brain.[2] High plasma IL-6 levels were found in a patient with ciHHV-6.[35] Elevated levels of IL-6, soluble tumor necrosis factor receptor (sTNFR)-1, and CSF IL-1 predict neurological sequele.[35,66] These alterations are similar to those found in acute encephalitis induced by other viruses and suggest a partially common pathway.[2,52,60,67,68]

RISK FACTORS

For HHV-6A, HHV-6B, and HHV-7, the etiological spectrum is not influenced by geography as these viruses are ubiquitous. Most primary HHV-6 infection occurs in the second six months of life, while HHV-7 primary infection is in the second or third year sequentially.[23,46] This temporal sequence does not apply in children with encephalitis or seizures, where the median age is one year for both. Absence of previous HHV-6 infecton predisposes individuals to HHV-7-induced neurological morbidity due to a lack of limited cross-reactivity between these viruses.[46] The majority of children and adults suffering from encephalitis due to primary HHV-7 infection are males.[45,48,60,69] The incidence of HHV-6 encephalitis in the general population is low,[1,22,33,56] but HHV-6 reactivation has been shown in 40 to 50% of HSCT recipients.[9,53,56] Most events have been due to HHV-6B; only 2 to 3% of events are caused by HHV-6A.[56,64] Younger age; underlying malignancy (leukemia or lymphoma); allogeneic transplants; unrelated, mismatched, or gender-mismatched donors; receipt of anti-T cell antibodies or steroids; and mainly unrelated cord blood cell transplantation (UCBT) might be risk factors for encephalitis. Males seem to be more frequently affected.[54,64] HHV-6-associated encephalitis or myelitis occurs more frequently in recipients who underwent two or more HSCTs compared to those who received only one HSCT.[50,54,55,70] An association of high HHV-6 levels in the blood and peripheral blood mononuclear cells (PBMCs) and CNS dysfunction with encephalitis has been documented.[22] HHV-6 was detected more frequently in the CSF of transplant recipients with encephalitis than in immunocompromised patients without encephalitis. Detection of HHV-6 DNA in the CSF of an HSCT recipient is potentially meaningful, and if these patients present with CNS symptoms antiviral therapy should be initiated.[56] Reactivation of CMV, HHV-6, and EBV differs with respect to risk factors and clinical outcome after HSCT.[71]

DIAGNOSIS

Infections of the CNS are a difficult diagnostic problem for both clinicians and microbiologists.

Clinical Manifestations and Other Relevant Information

The patient's clinical history is mandatory in the assessment of suspected encephalitis. Clinical diagnosis of ES or PR with concomitant neurological symptoms suggests an association with CNS disorders.[59] Geographical location, seasonal occurence, occupation, recent contact with animals, or insect bites can be relevant information for epidemic encephalitis.[3] Clinical parameters, ciHHV-6 status, immuncompromised conditions, use of recreational drugs, and medications known to affect the immune system or reactivate latent viruses have to be assessed.[4]

Fever, skin rash, and systemic symptoms can accompany or preclude HHV-6 or HHV-7 encephalitis.[9] Rheumatic symptoms might be associated with HHV-7 plus CMV infection and encephalitis.[48] Behaviorial, cognitive, and focal neurological signs; seizures; and autonomous and hypothalamic disturbances reflect abnormal CNS functions.[4] Unusual clinical presentations and inconclusive laboratory results are not uncommon in case reports.[69]

General Laboratory Procedures

Peripheral blood lymphocytosis is common in viral encephalitis.[4] In HHV-6 encephalitis following HSCT, CSF is not unusual, except that it is accompanied by elevated protein level. A minority of patients have been diagnosed with pleiocytosis, and leukopenia is frequent among transplanted patients.[1,2,20,54,56] Severe hyponatremia can be a prodromal or concomitant manifestation of HHV-6-PALE.[72] In HHV-7-associated encephalitis, laboratory results may not be consistent—for example, normal or elevated protein and glucose levels in the CSF, a high leukocyte count with lymphocyte predominance and neutrophils, normal or elevated peripheral white blood cell (WBC) count, differential counts, normal electrolyte and glucose levels, normal liver and renal function, and normal or elevated C-reactive protein (CRP) levels.[45,47,56] Simultaneous reactivation of HHV-6 and HHV-7[13] or HHV-7 and CMV[48] might modify laboratory results. Recent studies on the cytokine pattern alterations in both CSF and serum may demonstrate that they are important diagnostic critieria.[11,50,69]

Electroencephalography, Neuroimaging, and Histopathology

Electroencephalography (EEG) and a baseline quantitative EEG (QEEG) have been demonstrated to be objective means of measuring neurological disorders. EEG can show background abnormality prior to initial evidence of parenchyma involvement on neuroimaging. Several EEG abnormalities were described in acute, reactivated HHV-6 and HHV-7 infection, as well as ciHHV-6-associated encephalitis cases. Antiviral treatment and clinical improvement are often accompanied by normalized EEG.[20,35,45,56,69]

Characteristic neuroradiographic signs of HHV-6-associated encephalitis in both immunocompetent and immunocompromised patients have been identified by using computerized tomography (CT),[11,56,59] magnetic resonance imaging (MRI),[10,54,56]fluid-attenuated inversion recovery (FLAIR) MRI,[65] and/or acute-phase 2-deoxy-[^{18}F]fluoro-D-glucose positron emission tomography (FDG-PET).[73] HHV-6 encephalitis often involves both hippocampal and extrahippocampal structures, including the amygdala, entorhinal cortex, thalamus, hypothalamus, and deep forebrain structures, as well as the cerebellum and brainstem. Edema, necrotization, and sclerosis are frequently found.[52,59] In patients with HHV-7-associated encephalomyelitis[45,69] and simultaneous HHV-6 and HHV-7 reactivation-associated encephalitis,[13] CT and MRI scans have detected abnormalities in the cortex, cerebellum, and hippocampi. Improvement comes with antiviral therapy, but sequele have still been detected.

In autopsy specimens, histopathology has found high HHV-6 viral loads in the hippocampus, basal ganglia, insular cortex, temporal lobe, and cingular gyrus; astrogliosis and neuronal loss in regions of the hippocampus was probably due to HHV-6 protein expression.[59] Absence of viral inclusions has also been reported in fatal HHV-6 encephalitis.[33] Autopsy specimens obtained from an encephalitis case did not reveal macroscopic changes but did show multiple foci of microscopic neuronal degeneration, and HHV-7 was detected by PCR in the brain stem and CSF, suggesting that HHV-7 may induce brainstem encephalitis.[47]

Virological and Microbiological Investigations

Identifying clinically relevant HHV-6 or HHV-7 infections can be challenging due to their ubiquitous and persistent nature. Primary HHV-6 and HHV-7 infectons with acute febrile neurological illness occur at the time when routine childhood vaccines are administered. Neurological complications wrongly attributed to vaccination may in fact be due to coincident virus infection.[46] Detection of replicating HHV-6 or HHV-7 in the brain tissues would help prove etiology.[59,74] Brain biopsy is not used in routine clinical practice, but at autopsy brain material should be obtained for virus isolation, nucleic acid (NA) and antigen detection, immunochemistry, and *in situ* hybridization. The gold standard of diagnosis is considered to be virus isolation in cell culture; however, CSF or brain tissue samples rarely contain enough infectious materials, and autopsy materials do not contain infectious viruses.[4]

Cultivation is now replaced by the detection of specific NA by PCR-based technologies. The demonstration of viral DNA in blood or saliva alone should not be considered confirmatory in the diagnosis of encephalitis. Assays performed on CSF, CSF cells, brain biopsy, autopsy materials, tissue, blood, and/or PBMCs can give a more dependable diagnosis.[1,14,18,56] Detection of HHV-6 or HHV-7 NA from CSF depends upon when the sample was obtained. The highest yield is generally obtained when the virus appears transiently during the first week following the onset of symptoms.[1,46] PCR should be repeated after 3 to 7 days. Although HHV-6 DNA and mRNAs have been detected in normal brain tissue, these are found at a high frequency (42%) in the CSF of children with acute or past infection. The presence of ciHHV-6 must be considered when interpreting results.[56] ciHHV-6 is diagnosed in hair follicles.[35,37] Nested PCR (nPCR) with higher sensitivity is commonly used to detect HHV-6 and especially HHV-7[60] in CSF and autopsied brain materials.[3,39,45,47] Active infection such as reactivation or reinfection is best proven by quantitation of viral mRNA by reverse transcriptase (RT)-PCR.[6,46] Multiplex PCR (mPCR) assays are marketed for simultaneous detection of herpesviruses in a very small volume of CSF[1] and are able to distinguish among HHV-6A, HHV-6B, and HHV-7.[14,45] The use of quantitative PCR (qPCR) to measure virus loads before, during, and after the onset of encephalitis allows for a more comprehensive definition of the viral contribution to disease.[49] Low copy numbers of CSF HHV-6 DNA were detected in many of the patients with HHV-6 encephalitis in primary infection, whereas all tested CSF samples collected from posttransplant HHV-6 encephalitis patients contained high viral DNA copy numbers.[65] Real-time PCR makes it possible to establish viral DNA load in both CSF and serum in a shorter time.[46,60] The minimum detection level of most real-time PCR assays is extremely low (e.g., two copies per reaction), whereas the threshold level for development of HHV-6 encephalitis is around 10^4 copies per mL plasma.[11,50] Multiplex real-time PCR established four copies of HHV-6 and three copies of HHV-7 as the minimum quantitative detection level in encephalitis/encephalopathy patients.[75] Higher levels of HHV-6 DNA in peripheral blood are associated with the development of CNS dysfunction.[54] Patients with ciHHV-6 might show very high serum copy levels but low CSF levels.[35] Encephalitis and myelitis are associated with inflammation and alteration of the BBB; the detection of viral DNA in CSF by PCR *per se* might not discriminate between a CNS and a systemic infection.[48]

Antigens of HHV-6 or HHV-7 can be studied in brain cells by immunofluorescent (IF) or immunoperoxidase tests,[9] but these techniques seem unsatisfactory in diagnosis using CSF samples.[46] Intrathecal antibody production to HHV-6 or HHV-7 is strong evidence for etiology, but the detection of systemic serological responses should not be considered grounds for diagnosis. CSF is a convenient specimen and is recommended for neurological viral diagnosis. In a young child, serological tests are the key to diagnosis, as viral nucleic acid is not always detected in the CSF. Diagnostic methods must distinguish primary antibody response and pre-existing antibodies; primary infection may be verifyed by the use of antibody avidity test. In the immunocompromised patient, use of the avidity test is required as antibody titers may have decreased to the point of seronegativity and a secondary antibody response may therefore appear to be that of a primary antibody response. Assays have to distinguish between the antibody response to HHV-6A and to HHV-6B. Recently, a HHV-6A- and HHV-6B-specific immunoblotting assay has been developed.[76] Tests should be repeated after 2 to 4 weeks to show seroconversion or diagnostic increase in antibody levels.[4]

Differential Diagnosis

The etiological agents involved in sporadic encephalitis are difficult to establish, because the clinical symptoms and laboratory results can be very diverse and unusual. It is logical to conduct microbial investigations in many directions, including different herpesviruses, enteroviruses, influenza viruses, parainfluenza viruses, adenoviruses, and mycoplasma and chlamydophila viruses, to mention only a few.[59,69,74] Viral cultures from CSF, brain tissue, throat, and stool specimens and NA detection by PCR are commonly performed.[13,47] NA microarrays are suitable for detecting several microbes and their genotypes simultaneously.[4,6,61] Several of these microorganisms can infect a patient simultaneoeulsy; for example, in a child simultaneous acyclovir-resistant HSV-1 and adenovirus esophagitis, HHV-6-encephalitis, rotavirus gastroenteritis, and respiratory syncytial virus pneumonia were identified.[77] Antigen detection is not helpful in diagnosis using CSF samples.[4,60] Intrathecal but not serum antibodies might help sort out etiological agents. Antibody tests for measles, mumps, and rubella are occasionally necessary in countries with effective vaccination programs. Tests for arboviruses and zoonosis are useful in epidemic areas.[1] The simultaneous presence of several antibodies in the CSF suggests BBB breakdown. Different viruses induce different histological types of encephalitis. HSV, VZV, CMV, and HIV can induce panencephalitis, and RNA viruses elicit a patchy-nodular type of histology. The major target of the CNS by HHV-6 and HHV-7 is the brainstem. Acute disseminated encephalomyelitis (ADEM) follows febrile illnesses or immunization.[4,67] In pediatric patients, the exclusion of HSV-1, HSV-2, VZV, or CMV infection[5] or tuberculous meningitis[4] is important. In Japan, acute encephalopathy induced by influenza virus infection was the most common type followed by HHV-6 and rotavirus.[67] In another study, HSV-1, VZV, and enteroviruses were identified as the most common etiological agents.[1] In SOT and HSCT recipients, CMV causes the most severe disease.[14] The association between HHV-6 infection and the risk of bacterial and fungal infections is controversial, but persistent fungal and bacterial infections were found in patients with ciHHV-6.[14,35] In addition to reducing viral infection valaciclovir or ganciclovir prophylaxis can also reduce the risk for fungal infections.[13]

Prognosis, Therapy, and Prevention

All patients with acute encephalitis must be hospitalized, but in cases of HHV-6 or HHV-7 association, isolation is not required.[35] Primary infection by HHV-6 or HHV-7 usually does not require specific treatment. In the case of ES, due to high fever, prescribing the amoxicillin–clavulic acid combination may activate HHV-6 and therefore ought to be omitted.[18] In adolescents and adults with typical symptoms of PR, symptomatic therapy might contain steroids. In children with CNS diseases of unknown origin, prednisolone might be given.[45] Treatment with steroid derivates (e.g., hydrocortisone) enhances HHV-6 replication.[4,34,45] Patients with malignancies, and especially those who receive HSCT, undergo very intensive combined chemotherapy, intracranial or total body irradiation, and biological therapy.[20,22,65] These can activate HHV-6, HHV-7, and other viruses in the CNS.[74] HHV-6 and especially ciHHV-6 can be reactivated by common pharmaceuticals, among them histone deacetylase, an inhibitor used as an anticancer agent; aliphatic acid derivates used as anticonvulsants; analgesic, antipyretic, or antiemetic drugs; and anesthetic drugs (see references in Pellett et al.[18]). Because immunosuppression is vital for these patients and it seems impossible to alter treatment protocols, the use of particular drugs should be carefully considered and potential effects on the viruses should be monitored.[18,34,35,65] Therapy against hypercytokinemia might improve the physical state of the patients, but such attempts, including intensive steroid application, has resulted in the opposite effect.[22]

Clinical diagnosis of viral encephalitis before HHV-6 or HHV-7 identification might indicate HSV or VZV infection and acyclovir therapy, but acyclovir has little or no activity against HHV-6 or HHV-7.[4,47,49] When reactivation of HHV-6 or HHV-7 is recognized, patients must be treated immediately. Foscarnet, ganciclovir, and cidofovir have been shown to have *in vitro* inhibitory effects against HHV-6, although randomized, placebo-controlled trials are lacking.[13,49] Reports of successful treatment of several HSCT recipients and other patients have been published. A decrease of HHV-6 DNA in the CSF over time was documented, but some of the patients were left with lingering neurological compromise.[78] In both children and adults, symptomatic generalized epilepsy with seizures might follow PALE, which are resistant to multiple antiepileptic drugs.[57] Ganciclovir therapy resulted in gradual improvement in neurological symptoms of HHV-6B-related PRES.[65] The rate of decrease in HHV-6 DNA is slower in CSF than in blood. Clinical symptoms may not vanish until after several months of completion of antiviral therapy.[54] The potential emergence of ganciclovir-resistant mutants makes it necessary to monitor the virus carefully throughout therapy.[56,79] Low-dose foscarnet does not suppress HHV-6 reactivation and cannot prevent all cases of encephalitis.[50] HHV-7 is much less susceptible to ganciclovir treatment than HHV-6, although full recovery from encephalitis has been desribed.[6] Therapeutic failure after foscarnet[13] or ganciclovir[47] application has been reported. These drugs have serious side effects; therefore, they should be used only in life-threatening situations.[46]

Despite high toxicity, preemptive therapy by ganciclovir[50] and foscarnet[55] has been shown to be successful in preventing many but not all cases of HHV-6 encephalitis in transplant patients. So far, no concensus has been reached regarding appropiate preventive methods.[50] The recent guidelines from the American Society of Transplantation do not recommend antiviral prophylaxis for HHV-6 infection, except in the case of established disease; especially with encephalitis, intravenous ganciclovir and foscarnet are considered to be first-line agents.[14] Reactivated ciHHV-6 poses a high risk for CNS complications in recipients. Carrier persons should be banned from donation of cells, blood, or organs for transplantation.[18,20] Both foscarnet and valganciclovir therapy of ciHHV-6 patients with CNS disturbances has been reported to be successful with a rapid fall in CSF and blood viral copy numbers and no clinical sequelae.[35]

CONCLUSIONS

Human herpesvirus 6 and HHV-7 are emerging pathogens in both immunocompetent and immunocompromised individuals, especially in cases of transplantation, and may cause CNS diseases. When making the diagnosis of their infection, chromosomal integration should be considered, especially if persistently high levels of HHV-6 DNA are detected in the blood and CSF. Following successful chemotherapy and clinical improvement, relapse of symptoms should prompt a search for other co-pathogens. HHV-6 and HHV-7 can also reactivate consecutively or simultaneously. The scale of the contribution of HHV-6 or HHV-7 infection to serious neurological diseases including encephalitis has not been appreciated in the past, but such cases must now be fully investigated for these viruses. Particular attention must be paid to both donors and recipients who carry ciHHV-6, because ciHHV-6 readily reactivates under immunocompromised conditions; combined chemotherapy elicits severe CNS symptoms and it resists antiviral chemotherapy. Clinical symptoms and laboratory findings can be very diverse and vary from patient to patient; therefore, continuous monitoring of patients is very important. In the differential diagnosis of HHV-6A and HHV-6B, new technical reagents and applications have lately become available. There is a need for the development and use of rapid diagnostic tests for acute specimens. Ganciclovir, foscarnet, and cidofovir therapy of HHV-6- and especially HHV-7-induced encephalitis has been shown to be of limited value; toxicity, as well as morbidity and mortality, remain high. The emergence of chemotherapy-resistant virus mutants poses a new challenge and suggests that drug combinations will have to be used in the future. Several commonly used medications and immunosuppressants might reactivate latent HHV-6, HHV-7, or most importantly ciHHV-6; therefore, their application should be restricted to life-threatening conditions. One has to balance the benefit versus harm of any therapy of HHV-6- or HHV-7-induced encephalitis on an individual basis.

ACKNOWLEDGMENTS

The authors are thankful to Dr. Louise Chatlynne, Ms. Kristin Loomis, and Mr. Joshua Prichett (HHV-6 Foundation, Santa Barbara, CA) for their suggestions and corrections of the manuscript; to Dr. Károly Nagy (Director) for his encouragement and Mrs. Valéria Kövesdi for her many-sided and valuable clerical assistance through the preparation of the manuscript (Department of Medical Microbiology, Semmelweis University, Budapest).

REFERENCES

1. de Ory F, Avellón A, Echevarría JE, et al. Viral infections of the central nervous system in Spain: a prospective study. *J Med Virol* 2013;**85**:554–62.
2. Kawamura Y, Ohashi M, Asahito H, et al. Posterior reversible encephalopathy syndrome in a child with post-transplant HHV-6B encephalitis. *Bone Marrow Transplant* 2012;**47**:1381–2.
3. Michael BD, Solomon T. Seizures and encephalitis: clinical features, management and potential pathophysiologic mechanisms. *Epilepsia* 2012;**53**:63–71.
4. Steiner I, Budka H, Chaudhuri A, et al. Viral encephalitis: a review of diagnostic methods and guidelines for management. *Eur J Neurol* 2005;**12**:331–43.
5. Ibrahim AI, Obeid MT, Jouma MJ, et al. Prevalence of herpes simplex virus (types 1 and 2) varicella-zoster virus cytomegalovirus and human herpesvirus 6 and 7 DNA in cerebrospinal fluid of middle eastern patients with encephalitis. *J Clin Microbiol* 2005;**43**:4172–4.
6. Miranda CM, Torres TJP, Larrañaga LC, et al. Meningomyelitis associated with infection by human herpes virus 7: report of two cases. *Rev Med Child* 2011;**139**:1588–91.
7. Salahuddin SZ, Ablashi DV, Markham PD, et al. Isolation of a new virus HBLV in patients with lymphoproliferative disorders. *Science* 1986;**234**:596–601.
8. Frenkel N, Schirmer EC, Wyatt LS, et al. Isolation of a new herpesvirus from human CD4+ T cells. *Proc Natl Acad Sci USA* 1990;**87**:748–52.
9. Savolainen H, Lautenschlager I, Piiparinen H, et al. Human herpesvirus-6 and -7 in pediatric stem cell transplantation. *Pediatr Blood Cancer* 2005;**45**:820–5.
10. Crawford JR, Santi MR, Cornelison R, et al. Detection of human herpesvirus-6 in adult central nervous system tumors: predominance of early and late viral antigens in glial tumors. *J Neurooncol* 2009;**95**:49–60.
11. Kawabe S, Ito Y, Ohta R, et al. Comparison of the levels of human herpesvirus 6 (HHV-6) DNA and cytokines in the cerebrospinal fluid and serum of children with HHV-6 encephalopathy. *J Med Virol* 2010;**82**:1410–5.
12. Yoshikawa T, Asano Y. Central nervous system complications in human herpesvirus-6 infection. *Brain Dev* 2000;**22**:307–14.

13. Holden SR, Vas AL. Severe encephalitis in a haematopoietic stem cell transplant recipient caused by reactivation of human herpesvirus 6 and 7. *J Clin Virol* 2007;**40**:245–7.
14. Lautenschlager I, Razonable RR. Human herpesvirus-6 infections in kidney, liver lung and heart transplantation: review. *Transpl Int* 2012;**25**:493–502.
15. Ongrádi J, Kövesdi V, Nagy K, Matteoli B, Ceccherini-Nelli L, Ablashi D. *In vitro* and *in vivo* transactivation of HIV by HHV-6. In: Chang TL, editor. *HIV–Host Interactions*. Rijeca, Croatia: InTech; 2011. p. 257–98.
16. Tai AK, Luka J, Ablashi D, et al. HHV-6A infection induces expression of HERV-K18-encoded superantigen. *J Clin Virol* 2009;**46**:47–8.
17. Turcanova VL, Bundgaard B, Höllsberg P. Human herpesvirus-6B induces expression of the human endogenous retrovirus K18-encoded superantigen. *J Clin Virol* 2009;**46**:15–19.
18. Pellett PE, Ablashi DV, Ambros PF, et al. Chromosomally integrated human herpesvirus 6: questions and answers. *Rev Med Virol* 2012;**22**:144–55.
19. Morissette G, Flamand L. Herpesviruses and chromosomal integration. *J Virol* 2010;**84**(12):100–9.
20. Wittekindt B, Berger A, Porto L, et al. Human herpes virus-6 DNA in cerebrospinal fluid of children undergoing therapy for acute leukaemia. *Brit J Haematol* 2009;**145**:542–5.
21. Troy SB, Blackburn BG, Yeom K, et al. Severe encephalomyelitis in an immunocompetent adult with chromosomally integrated human herpesvirus 6 and clinical response to treatment with foscarnet plus ganciclovir. *Clin Dis Inf* 2008;**47**:93–6.
22. Ogata M, Satou T, Kawano R, et al. Correlations of HHV-6 viral load and plasma IL-6 concentration with HHV-6 encephalitis in allogeneic stem cell transplant recipients. *Bone Marrow Transplant* 2010;**45**:129–36.
23. Ongrádi J, Csiszár A, Maródi C, et al. Onset of antibodies to human herpesvirus types 6 and 7 in Hungarian children. *Orv Hetil* 1999;**140**:935–40.
24. Bates M, Monze M, Bima H, et al. Predominant human herpesvirus 6 variant A infant infections in an HIV-1 endemic region of Sub-Saharan Africa. *J Med Virol* 2009;**8**:779–89.
25. Alvarez-Lafuente R, de las Heras V, Garcia-Montojo M, et al. Human herpesvirus-6 and multiple sclerosis: relapsing-remitting versus secondary progrssive. *Mult Scler* 2007;**13**:78–83.
26. Hall CB, Caserta MT, Schnabel KC, et al. Persistence of human herpesvirus 6 according to site and variant: possible greater neurotropism of variant A. *Clin Inf Dis* 1998;**26**:132–7.
27. Portolani M, Tamassia MG, Gennari W, et al. Post-mortem diagnosis of encephalitis in a 75-year-old man associated with human herpesvirus-6 variant A. *J Med Virol* 2005;**77**:244–8.
28. Ongrádi J, Ahmad A, Menezes J. *In vitro* cytokine induction by HHV-7 in leukocytes. *Orv Hetil* 1999;**140**:1935–9.
29. Flamand L, Stefanescu I, Ablashi DV, Menezes J. Activation of Epstein–Barr virus replicatice cycle by human herpesvirus 6. *J Virol* 1993;**67**:6768–77.
30. De Bolle L, Naesens L, De Clercq E. Update on human herpesvirus 6 biology clinical features and therapy. *Clin Microbiol Rev* 2005;**18**:217–45.
31. Yamanishi K, Okuno T, Shiraki K, Takahashi M, et al. Identification of human herpesvirus-6 as a causal agent for exanthem subitum. *Lancet* 1988;**14**:1065–7.
32. Harberts E, Yao K, Wohler JE, et al. Human herpesvirus-6 entry into the central nervous system through the olfactory pathway. *Proc Natl Acad Sci USA* 2011;**108**:13734–9.
33. Drobyski WR, Knox KK, Majewski D, Carrigan D, et al. Brief report: fatal encephalitis due to variant B human herpesvirus-6 infection in a bone marrow-transplant recipient. *N Engl J Med* 1994;**330**:1356–60.
34. Arbuckle JH, Medveczky MM, Luka J, et al. The latent human herpesvirus-6A genome specifically integrates in telomeres of human chromosomes *in vivo* and *in vitro*. *Proc Natl Acad Sci USA* 2010;**107**:5563–8.

35. Montoya JG, Neely MN, Gupta S, et al. Antiviral therapy of two patients with chromosomally integrated human herpesvirus-6A presenting with cognitive dysfunction. *J Clin Virol* 2012;**55**:40–5.
36. Hall CB, Caserta MT, Schnabel KC, et al. Congenital infections with human herpesvirus 6 (HHV6) and human herpesvirus 7 (HHV7). *J Pediatr* 2004;**145**:472–7.
37. Ward KN, Leong HN, Nacheva EP, et al. Human herpesvirus 6 chromosomal integration in immunocompetent patients results in high levels of viral DNA in blood sera and hair follicles. *J Clin Microbiol* 2006;**44**:1571–4.
38. Amirian ES, Scheurer ME. Chromosomally integrated human herpesvirus 6 in familial glioma etiology. *Med Hypotheses* 2012;**79**:193–6.
39. Yoshikawa T, Ihira M, Suzuki K, et al. Invasion by human herpesvirus 6 and human herpesvirus 7 of the central nervous system in patients with neurological signs and symptoms. *Arch Dis Child* 2000;**83**:170–1.
40. Drago F, Ranieri E, Malaguti F, et al. Human herpesvirus 7 in patients with pityriasis rosea. Electron microscopy investigations and polymerase chain reaction in mononuclear cells plasma and skin. *Dermatology* 1997;**195**:374–8.
41. Vág T, Sonkoly E, Kárpáti S, et al. Avidity of antibodies to human herpesvirus 7 suggests primary infection in young adults with pityriasis rosea. *J Eur Acad Dermatol Venereol* 2004;**18**:738–40.
42. Asano Y, Yoshikawa T, Kajita Y, et al. Fatal encephalitis/encephalopathy in primary human herpesvirus-6 infection. *Arch Dis Child* 1992;**67**:1484–5.
43. Yoshikawa T, Ohashi M, Miyake F, et al. Exanthem subitum-associated encephalitis: nationwide survey in Japan. *Pediatr Neurol* 2009;**41**:353–8.
44. Bozzola E, Krzysztofiak A, Bozzola M, Calcaterra V, et al. HHV6 meningoencephalitis sequelae in previously healthy children. *Infection* 2012;**40**:563–6.
45. Pohl-Koppe A, Blay M, Jäger G, et al. Human herpes virus type 7 DNA in the cerebrospinal fluid of children with central nervous system diseases. *Eur J Pediatr* 2001;**160**:351–8.
46. Ward KN. The natural history and laboratory diagnosis of human herpesviruses-6 and -7 infections in the immunocompetent. *J Clin Virol* 2005;**32**:183–93.
47. Chan PK, Chik KW, To KF, Li CK, et al. Case report: human herpesvirus 7 associated fatal encephalitis in a peripheral blood stem cell transplant recipient. *J Med Virol* 2002;**266**:493–6.
48. Ginanneschi F, Donati D, Moschettini D, et al. Encephaloradiculomyelitis associated to HHV-7 and CMV co-infection in immunocompetent host. *Clin Neurol Neurosurg* 2007;**109**:272–6.
49. Ljungman P. Beta-herpesvirus challenges in the transplant recipient. *J Infect Dis* 2002;**186**:99–109.
50. Ogata M, Satou T, Inoue Y, et al. Foscarnet against human herpesvirus (HHV)-6 reactivation after allo-SCT: breakthrough HHV-6 encephalitis following antiviral prophylaxis. *Bone Marrow Transplant* 2013;**48**:257–64.
51. Dockrell DH, Mendez JC, Jones M, et al. Human herpesvirus 6 seronegativity before transplantation predicts the occurrence of fungal infection in liver transplant recipients. *Transplantation* 1999;**67**:399–403.
52. Singh N, Paterson DL. Encephalitis caused by human herpesvirus-6 in transplant recipients: relevance of a novel neurotropic virus. *Transplantation* 2000;**69**:2474–9.
53. Yoshikawa T, Asano Y, Ihira M, et al. Human herpesvirus 6 viremia in bone marrow transplant recipients: clinical features and risk factors. *J Infect Dis* 2002;**185**:847–53.
54. Hirabayashi K, Nakazawa Y, Katsuyama Y, et al. Successful ganciclovir therapy in a patient with human herpesvirus-6 encephalitis after unrelated cord blood transplantation: usefulness of longitudinal measurements of viral load in cerebrospinal fluid. *Infection* 2013;**41**:219–23.

55. Ishiyama K, Katagiri T, Ohata K, et al. Safety of pre-engraftment prophylactic foscarnet administration after allogeneic stem cell transplantation. *Transpl Infect Dis* 2012;**14**:33–9.
56. Zerr M. Human herpesvirus 6 and central nervous system disease in hematopoietic cell transplantation. *J Clin Virol* 2006;**37**:52–6.
57. Howell KB, Tiedemann K, Haeusler G, et al. Symptomatic generalized epilepsy after HHV-6 posttransplant acute limbic encephalitis in children. *Epilepsia* 2012;**53**:e122–6.
58. Tsukamoto S, Sakai S, Takeda Y, et al. Posterior reversible encephalopathy syndrome in an adult patient with acute lymphoblastic leukemia after remission induction chemotherapy. *Int J Hematol* 2012;**95**:204–8.
59. Yao K, Crawford JR, Komaroff AL, et al. Review part 2: human herpesvirus-6 in central nervous system diseases. *J Med Virol* 2010;**82**:1669–78.
60. Kawada J, Kimura H, Hara S, et al. Absence of associations between influenza-associated encephalopathy and human herpesvirus 6 or human herpesvirus 7. *Pediatr Infect Dis J* 2003;**22**:115–9.
61. Boriskin YS, Rice PS, Stabler RA, et al. DNA microarrays for virus detection in cases of central nervous system infection. *J Clin Microbiol* 2004;**42**:5811–8.
62. Donati D, Akhyani N, Fogdell-Hahn A, et al. Detection of human herpesvirus-6 in mesial temporal lobe epilepsy surgical brain resections. *Neurology* 2003;**61**:1405–11.
63. Yoshikawa T, Asano Y, Akimoto S, et al. Latent infection of human herpesvirus 6 in astrocytoma cell line and alteration of cytokine synthesis. *J Med Virol* 2002;**66**:497–550.
64. Sakai R, Kanamori H, Motohashi K, et al. Long-term outcome of human herpesvirus-6 encephalitis after allogeneic stem cell transplantation. *Biol Blood Marrow Transplant* 2011;**17**:1389–94.
65. Kawamura Y, Sugata K, Ihira M, et al. Different characteristics of human herpesvirus 6 encephalitis between primary infection and viral reactivation. *J Clin Virol* 2011;**51**:2–19.
66. Ichiyama T, Ito Y, Kubota M, et al. Serum and cerebrospinal fluid levels of cytokines in acute encephalopathy associated with human herpesvirus-6 infection. *Brain Dev* 2009;**31**:731–8.
67. Hoshino A, Saitoh M, Oka A, et al. Epidemiology of acute encephalopathy in Japan with emphasis on the association of viruses and syndromes. *Brain Dev* 2012;**34**:337–43.
68. Wang SM, Lei HY, Yu CK, et al. Acute chemokine response in the blood and cerebrospinal fluid of children with enterovirus 71-associated brainstem encephalitis. *J Infect Dis* 2008;**19**:1002–6.
69. Ward KN, Kalima P, MacLeod KM, et al. Neuroinvasion during delayed primary HHV-7 infection in an immunocompetent adult with encephalitis and flaccid paralysis. *J Med Virol* 2002;**67**:538–41.
70. Shimazu Y, Kondo T, Ishikawa T, et al. Human herpesvirus-6 encephalitis during hematopoietic stem cell transplantation leads to poor prognosis. *Transpl Infect Dis* 2013;**15**(2):195–201.
71. Jaskula E, Dlubek D, Sedzimirska M, et al. Reactivations of cytomegalovirus, human herpes virus 6 and Epstein–Barr virus differ with respect to risk factors and clinical outcome after hematopoietic stem cell transplantation. *Transplant Proc* 2010;**42**:3273–6.
72. Kawaguchi T, Takeuchi M, Kawajiri C, et al. Severe hyponatremia caused by syndrome of inappropriate secretion of antidiuretic hormone developed as initial manifestation of human herpesvirus-6-associated acute limbic encephalitis after unrelated bone marrow transplantation. *Transpl Infect Dis* 2013;**15**:E54–7.
73. Hubele F, Bilger K, Kremer S, et al. Sequential FDG PET and MRI findings in a case of human herpes virus 6 limbic encephalitis. *Clin Nucl Med* 2012;**37**:716–7.
74. Yao K, Gagnon S, Akhyani N, et al. Reactivation of human herpesvirus-6 in natalizumab treated multiple sclerosis patients. *PLoS One* 2008;**3**(4):e2028.

75. Wada K, Mizoguchi S, Ito Y, et al. Multiplex real-time PCR for the simultaneous detection of herpes simplex virus, human herpesvirus 6, and human herpesvirus 7. *Microbiol Immunol* 2009;**53**:22–9.
76. Higashimoto Y, Ohta A, Nishiyama Y, Ihira M, Sugata K, Asano Y, et al. Development of a human herpesvirus 6 species-specific immunoblotting assay. *J Clin Microbiol* 2012;**50**:1245–51.
77. Ernst J, Sauerbrei A, Krumbholz A, Egerer R, et al. Multiple viral infections after haploidentical hematopoietic stem cell transplantation in a child with acute lymphoblastic leukemia. *Transpl Infect Dis* 2012;**14**:82–8.
78. Astriti M, Zeller V, Boutolleau D, et al. Fatal HHV-6 associated encephalitis in an HIV-1 infected patient treated with cidofovir. *J Infect* 2006;**52**:237–42.
79. Baldwin K. Ganciclovir-resistant human herpesvirus-6 encephalitis in a liver transplant patient: a case report. *J Neurovirol* 2011;**17**:193–5.

Chapter 6

Cognitive Dysfunction from HHV-6A and HHV-B

Danielle M. Zerr[a] and Anthony L. Komaroff[b]
[a]*University of Washington, Seattle Children's Hospital, Seattle, Washington,* [b]*Harvard Medical School, Brigham and Women's Hospital, Boston, Massachusetts*

INTRODUCTION

As discussed elsewhere in this volume, *in vitro* studies have demonstrated that human herpesvirus 6 (HHV-6) is a neurotropic virus. Indeed, immunohistochemistry and *in situ* cytohybridization studies indicate that glial cells in the central nervous system (CNS) may be *in vivo* reservoirs for the virus.[1–3] HHV-6 is likely carried into the CNS by leukocytes, which are a primary reservoir. It also may reach the brain via the olfactory pathway.[4] Much of the literature on HHV-6 was generated before the important differences between HHV-6A and HHV-6B were recognized and before technology for distinguishing the two viruses was widely available. When a report clearly distinguished these viruses, we will so indicate. When the distinction was not made, we will simply refer to "HHV-6."

As reviewed elsewhere in this volume, HHV-6 has been associated with multiple neurological diseases that can cause cognitive dysfunction. HHV-6 (particularly HHV-6B) is a cause of roseola infantum (exanthem subitum).[5] Since the first description of roseola in 1913,[6] encephalitis has been recognized as a rare complication. In this chapter, we focus primarily on the literature linking HHV-6A and HHV-6B to encephalitis, a subject also considered elsewhere in this book (see Chapter 5). We also discuss evidence linking HHV-6 to chronic fatigue syndrome—an illness in which cognitive dysfunction has been documented.

We think there is strong evidence that HHV-6 causes encephalitis in immunocompromised subjects and in rare cases of primary infection (which occur predominantly in infants and young children). There is also reasonably strong evidence that it can cause encephalitis, presumably during reactivated infection, in immunocompetent older children and adults. Having said this, several factors make it difficult to conclude definitively that HHV-6 is a cause of encephalitis.

Human Herpesviruses HHV-6A, HHV-6B & HHV-7.

THE DIFFICULTY OF ATTRIBUTING A CAUSAL ROLE TO HHV-6 IN ENCEPHALITIS

The Definition of Encephalitis

The term "encephalitis" has been interpreted differently by different authors in studies of HHV-6. The Infectious Diseases Society of America defines it as "the presence of an inflammatory process of the brain in association with clinical evidence of neurologic dysfunction."[7] Without evidence of an inflammatory process in the brain, the clinical phenotype is called "encephalopathy." We are unaware of any consensus on what constitutes evidence of "an inflammatory process in the brain." Just the presence of a fever or of peripheral markers of inflammation (such as elevated erythrocyte sedimentation rate or C-reactive protein in blood) would not seem to qualify. On the other hand, the presence of elevated levels of proinflammatory cytokines, or pleocytosis, in the cerebrospinal fluid (CSF) or electroencephalographic or neuroimaging findings consistent with encephalitis would seem to qualify. Biopsy evidence of an inflammatory response in brain parenchyma would also seem to qualify, but biopsies often are not obtained in cases of suspected encephalitis.

The reports of encephalitis that we include in this review all include evidence of acutely altered mental status (disorientation or confusion, unresponsiveness) or speech or movement disorders along with at least one of the markers of CNS inflammation listed above. As we have defined encephalitis, a single uncomplicated generalized febrile seizure in a child does not constitute "neurologic dysfunction" sufficient to say a child has encephalitis, but repeated seizures, focal seizures, or status epilepticus are taken as evidence of encephalitis.

Attributing Pathogenicity to What Is Typically a Ubiquitous Commensal

Human infection with HHV-6 is both ubiquitous and lifelong and begins at a very early age. Yet, many humans never develop any recognized illness from this chronic infection. Attributing a causal role for the virus in encephalitis requires more evidence than simply detecting the presence of the virus in a patient with encephalitis. The epidemiologist Bradford Hill put forward well-accepted criteria for evaluating causality: the strength of association, consistency, specificity, temporality, biological gradient (dose–response relationship), biological plausibility, coherence, analogy, and experimental evidence.[8,9] We shall use those as we review the literature.

Understandable Deficiencies in the Literature

There are several practical problems in interpreting the literature. Few, if any, of the reports of HHV-6 and encephalitis have systematically examined all known viral, bacterial, fungal, and parasitic causes of encephalitis. Few have systematically obtained simultaneous and repeated samples of brain, CSF, and blood. Few have systematically assessed noninfectious causes of encephalopathy, such as various toxins, metabolic encephalopathy (e.g., diabetic ketoacidosis), high doses of CNS-active medications, several autoimmune diseases (e.g., lupus cerebritis), cerebrovascular accidents, neoplastic diseases of the CNS, or some neurodegenerative disorders. Such causes of encephalopathy are mistaken as infectious encephalitis remarkably often.[10] In addition, many older reports used diagnostic technologies that lacked the sensitivity and specificity of contemporary assays.

These observations are not made to criticize. Much of the literature on HHV-6 encephalitis consists of case reports and case series of unusual patients. The subjects had not been enrolled in research studies of encephalitis. Conventional, widely available diagnostic technologies, not state-of-the-art techniques, were employed. Moreover, the possibility of HHV-6 as an etiologic agent often arose only relatively late in the clinical course—too late to collect data and specimens at an earlier stage of the illness. Even more systematic studies of large numbers of subjects with possible HHV-6 encephalitis have often been conducted retrospectively, using existing clinical information and laboratory samples. With a few exceptions, reported here, they have not involved the prospective collection of specified clinical information and biological samples in subjects with apparent encephalitis. In addition, studies employing state-of-the-art diagnostic technology in cases of suspected HHV-6-associated encephalitis have demonstrated considerable variability between patients in the viral load detected in blood or CSF specimens, as well as instances in which this cell-associated virus is clearly present in the brain but undetectable in the blood or CSF.

Moreover, even when HHV-6 is present in a case of encephalitis, and a systematic search for other known causes of encephalitis has revealed no other explanation, it still can be argued that the detection of HHV-6 in the CNS may be an epiphenomenon—that HHV-6 has been reactivated during the course of CNS infection with another as-yet-undiscovered pathogen that is the true cause of the encephalitis. That possibility is not farfetched, as the most systematic search for etiologic agents in encephalitis, the California Encephalitis Project, failed to find any known infectious pathogen or noninfectious cause in 62% of 334 patients.[10] Identifying HHV-6 in the CNS of a patient with encephalitis, in the absence of other known pathogens, surely constitutes strong evidence that the virus is the cause of the encephalitis, but it does not constitute absolute proof. Thus, even with the best available diagnostic technology, etiologic inferences are difficult.

Inherited, Chromosomally Integrated Virus

A final problem in diagnosing HHV-6 encephalitis involves inherited, chromosomally integrated virus. As discussed elsewhere in this volume (see Chapter 15), in contrast to what is typical for the other human herpesviruses, HHV-6 presumably achieves latency not through the formation of episomes but by integrating its genome into the host chromosome.[11] Some millennia ago, the virus integrated its genome into that of a germ cell. As a consequence, approximately 0.2 to 0.8% of people are born with the HHV-6 genome in every cell in their bodies.[12,13]

Such inherited, chromosomally integrated HHV-6 can present challenges when attempting to diagnose HHV-6-associated disease. High levels of HHV-6 DNA are detectable in blood, CSF, and tissue samples of affected individuals, falsely suggesting an etiologic role for the virus in disease. The possibility of chromosomally integrated or inherited HHV-6 should be considered in the setting of high and persistent levels of HHV-6 DNA in the peripheral blood, particularly if the patient has no clinical signs of encephalitis and/or the level of viremia does not decline following initiation of antiviral therapy with activity against HHV-6.

HHV-6 ENCEPHALITIS AT THE TIME OF PRIMARY INFECTION

Case Reports

Since the discovery of HHV-6 in 1986,[14] there have been many scattered case reports linking it to encephalitis in infants and young children.[15–39] Some of these reports have provided strong evidence of primary infection, such as immunoglobulin M (IgM) antibodies combined with IgG seroconversion and HHV-6 cultured from the blood. In many reports, however, primary infection is simply assumed because of the child's age. Because most children have been infected by HHV-6 by age 2 years, and because some of the children in these case reports were older than 2 years at the time they became ill, it is possible that some of these cases represented reactivated infection.

The clinical presentations described in these case reports are varied. In some of the cases, the patient had roseola. In many other cases, fever and a rash were present, but the phenotype did not fully meet the description of roseola. In a few cases, no fever and rash were present, just clear neurologic dysfunction often preceded by days of malaise and mild headache. Seizures occurred in the majority of cases. Often the neurological symptoms and signs preceded the appearance of a rash. In some cases, neuroimaging or brain biopsy revealed areas of demyelinization, whereas in other cases only a cytodestructive process was observed. A few patients recovered without clinically apparent neurological sequelae, but most were left with some sequelae. In some cases, the illness was fatal. In some cases, clinical improvement was seen following steroid therapy, but in other cases it was not. Antiviral medications with potent *in vitro* activity against HHV-6 were not employed in any cases, typically because they were unavailable at the time of the illness. In all of the cases reported, several other causes of encephalitis had been pursued and ruled out; however, in few of the cases an exhaustive search for all known etiologic agents had been conducted.

Case Series and Large Systematic Surveys

In 1994, Hall and her colleagues[40] published a landmark study of primary HHV-6 in the emergency department setting, a study that provided the context for considering the relationship of HHV-6 to encephalitis in children. The team studied 2587 children under age 3 presenting to the emergency department of a large urban hospital with symptoms and signs suggesting infection, as well as 352 healthy children of similar age. Primary HHV-6 infection confirmed by viremia and seroconversion was found in 160 (6.2%) of the children, all of whom were febrile. Only a minority had roseola. Primary HHV-6 infection was found in 9.7% of all febrile children, in 20% of all febrile children between the ages of 6 and 12 months, and in a third of all children with febrile seizures. In striking contrast, primary HHV-6 infection was never identified in afebrile cases or in the healthy control subjects.

Children with primary infection were followed closely for another two years. The HHV-6 genome was detectable in peripheral blood mononuclear cells in 37 of 56 children (66%) after primary infection. Reactivation of the virus, defined serologically, sometimes with recurrent fevers, was noted in 30 cases. This study indicates that primary HHV-6 infection is often associated with febrile seizures in the emergency department setting. The study did not report that any subject developed severe cognitive, sensory, or motor neurologic abnormalities; such devastating consequences of HHV-6 infection as described in case reports are likely to be very uncommon. A subsequent report of these patients by this team found HHV-6 DNA in the CSF of 72 of 487 (14.8%) of children with fever, "signs of sepsis," or seizures.[41] Of particular interest, the investigators found that HHV-6 often persisted in the CSF when it no longer could be isolated or identified by nucleic acid amplification testing in blood. These data suggest that detection of HHV-6 DNA in the CSF of a young child could potentially be explained by previous HHV-6 infection.

Other studies before and since Hall's publication have evaluated the frequency of HHV-6-associated encephalitis in patients with roseola. In 1993, a Japanese series of 21 infants (mean age, 9 months) with roseola and CNS complications was reported.[42] In all cases, HHV-6 had been isolated from blood. In six cases, nucleic acid amplification testing of CSF revealed HHV-6. In all cases, the neurologic complications (typically repeated seizures) preceded the appearance of rash. Some of the infants recovered without neurologic sequelae, others were left with sequelae, and one died. A survey of all hospitals in Japan identified 86 cases of roseola-associated encephalitis; many were found to have HHV-6, typically by serologic study (53 patients) or by nucleic acid amplification of plasma (34 patients).[43] Careful examination of the clinical information indicated that only 10% of the patients had the full phenotype of roseola. Approximately half of the patients suffered permanent neurologic sequelae, and the disease was fatal in two.

Other studies have examined the role of HHV-6 in all cases of encephalitis not specifically associated with roseola and have found varying frequencies of HHV-6. A summary of the 256 cases of encephalitis in children reported from Japan in 1993 identified two cases of HHV-6 infection.[44] However, the study did not prospectively collect clinical data and laboratory specimens and did not test for HHV-6 in most instances. Thus, the study could not estimate the prevalence of HHV-6 encephalitis. A study of 106 children with encephalitis or meningitis (mean age, 4 years) in Syria found HHV-6 by nucleic acid amplification assay in the CSF of 3 (2.8%).[45] In contrast, HSV-1 DNA was found in the CSF of 30% of cases. Similarly, investigators at one large hospital in China surveyed 405 children under age 2 years with suspected encephalitis.[46] CSF from every child was tested with nucleic acid amplification assay; 23 specimens (5.7%) were positive for HHV-6 DNA, including 20 cases of HHV-6B and 3 cases of HHV-6A. The report does not clearly describe whether other potential CNS pathogens were ruled out. Finally, another large survey of cases of acute encephalopathy recently was conducted in Japan, by a different research team than described previously.[43] The new survey involved 983 cases in children with a median age of 3 years.[47] The basis in each case for linking a particular pathogen with encephalopathy is not clearly described in the report, and criteria might well have been different from one hospital to another. Influenza virus was incriminated in 263 cases (26.8%), HHV-6 in 168 cases (17.1%), rotavirus in 40 cases (4.1%), and respiratory syncytial virus in 17 cases (1.7%).

British investigators conducted a 3-year prospective study of 205 immunocompetent children ages 2 to 23 months who were hospitalized with suspected encephalitis.[48] Of 156 in whom serologic studies on paired samples could be conducted, 13 (8.3%) had evidence of primary infection with HHV-6. In a subsequent study involving both immunocompetent children and adults with encephalitis, these investigators performed viral load measurements of HHV-6 in CSF, serum, and whole blood and also distinguished primary infection from reactivated infection serologically.[49] The investigators assumed that subjects with extraordinarily high viral loads in serum or whole blood had inherited chromosomally integrated HHV-6. Among children under age 2 with suspected encephalitis, the investigators found that the prevalence of HHV-6 DNA in the CSF associated with primary infection was 2.5%, while the prevalence of HHV-6 detection associated with inherited HHV-6 was 2.0%. The investigators argue that, before concluding that HHV-6 DNA in the CSF represents CNS infection, future investigators should be sure that the subject did not have inherited HHV-6. However, the report does not provide the data needed to address the question that it raises: Did the children with clinical evidence of encephalitis and inherited HHV-6 have encephalitis caused by HHV-6, or not? For example, 20 of the 21 patients with clinical encephalitis and inherited HHV-6 (95%) had HHV-6 detected in the CSF. Were some or all of those results false positives caused by leukocytes in the CSF—leukocytes that reflected the immune response to a pathogen other than HHV-6? Or were they true positives? Some clinical[50,51] and laboratory[11] evidence indicates that inherited HHV-6 DNA may be transcriptionally active and capable of producing full virions.

A multicenter prospective study recently reported its evaluation of the roles of HHV-6 and HHV-7 in febrile status epilepticus, which, as stated earlier, we take to be a possible indicator of encephalitis.[52] Cases were enrolled within 72 hours of the diagnosis. The ages of the children ranged from 1 month to 5 years. Medical history items were collected via a standardized study questionnaire. Complete physical and neurologic examinations were performed. Blood specimens at enrollment and at one month were collected from 169 children. Magnetic resonance imaging (MRI) and electroencephalography (EEG) were performed in every case. Lumbar punctures were performed based on clinical indication in 74% of cases, and CSF was obtained for testing in each instance. HHV-6B viremia was found in 54 children (32%), 38 of whom had primary infection and 16 of whom had reactivated infection. No HHV-6A infections were noted. HHV-7 infection was observed in 12 children (7.1%), 8 of whom had primary infection and 4 of whom had reactivated infection. Neither HHV-6 nor HHV-7 DNA was detected in CSF, including in those subjects with HHV-6 or HHV-7 viremia. The report does not indicate that other potential pathogens were evaluated. Thus, while the report strongly suggests that HHV-6 plays an etiologic role in some cases of febrile status epilepticus, it does not clearly establish the frequency with which this occurs.

In an effort to more definitively link HHV-6 with CNS findings, Japanese investigators measured the levels of various cytokines in blood and CSF in 15 infants with acute encephalopathy and 12 with febrile seizures, in which primary HHV-6 infection was identified by isolation of HHV-6 from the throat and/or an antibody rise on paired sera.[53] Cytokine levels were compared to normative levels in healthy infants; no measurements were taken of levels in CSF or blood from healthy control subjects, and the case and control assays were run in parallel. In infants with acute encephalopathy, elevated levels of serum interleukin 6 (IL-6) were found, along with elevated levels of CSF IL-6. Thus, the report provides evidence of an inflammatory response in infants with apparent HHV-6-associated encephalitis.

Most of the studies did not report neurological outcomes in cases of HHV-6-associated encephalitis. There clearly were cases of permanent neurologic damage.[43]

HHV-6 ENCEPHALITIS IN IMMUNOCOMPETENT ADULTS

Case Reports

There have been many scattered case reports linking HHV-6 to encephalitis in immunocompetent adults.[54–75] Some of these cases of encephalitis occurred along with other conditions that also have been linked to HHV-6, including hepatitis, severe hypersensitivity syndrome associated with the use of anticonvulsants, and Stevens–Johnson syndrome. In one case report, the encephalitic episode occurred in a patient who died from a demyelinating disease that had been diagnosed as multiple sclerosis but that was identified at postmortem examination to be subacute leukoencephalitis.[57] Another report of autopsy specimens from 15 patients who died from acute hemorrhagic leukoencephalitis found HHV-6 DNA in one case.[64] In one report, a patient had a recurrent attack of HHV-6-associated encephalitis, with the two episodes separated by 3 years.[71] Finally, one patient with HHV-6-associated limbic encephalitis developed a syndrome of inappropriate secretion of antidiuretic hormone.[72]

Case Series and Large Systematic Surveys

The first systematic study of HHV-6-associated encephalitis was reported in 1995.[76] Blood and CSF samples saved from two earlier studies were tested by nucleic acid amplification assays for HHV-6. The first group of 37 patients came from a multicenter study of patients with possible herpes simplex encephalitis. The second group included 101 patients evaluated for possible herpes simplex encephalitis at the University of Alabama at Birmingham. In all 138 patients, infection with *Herpes simplex* had been ruled out. It is unclear to what extent other potential pathogens had been evaluated. HHV-6 DNA was identified in the CSF of nine (6.5%) of all patients. There were no significant differences in clinical presentation, laboratory findings, or neurodiagnostic studies between the HHV-6-positive patients versus HHV-6-negative patients, although the small number of subjects who were HHV-6 positive limited the power of the study to detect differences.

In a study of 1482 patients hospitalized for symptoms of encephalitis or meningitis, a sensitive nucleic acid amplification test was used to look for HHV-6 DNA in various specimens—predominantly (95%) CSF.[77] Positive results were found in six children between the ages of 4 and 20, and eight adults between 21 and 81 years of age.

Perhaps the most exhaustive attempt to identify pathogens responsible for encephalitis in immunocompetent adults has been the California Encephalitis Project, first reported in 2003.[10] Physicians throughout the state were notified about the project and were asked to refer potential cases to the study team. Patients were enrolled if they were older than 6 months of age, had encephalopathy (as defined above), required hospitalization, and had one or more of the following: fever, seizure, focal neurological findings, pleocytosis, or EEG or neuroimaging findings. In total, 334 patients were enrolled. Although the study enrolled children as young as 6 months, the data indicate that few young children were enrolled (the report does not present the mean age of the subjects enrolled nor does it report the results by age strata). CSF was obtained from 96% of subjects, acute phase sera from 98%, and convalescent phase sera from 63%.

Either serological or nucleic acid amplification tests were employed for 26 viral or bacterial agents. However, each of these tests was performed only if the patient met certain prespecified clinical criteria suggesting that an agent was a plausible cause of the encephalitis. As stated earlier, the investigators could not identify an infectious or noninfectious cause of the encephalitis in 62% of cases. The original report[10] does not identify HHV-6 as one of the infectious agents that was considered, yet it states that one patient appeared to have primary infection with HHV-6 on the basis of unspecified serologic criteria.

Investigators at the National Institute of Neurological Disorders and Stroke studied CSF from 35 patients referred to the California Encephalitis Project in 2006 in whom no cause of encephalitis had been determined.[78] The team employed highly sensitive and specific nucleic acid amplification tests and assays for both IgG and IgM antibody to HHV-6. CSF from several comparison groups was studied with the same assays: 18 subjects with relapsing-remitting multiple sclerosis (RRMS), 8 from post-hematopoietic cell transplant patients with neurological complications that had been associated with HHV-6, and 14 with other neurological diseases (including West Nile virus encephalitis, epilepsy, lymphoma, stroke, and schizophrenia). HHV-6 IgG levels in CSF were significantly higher in the encephalitis patients than in the comparison groups, although elevations were also seen in the RRMS group. HHV-6 IgM levels were significantly higher in the encephalitis patients than in the comparison groups. Most persuasive, HHV-6 DNA was detected in *cell-free* CSF from 40% of subjects with encephalitis compared to none of the controls. Levels of HHV-6 DNA correlated significantly with levels of HHV-6 IgG. Because it could be argued that HHV-6 DNA detected in CSF might have come from mononuclear cells recruited to the brain during encephalitis caused by any agent, the finding of DNA in cell-free CSF is particularly strong evidence of HHV-6 infection of the CNS.

HHV-6 ENCEPHALITIS IN INHERITED, CHROMOSOMALLY INTEGRATED VIRUS

As mentioned earlier, subjects with encephalitis who inherit chromosomally integrated HHV-6 DNA can have positive nucleic acid amplification assays in blood and CSF that falsely suggest HHV-6 is the cause of their encephalitis. Indeed, such false-positive results may occur almost as often as true reactivation of HHV-6 in patients with encephalitis.[49] Whether individuals with integrated HHV-6 can also reactivate HHV-6 and suffer complications such as encephalitis or CNS dysfunction is unknown. A case of encephalitis occurring in an immunocompetent patient with integrated HHV-6 has been reported.[50] Although a thorough evaluation was performed and an alternative etiology was not identified, causality remains uncertain. In addition, another case report describes two siblings with integrated HHV-6 and cognitive dysfunction. Both patients appeared to respond to antiviral therapy.[51]

HHV-6 ENCEPHALITIS IN HEMATOPOIETIC CELL TRANSPLANTATION RECIPIENTS

Epidemiology of HHV-6 Encephalitis in the HCT Recipient

Primary infection due to HHV-6 occurs during early childhood. Thus, most recipients of hematopoietic cell transplantation (HCT) have already been infected with HHV-6 prior to transplantation. With the transplantation process, however, HHV-6 becomes active again or "reactivates" in 30 to 50% of patients between 2 and 4 weeks after transplantation.[79–83] A small proportion of patients with HHV-6 reactivation will develop encephalitis. HHV-6B accounts for most reactivation events, with HHV-6A accounting for fewer than 3% of cases.[81,84,85] Similarly, of the reported cases of HHV-6 encephalitis where the virus was typed, approximately 90% of them have been due to HHV-6B.[86–89]

As discussed earlier, it is challenging to try to link a ubiquitous and persistent virus such as HHV-6 to disease, including in the immunocompromised host; however, the accumulated evidence supports a causal association between HHV-6 and encephalitis in HCT recipients. For example, fulfilling Bradford Hill's criteria for the strength and consistency of the association, multiple case reports[8,86–89] and observational studies[79,81,83,90–93] have linked HHV-6 with encephalitis in HCT recipients. In fact, HHV-6 has been observed to be the most common cause of encephalitis in this setting, demonstrating some level of specificity.[94] Regarding temporality, given the early age of primary infection, it is likely that HCT recipients with HHV-6 encephalitis were first infected with the virus well before the episode of encephalitis. In addition, a few studies have also demonstrated the presence of HHV-6 reactivation before the onset of encephalitis.[92,95] Biological plausibility is met in that the concept that HHV-6 can cause encephalitis is consistent with existing biological and medical knowledge about HHV-6 and encephalitis (HHV-6 can infect the cells of the CNS, etc.). In addition, the proposed causal association does not conflict with what is known about HHV-6 and encephalitis (coherence). Finally, an analogy exists in that other similar viruses (herpesviruses) are known to cause encephalitis. The one criterion that has not been clearly demonstrated is experimental evidence that the condition can be prevented or ameliorated by an appropriate antiviral regimen. One small, uncontrolled study suggests that may be the case.[95]

As mentioned above, several observational studies of HCT recipients have demonstrated associations between systemic HHV-6 reactivation and encephalitis or CNS dysfunction.[79,81,83,90–93] In a study of 50 allogeneic HCT recipients, HHV-6 reactivation was detected in 24 patients (48%), and 4 patients (8%) developed HHV-6 encephalitis.[83] The HHV-6 DNA levels were significantly higher in the plasma of those who developed encephalitis compared with those who did not. Similarly, in a study of 111 HCT recipients conducted by the same group, 60 (54%) patients were found to have HHV-6 reactivation.[92] Within these 60 patients, high-level plasma HHV-6 DNA levels ($\geq 10^4$ copies per mL) was associated with encephalitis; none of 36 patients with low-level HHV-6 reactivation developed encephalitis compared to 8 of 24 patients with high-level HHV-6 reactivation ($p=0.0003$). Another study of 315 allogeneic HCT recipients demonstrated an independent association between HHV-6 reactivation and subsequent delirium, as indicated by neuropsychiatric screening.[91] HHV-6 reactivation was also independently associated with subsequent neurocognitive decline at approximately 3 months after HCT, as measured by baseline and follow-up neurocognitive testing. An important limitation of this study is that CSF samples were obtained from only 4 of the 19 patients with HHV-6-associated delirium, 2 of whom had HHV-6 DNA detected by polymerase chain reaction. In addition, brain MRI was obtained in only 9 of these 19 patients, and none had the typical abnormalities associated with HHV-6 encephalitis. This study suggests that HHV-6 may lead to CNS dysfunction in the absence of encephalitis, but further studies are necessary to prove this.

Published reports of HHV-6 encephalitis provide information regarding characteristics of the patient population most at risk of developing HHV-6 encephalitis.[86–89] Approximately 85% of reported HHV-6 encephalitis cases have occurred in recipients of unrelated or HLA-mismatched related allogeneic HCT.[89] Recipients of umbilical cord transplantations also make up a significant proportion of case reports.[89] Indeed, several studies provide relatively strong evidence that umbilical cord transplantation is a risk factor for HHV-6 encephalitis.[93,96–98] In this single-center study, the incidence of HHV-6 encephalitis in allogeneic HCT recipients was 0.7% for adult-donor transplantations and 9.9% for umbilical cord blood transplantations.

Clinical Presentation of HHV-6 Encephalitis in HCT Recipients

As discussed above, numerous case reports and case series describing HHV-6 encephalitis in allogeneic HCT recipients have been published.[86–89,99,100] Typically, in such cases, HHV-6 was identified in the CSF without another cause identified to explain the CNS abnormalities. In these reports, the symptoms and signs of encephalitis typically presented between 2 and 6 weeks after transplantation and were characterized by confusion and anterograde amnesia. Overt seizures have been reported in 40 to 70% of patients,[89,99] and when electroencephalograms (EEGs) are used an even higher proportion of patients are observed to have seizures.[87,88,101] Among patients with clinically apparent seizures, some have generalized seizures, whereas others have partial seizures.[87]

One group has termed the encephalitis associated with HHV-6 in HCT recipients *post-transplant acute limbic encephalitis* (PALE), describing it as a distinct syndrome of anterograde amnesia, a syndrome of inappropriate antidiuretic hormone, mild pleocytosis, temporal electroencephalogram abnormalities often reflecting clinical or subclinical seizures, and magnetic resonance imaging hyperintensities in the limbic system.[87] In this series of nine patients, engraftment preceded the onset of the neurologic syndrome in all patients, and the median time to onset of neurologic symptoms was 29 days following HCT (range, 14 to 61 days).[87] The syndrome began as confusion in most patients and progressed to dense anterograde amnesia, with patchy retrograde amnesia occurring within several days in all patients. Fever was present in only two patients and could have been due to other processes that were present, such as graft-versus-host disease or concurrent infection.

Cerebrospinal fluid findings are typically normal or only mildly abnormal in patients with HHV-6 encephalitis.[89] The most common abnormalities are elevations in CSF protein concentration and white blood cell count. In a series of nine allogeneic HCT recipients with HHV-6 encephalitis, the median CSF white blood cell count was 5 leukocytes per mm^3 (range, 1 to 41) with lymphocyte predominance; the median protein concentration was 48 mg/dL (range, 19 to 189).[87] The glucose concentration was normal in all patients.

Abnormal findings are noted on brain MRI in most patients with HHV-6 encephalitis.[89,99] Abnormalities typically involve the medial temporal lobes, particularly the amygdala and hippocampus; hyperintensities in these regions are visualized on T2, fluid-attenuated inversion recovery (FLAIR), and diffusion-weighted imaging (DWI) sequences.[87,93,99,102] Patients with HHV-6 encephalitis may also have abnormalities in limbic structures outside the medial temporal lobes.[103] Computed tomography (CT) of the brain, especially when obtained early in the course of illness, is often normal.[102]

As noted above, seizures are common in patients with HHV-6 encephalitis. On EEG, focal abnormalities are often observed over the temporal or fronto-temporal regions.[87] Abnormalities may include epileptiform activity, including electrographic seizures, periodic lateralized epileptiform discharges, and/or sporadic interictal discharges. Diffuse slowing can also occur.

Autopsy studies of patients who died with HHV-6 encephalitis demonstrate lesions involving the white and/or grey matter and injuries characterized by necrosis, neuronal loss, demyelination, and astrogliosis.[86–88,104,105] Earlier in the course of infection, edema and inflammation may be seen.[106] Consistent with the clinical findings of memory impairment and focal findings on brain imaging, the hippocampus is the area most commonly involved on pathology,[86–88,104,105] but other areas of the brain may also be involved.[105,106] Few published studies have attempted to correlate the detection of HHV-6 with pathology. In the studies that have tried to do this, high levels of HHV-6 mRNA and HHV-6 antigen have been documented in diseased areas of the brain, with astrocytes being the predominant cell involved.[86,105] A more extensive discussion of HHV-6 encephalitis in HCT is presented in Chapter 13 of this book.

LABORATORY DIAGNOSIS IN ENCEPHALITIS

As described in more detail elsewhere in this book (see Chapter 2), direct detection of HHV-6 in CSF using nucleic acid amplification assays is the preferred method for diagnosis for HHV-6 encephalitis, and in the absence of another identified etiology detection of HHV-6 in the CSF is generally considered diagnostic of HHV-6 encephalitis. The HHV-6 viral load from the CSF varies widely in HCT recipients with HHV-6 encephalitis. For example, in a series of eight patients with HHV-6 encephalitis, the peak median CSF viral load was 16,600 copies per mL and ranged from 600 to 288,975 copies per mL.[100] In another series that included two patients with HHV-6 encephalitis who underwent quantitative nucleic acid amplification of the CSF at the time of onset, the HHV-6 DNA levels were 203,000 copies per mL in one patient and >999,000 copies per mL in the other.[87] Unfortunately, with the exception of one report,[49] HHV-6 viral load studies have not been reported in cases of HHV-6-associated encephalitis during primary infection, nor in immunocompetent older children and adults.

It should be noted that young children have detectable HHV-6 in their CSF around the age when primary HHV-6 infection occurs even in the absence of encephalitis.[107] Thus, when evaluating young children (under the age of 3 years), the possibility of an unrelated relatively recent primary infection should be considered.

Few studies have investigated the viral dynamics in patients with HHV-6 encephalitis. In a study of 11 HCT recipients with HHV-6 detected by nucleic acid amplification in the CSF and serum (8 of whom had HHV-6 encephalitis) who were treated with foscarnet and/or ganciclovir, the concentration of HHV-6 in the CSF declined more slowly than in the serum.[100] This demonstrates that the inability to detect HHV-6 DNA in a blood specimen should not rule out HHV-6 encephalitis. Furthermore, it has been demonstrated in three patients who died with or after HHV-6 encephalitis after HCT that active HHV-6 infection can still be detected in the brain tissue even after HHV-6 DNA has become undetectable in the serum and CSF,[86] suggesting that the absence of serum or CSF HHV-6 DNA should not necessarily rule out HHV-6 encephalitis, particularly if samples are obtained days or weeks after presentation.

Nucleic acid amplification of a blood specimen may provide supportive evidence for the presence of HHV-6. Detection of HHV-6 DNA in plasma or serum correlates well with viremia and seroconversion[108,109] and is commonly used to detect primary infection or reactivation. HHV-6 DNA detection in whole blood has also been used to identify HHV-6 reactivation after HCT, with similar frequencies of detection and similar clinical associations as seen in studies using plasma or serum.[100] In contrast, detection of viral DNA in peripheral blood mononuclear cells by nucleic amplification assays can be difficult to interpret because the mononuclear cell is a site of virus latency.

Inherited Chromosomally Integrated HHV-6

Distinguishing inherited HHV-6 from HHV-6-associated disease can be challenging, especially when dealing with a single positive result. As noted above, high levels of HHV-6 DNA can be detected in the CSF in individuals with chromosomal integration. For example, among 21 patients with presumed chromosomal integration, HHV-6 DNA was detected in the CSF with a mean level of $4.0 \log_{10}$ copies per mL (95% CI, 3.5–4.5).[49] This was higher than the mean level detected in immunocompetent children with primary HHV-6 infection ($2.4 \log_{10}$ copies per mL; 95% CI, 1.0–3.7), but similar to levels detected in HCT recipients with HHV-6 encephalitis.

Despite these challenges, there are strategies to help distinguish the integrated state from reactivation. Fluorescent *in situ* hybridization (FISH) with a specific HHV-6 probe performed on metaphase chromosome preparations from peripheral blood will demonstrate integrated HHV-6.[111,112] In addition, in contrast to individuals with latent HHV-6 infection, individuals with integrated HHV-6 have detectable HHV-6 in their hair follicles.[113] Thus, if FISH is not available, nucleic acid amplification testing of hair follicles for HHV-6 DNA can be performed to identify an individual with integrated HHV-6. This would mean testing the donor's hair follicles in the case of an HCT recipient whose donor is suspected to have integrated HHV-6. When there is concern for integrated HHV-6 in the HCT donor cells, testing of donor serum or whole blood using a nucleic acid amplification test may provide supporting evidence for integrated HHV-6.[113,114] Both HHV-6A and HHV-6B can be integrated and inherited. Because the vast majority of cases of encephalitis associated with primary infection as well as those associated with reactivation events following HCT are due to HHV-6B, detection of HHV-6A can be a clue to chromosomally integrated HHV-6.[81,84,85] The possibility of chromosomally integrated or inherited HHV-6 should be considered in the setting of high and persistent levels of HHV-6 DNA in the peripheral blood, particularly if the patient has no clinical signs of encephalitis and/or the level of viremia does not decline following initiation of antiviral therapy with activity against HHV-6.

TREATMENT OF ENCEPHALITIS

No antiviral agent has been approved by the U.S. Food and Drug Administration for the treatment of HHV-6 infection; however, *in vitro* studies demonstrate that foscarnet, ganciclovir, and cidofovir have antiviral activity against HHV-6.[115] There are no controlled trials of antiviral treatment in cases of encephalitis associated with primary HHV-6 infection. There also are no such trials in immunocompetent adults with encephalitis associated with HHV-6 infection. In one such patient, treatment with cidofovir was associated with clinical recovery.[66] Another such patient was treated with foscarnet plus ganciclovir and also recovered.[50] In immunocompromised patients, there are small case series suggesting potential anti-HHV-6 effect of foscarnet and ganciclovir. Studies of patients with HHV-6 encephalitis demonstrate an association between foscarnet or ganciclovir treatment and a reduction of viral DNA levels in serum and CSF.[100] However, these studies did not have a comparison group of patients with HHV-6 encephalitis who were not treated with these antivirals. Although the optimal therapy is unknown, based on the available data, foscarnet or ganciclovir is recommended for the treatment of HHV-6 encephalitis in HCT recipients.[7,116,117] Cidofovir has been proposed as a second-line agent, but it is avoided as a first-line agent because it is highly nephrotoxic.[116]

OUTCOMES IN ENCEPHALITIS

As summarized above, outcomes of apparent HHV-6-associated encephalitis in cases of primary infection (typically, in infants and young children) and of reactivated infection in older children and adults range widely. Larger, more population-based studies have found that many patients recover completely, although many retain neurologic sequelae, and some fatalities are seen. Outcomes of HHV-6 encephalitis have been difficult to evaluate in HCT recipients, as they often have complex comorbidities. It appears that outcomes vary widely, with some patients recovering full neurologic function and other patients being left with residual neurologic deficits. Similarly, some patients with HHV-6 encephalitis have appeared to die from encephalitis, but many have died of other identifiable causes.[86–88,104,105] In a review of case reports and case series of patients with HHV-6 encephalitis, information was provided on the neurologic course and outcomes in 44 cases.[89] Although difficult to determine with certainty, 11 patients (25%) had a progressive course and died within 1 to 4 weeks of diagnosis. Eight patients (18%) improved but were left with residual neurologic compromise. Nineteen patients (43%) made a full recovery, although for some the recovery process lasted several weeks and required rehabilitation services. Six patients (14%) initially showed improvement but then succumbed to respiratory failure, multiorgan failure, or other documented infections. Whether these conditions were directly or indirectly a result of the HHV-6 infection remains unclear. Similar observations were made in a surveillance study of HHV-6 encephalitis in HCT recipients in Japan.[99]

CHRONIC FATIGUE SYNDROME

Chronic fatigue syndrome (CFS) is a chronic illness defined entirely by a constellation of symptoms: profound fatigue, impaired memory and concentration, headaches, muscle and joint pain, and post-exertional malaise.[118] The central nervous system is involved in CFS, as indicated by multiple studies involving neuroendocrine studies, magnetic resonance imaging, *in vivo* magnetic resonance spectroscopy, single photon emission computed tomography, positron emission tomography, electroencephalography, autonomic nervous system testing, and spinal fluid studies, as summarized in detail elsewhere.[119]

Patients with CFS complain of cognitive impairment, and neuropsychological testing has revealed abnormalities in patients with CFS,[120–133] abnormalities not explained by a coexisting depression.[122] The most commonly reported deficits involve attention, memory, and reaction time, as summarized in a recent meta-analysis of 50 eligible studies.[134] Another recent study directly compared cognitive function in 25 patients with CFS, 25 with major depression, and 25 healthy controls. Patients with CFS were found to have impairment in attention and working, visual, and verbal episodic memory; these deficits were not found in the comparison groups.[135]

Many studies have found evidence of reactivated HHV-6 in patients with CFS. The first large study to suggest a link included 259 patients with a "CFS-like" illness (the case definition had not yet been developed). Primary culture of lymphocytes showed active replication of HHV-6 in 70% of the patients in contrast to 20% of the age- and gender-matched healthy control subjects ($p<10^{-8}$).[136] Other studies employing assays that can detect active infection—IgM early antigen antibodies, PCR of serum or plasma, and primary cell culture—also have found an association between CFS and active HHV-6 infection.[136–143] Not surprisingly, given the ubiquity of HHV-6 infection in humans, studies employing assays that detect only latent infection have not consistently found differences between patients with CFS and healthy control subjects.

Thus, there is considerable evidence that CFS involves pathology in the central and autonomic nervous system, that patients with CFS often have documented cognitive dysfunction, and that reactivated infection with HHV-6 can be found in peripheral blood. As indicated in other chapters in this volume, HHV-6 is a plausible candidate to cause some cases of CFS, as it is a central nervous system pathogen and has been linked to various neurological disorders that can cause cognitive dysfunction. However, there currently is no persuasive evidence that HHV-6 is an etiologic agent of CFS, or that the virus causes the cognitive dysfunction seen in CFS.

CONCLUSION

Human herpesvirus 6 is an established cause of encephalitis, and presumably of the cognitive dysfunction associated with it. There is strong evidence that HHV-6 can cause encephalitis during primary infection, but given that virtually all children experience primary infection at a very young age HHV-6 encephalitis leading to severe symptoms or residual neurological damage must be a very unusual event. The same can be said about reactivated HHV-6 infection causing encephalitis in immunocompetent older children and adults. In contrast, HHV-6 is perhaps the most common cause of encephalitis in immunosuppressed subjects, particularly following hematopoietic cell transplantation. Although reactivated HHV-6 infection in the peripheral blood often is seen in patients with chronic fatigue syndrome and many patients with that illness have documented abnormalities of cognition, there is no strong evidence that HHV-6 infection in the peripheral blood or in the central nervous system is the cause of either the illness or the cognitive dysfunction seen in patients with the illness.

ACKNOWLEDGMENTS

The authors thank Jessica L. Erickson, Bailey Bonura, and Amanda Adler for their help in retrieving and organizing the literature reviewed.

REFERENCES

1. Challoner PB, Smith KT, Parker JD, et al. Plaque-associated expression of human herpesvirus 6 in multiple sclerosis. *Proc Natl Acad Sci USA* 1995;**92**:7440–4.
2. Donati D, Akhyani N, Fogdell-Hahn A, et al. Detection of human herpesvirus-6 in mesial temporal lobe epilepsy surgical brain resections. *Neurology* 2003;**61**:1405–11.
3. Fotheringham J, Donati D, Akhyani N, et al. Association of human herpesvirus-6B with mesial temporal lobe epilepsy. *PLOS Med* 2007;**4**:848–57.
4. Harberts E, Yao K, Wohler JE, et al. Human herpesvirus-6 entry into the central nervous system through the olfactory pathway. *Proc Natl Acad Sci USA* 2011;**108**:13734–9.
5. Yamanishi K, Okuno T, Shiraki K, et al. Identification of human herpesvirus-6 as a causal agent for exanthem subitum. *Lancet* 1988;**1**:1065–7.
6. Zahorsky J. Roseola infantum. *JAMA* 1913;**61**:1446–50.
7. Tunkel AR, Glaser CA, Bloch KC, et al. The management of encephalitis: clinical practice guidelines by the Infectious diseases society of america. *Clin Infect Dis* 2008;**47**:303–27.
8. Hill AB. The environment and disease: association or causation? *Proc R Soc Med* 1965;**58**:295–300.
9. Weed DL. On the use of causal criteria. *Int J Epidemiol* 1997;**26**:1137–41.
10. Glaser CA, Gilliam S, Schnurr D, et al. In search of encephalitis etiologies: diagnostic challenges in the California Encephalitis Project, 1998–2000. *Clin Infect Dis* 2003;**36**:731–42.
11. Arbuckle JH, Medveczky MM, Luka J, et al. The latent human herpesvirus-6A genome specifically integrates in telomeres of human chromosomes *in vivo* and *in vitro*. *Proc Natl Acad Sci USA* 2010;**107**:5563–8.
12. Tanaka-Taya K, Sashihara J, Kurahashi H, et al. Human herpesvirus 6 (HHV-6) is transmitted from parent to child in an integrated form and characterization of cases with chromosomally integrated HHV-6 DNA. *J Med Virol* 2004;**73**:465–73.
13. Leong HN, Tuke PW, Tedder RS, et al. The prevalence of chromosomally integrated human herpesvirus 6 genomes in the blood of UK blood donors. *J Med Virol* 2007;**79**:45–51.
14. Salahuddin SZ, Ablashi DV, Markham PD, et al. Isolation of a new virus, HBLV, in patients with lymphoproliferative disorders. *Science* 1986;**234**:596–601.
15. Yoshikawa T, Nakashima T, Suga S, et al. Human herpesvirus-6 DNA in cerebrospinal fluid of a child with exanthem subitum and meningoencephalitis. *Pediatrics* 1992;**89**:888–90.
16. Asano Y, Yoshikawa T, Kajita Y, et al. Fatal encephalitis/encephalopathy in primary human herpesvirus-6 infection. *Arch Dis Child* 1992;**67**:1484–5.
17. Vinters HV, Wang R, Wiley CA. Herpesviruses in chronic encephalitis associated with intractable childhood epilepsy. *Hum Pathol* 1993;**24**:871–9.
18. Jones CM, Dunn HG, Thomas EE, Cone RW, Weber JM. Acute encephalopathy and status epilepticus associated with human herpes virus 6 infection. *Dev Med Child Neurol* 1994;**36**:646–50.
19. Kamei A, Ichinohe S, Onuma R, Hiraga S, Fujiwara T. Acute disseminated demyelination due to primary human herpesvirus-6 infection. *Eur J Pediatr* 1997;**156**:709–12.
20. Webb DW, Bjornson BH, Sargent MA, Hukin J, Thomas EE. Basal ganglia infarction associated with HHV-6 infection. *Arch Dis Child* 1997;**76**:362–4.

21. Matsumoto K, Kato A, Inagi R, Yamanishi K, Kawano K. Herpesvirus-6 (HHV-6) encephalitis in a child presenting as a focal necrotic lesion. *Acta Neurochir (Wien)* 1999;**141**:439–40.
22. Rantala H, Mannonen L, Ahtiluoto S, et al. Human herpesvirus-6 associated encephalitis with subsequent infantile spasms and cerebellar astrocytoma. *Dev Med Child Neurol* 2000;**42**:418–21.
23. Ahtiluoto S, Mannonen L, Paetau A, et al. *In situ* hybridization detection of human herpesvirus 6 in brain tissue from fatal encephalitis. *Pediatrics* 2000;**105**:431–3.
24. Poppe M, Brück W, Hahn G, et al. Fulminant course in a case of diffuse myelinoclastic encephalitis—a case report. *Neuropediatrics* 2001;**32**:41–4.
25. Kato Z, Kozawa R, Teramoto T, Hashimoto K, Shinoda S, Kondo N. Acute cerebellitis in primary human herpesvirus-6 infection. *Eur J Pediatr* 2003;**162**:801–3.
26. Akasaka M, Sasaki M, Ehara S, Kamei A, Chida S. Transient decrease in cerebral white matter diffusivity on MR imaging in human herpes virus 6–encephalopathy. *Brain Dev* 2005;**27**:30–3.
27. Murakami A, Morimoto M, Adachi S, Ishimaru Y, Sugimoto T. Infantile bilateral striatal necrosis associated with human herpes virus-6 (HHV-6) infection. *Brain Dev* 2005;**27**:527–30.
28. Enoki H, Takeda S, Matsubayashi R, Matsubayashi T. Steroid therapy in an infant with human herpesvirus 6 encephalopathy. *Brain Dev* 2006;**28**:597–9.
29. Takaya J, Araki A, Mori K, Kaneko K. Usefulness of diffusion-weighted MRI in human herpesvirus-6 encephalitis. *Acta Paediatr* 2007;**96**:137–8.
30. Takanashi J, Barkovich AJ, Tada H, Takada N, Fuji K, Kohno Y. Cortical liquefaction in severe human herpesvirus 6 encephalopathy. *Neurology* 2006;**66**:452–3.
31. Ohsaka M, Houkin K, Takigami M, Koyanagi I. Acute necrotizing encephalopathy associated with human herpesvirus-6 infection. *Pediatr Neurol* 2006;**34**:160–3.
32. Nagasawa T, Kimura I, Abe Y, Oka A. HHV-6 encephalopathy with cluster of convulsions during eruptive stage. *Pediatr Neurol* 2007;**36**:61–3.
33. Crawford JR, Kadom N, Santi MR, Mariani B, Lavenstein BL. Human herpesvirus 6 rhombencephalitis in immunocompetent children. *J Child Neurol* 2007;**22**:1260–8.
34. Vianello F, Barbaro F, Cogo P, Furlan A, Trevenzoli M, Sgarabotto D. Co-infection with *Mycoplasma pneumoniae* and human herpesvirus 6 (HHV-6) in an immunocompetent child with meningoencephalitis: a random association? *Infection* 2007;**36**:174–6.
35. Skelton BW, Hollingshead MC, Sledd AT, Phillips CD, Castillo M. Acute necrotizing encephalopathy of childhood: typical findings in an atypical disease. *Pediatr Radiol* 2008;**38**:810–3.
36. Okumura A, Suzuki M, Kidokoro H, et al. The spectrum of acute encephalopathy with reduced diffusion in the unilateral hemisphere. *Eur J Paediatri Neurol* 2009;**13**:154–9.
37. Olli-Lähdesmäki T, Haataja L, Parkkola R, Waris M, Bleyzac N, Ruuskanen O. High-dose ganciclovir in HHV-6 encephalitis of an immunocompetent child. *Pediatr Neurol* 2010;**43**:53–6.
38. Kawamura Y, Sugata K, Ihira M, et al. Different characteristics of human herpesvirus 6 encephalitis between primary infection and viral reactivation. *J Clin Virol* 2011;**51**:12–19.
39. Matsumoto H, Hatanaka D, Ogura Y, Chida A, Nakamura Y, Nonoyama S. Severe human herpesvirus 6-associated encephalopathy in three children: analysis of cytokine profiles and the carnitine palmitoyltransferase 2 gene. *Pediatr Infect Dis* 2011;**30**:999–1001.
40. Hall CB, Long CE, Schnabel KC, et al. Human herpesvirus-6 infection in children. A prospective study of complications and reactivation. *N Engl J Med* 1994;**331**:432–8.
41. Caserta MT, Hall CB, Schnabel K, et al. Neuroinvasion and persistence of human herpesvirus 6 in children. *J Infect Dis* 1994;**170**:1586–9.

42. Suga S, Yoshikawa T, Asano Y, et al. Clinical and virological analyses of 21 infants with exanthem subitum (roseola infantum) and central nervous system complications. *Ann Neurol* 1993;**33**:597–603.
43. Howell KB, Tiedemann K, Haeusler G, et al. Symptomatic generalized epilepsy after HHV6 posttransplant acute limbic encephalitis in children. *Epilepsia* 2012;**53**:e122–6.
44. Ishikawa T, Asano Y, Morishima T, et al. Epidemiology of acute childhood encephalitis. Aichi Prefecture, Japan, 1984–1990. *Brain Dev* 1993;**15**:192–7.
45. Ibrahim AI, Obeid MT, Jouma MJ, Roemer K, Mueller-Lantzsch N, Gärtner BC. Prevalence of herpes simplex virus (types 1 and 2), varicella-zoster virus, cytomegalovirus, and human herpesvirus 6 and 7 DNA in cerebrospinal fluid of Middle Eastern patients with encephalitis. *J Clin Microbiol* 2005;**43**:4172–4.
46. Lou J, Wu Y, Cai M, Wu X, Shanq S. Subtype-specific, probe-based, real-time PCR for detection and typing of human herpesvirus-6 encephalitis from pediatric patients under the age of 2 years. *Diagn Microbiol Infect Dis* 2011;**70**:223–9.
47. Hoshino A, Saitoh M, Oka A, et al. Epidemiology of acute encephalopathy in Japan, with emphasis on the association of viruses and syndromes. *Brain Dev* 2012;**34**:337–43.
48. Ward KN, Andrews NJ, Verity CM, Miller E, Ross EM. Human herpesviruses-6 and -7 each cause significant neurological morbidity in Britain and Ireland. *Arch Dis Child* 2005;**90**:619–23.
49. Ward KN, Leong HN, Thiruchelvam AD, Atkinson CE, Clark DA. Human herpesvirus 6 DNA levels in cerebrospinal fluid due to primary infection differ from those due to chromosomal viral integration and have implications for diagnosis of encephalitis. *J Clin Microbiol* 2007;**45**:1298–304.
50. Troy SB, Blackburn BG, Yeom K, Caulfield AK, Bhangoo MS, Montoya JG. Severe encephalomyelitis in an immunocompetent adult with chromosomally integrated human herpesvirus 6 and clinical response to treatment with foscarnet plus ganciclovir. *Clin Infect Dis* 2008;**47**:e93–6.
51. Montoya JG, Neely MN, Gupta S, et al. Antiviral therapy of two patients with chromosomally-integrated human herpesvirus-6A presenting with cognitive dysfunction. *J Clin Virol* 2012;**55**:40–5.
52. Epstein LG, Shinnar S, Hesdorffer DC, et al. Human herpesvirus 6 and 7 in febrile status epilepticus: the FEBSTAT study. *Epilepsia* 2012;**53**:1481–8.
53. Ichiyama T, Ito Y, Kubota M, Yamazaki T, Nakamura K, Furukawa S. Serum and cerebrospinal fluid levels of cytokines in acute encephalopathy associated with human herpesvirus-6 infection. *Brain Dev* 2009;**31**:731–8.
54. Rawlinson WD, Hueston LC, Irving WL, Cunningham AL. Cytomegalovirus meningoencephalitis in healthy adults with coincident infection by human herpesvirus type 6. *Aust NZ J Med* 1992;**22**:504–5.
55. Sloots TP, Mackay IM, Carroll P. Meningoencephalitis in an adult with human herpesvirus-6 infection. *Med J Aust* 1993;**159**:838.
56. Sloots TP, Mackay IM, Pope JH. Diagnosis of human herpesvirus-6 infection in two patients with central nervous system complications. *Clin Diagn Virol* 1995;**3**:333–41.
57. Carrigan DR, Harrington D, Knox KK. Subacute leukoencephalitis caused by CNS infection with human herpesvirus-6 manifesting as acute multiple sclerosis. *Neurology* 1996;**47**:145–8.
58. Merelli E, Sola P, Barozzi P, Torelli G. An encephalitic episode in a multiple sclerosis patient with human herpesvirus 6 latent infection. *J Neurol Sci* 1996;**137**:42–6.
59. Novoa LJ, Nagra RM, Nakawatase T, Edwards-Lee T, Tourtellotte WW, Cornford ME. Fulminant demyelinating encephalomyelitis associated with productive HHV-6 infection in an immunocompetent adult. *J Med Virol* 1997;**52**:301–8.

60. Ikusaka M, Ota K, Honma Y, Shibata K, Uchiyama S, Iwata M. Meningoencephalitis associated with human herpesvirus-6 in an adult. *Intern Med* 1997;**36**:157.
61. Torre D, Speranza F, Martegani R, et al. Meningoencephalitis caused by human herpesvirus-6 in an immunocompetent adult patient: case report and review of the literature. *Infection* 1998;**26**:402–4.
62. Portolani M, Pecorari M, Tamassia MG, Gennari W, Beretti F, Guaraldi G. Case of fatal encephalitis by HHV-6 variant A. *J Med Virol* 2001;**65**:133–7.
63. Beovic B, Pecaric-Meglic N, Marin J, Bedernjak J, Muzlovic I, Cizman M. Fatal human herpesvirus 6-associated multifocal meningoencephalitis in an adult female patient. *Scand J Infect Dis* 2001;**33**:942–4.
64. An SF, Groves M, Martinian L, Kuo LT, Scaravilli F. Detection of infectious agents in brain of patients with acute hemorrhagic leukoencephalitis. *J Neurovirol* 2002;**8**:439–46.
65. Fujino Y, Nakajima M, Inoue H, Kusuhara T, Yamada T. Human herpesvirus 6 encephalitis associated with hypersensitivity syndrome. *Ann Neurol* 2002;**51**:771–4.
66. Denes E, Magy L, Pradeau K, Alain S, Weinbreck P, Ranger-Rogez S. Successful treatment of human herpesvirus 6 encephalomyelitis in immunocompetent patient. *Emerg Infect Dis* 2004;**10**:729–31.
67. Torre D, Mancuso R, Ferrante P. Pathogenic mechanisms of meningitis/encephalitis caused by human herpesvirus-6 in immunocompetent adult patients. *Clin Infect Dis* 2005;**41**:422–3.
68. Portolani M, Tamassia MG, Gennari W, et al. Post-mortem diagnosis of encephalitis in a 75-year-old man associated with human herpesvirus-6 variant A. *J Med Virol* 2005;**77**:244–8.
69. Birnbaum T, Padovan CS, Sporer B, et al. Severe meningoencephalitis caused by human herpesvirus 6 type B in an immunocompetent woman treated with ganciclovir. *Clin Infect Dis* 2005;**40**:887–9.
70. Isaacson E, Glaser CA, Forghani B, et al. Evidence of human herpesvirus 6 infection in 4 immunocompetent patients with encephalitis. *Clin Infect Dis* 2005;**40**:890–3.
71. Sawada J, Nakatani-Enomoto S, Aizawa H, et al. An adult case of relapsing human herpesvirus-6 encephalitis. *Intern Med* 2007;**46**:1617–20.
72. Sakuma K, Kano Y, Fukuhara M, Shiohara T. Syndrome of inappropriate secretion of antidiuretic hormone associated with limbic encephalitis in a patient with drug-induced hypersensitivity syndrome. *Clin Exp Dermatol* 2008;**33**:287–90.
73. Pot C, Burkhard PR, Villard J, et al. Human herpesvirus-6 variant A encephalomyelitis. *Neurology* 2008;**70**:974–6.
74. Peppercorn AF, Miller MB, Fitzgerald Weber DJ, Groben PA, Cairns BA. High-level human herpesvirus-6 viremia associated with onset of Stevens–Johnson syndrome: report of two cases. *J Burn Care Res* 2010;**31**:365–8.
75. Niehusmann P, Mittelstaedt T, Bien CG, et al. Presence of human herpes virus 6 DNA exclusively in temporal lobe epilepsy brain tissue of patients with history of encephalitis. *Epilepsia* 2010;**51**:2478–83.
76. McCullers JA, Lakeman FD, Whitley RJ. Human herpesvirus 6 is associated with focal encephalitis. *Clin Infect Dis* 1995;**21**:571–6.
77. Tavakoli NP, Nattanmai S, Hull R, et al. Detection and typing of human herpesvirus 6 by molecular methods in specimens from patients diagnosed with encephalitis or meningitis. *J Clin Microbiol* 2007;**45**:3972–8.
78. Yao K, Honarmand S, Espinoza A, Akhyani N, Glaser C, Jacobson S. Detection of human herpesvirus-6 in cerebrospinal fluid of patients with encephalitis. *Ann Neurol* 2009;**65**:257–67.

79. Ljungman P, Wang FZ, Clark DA, et al. High levels of human herpesvirus 6 DNA in peripheral blood leucocytes are correlated to platelet engraftment and disease in allogeneic stem cell transplant patients. *Br J Haematol* 2000;**111**:774–81.
80. Yoshikawa T, Asano Y, Ihira M, et al. Human herpesvirus 6 viremia in bone marrow transplant recipients: clinical features and risk factors. *J Infect Dis* 2002;**185**:847–53.
81. Zerr DM, Corey L, Kim HW, Huang M-L, Nguy L, Boeckh M. Clinical outcomes of human herpesvirus 6 reactivation after hematopoietic stem cell transplantation. *Clin Infect Dis* 2005;**40**:932–40.
82. Hentrich M, Oruzio D, Jäger G, et al. Impact of human herpesvirus-6 after haematopoietic stem cell transplantation. *Br J Haematol* 2005;**128**:66–72.
83. Ogata M, Kikuchi H, Satou T, et al. Human herpesvirus 6 DNA in plasma after allogeneic stem cell transplantation: incidence and clinical significance. *J Infect Dis* 2006;**193**:68–79.
84. Reddy S, Manna P. Quantitative detection and differentiation of human herpesvirus 6 subtypes in bone marrow transplant patients by using a single real-time polymerase chain reaction assay. *Biol Blood Marrow Transplant* 2005;**11**:530–41.
85. Wang LR, Dong LJ, Zhang MJ, Lu DP. The impact of human herpesvirus 6B reactivation on early complications following allogeneic hematopoietic stem cell transplantation. *Biol Blood Marrow Transplant* 2006;**12**:1031–7.
86. Fotheringham J, Akhyani N, Vortmeyer A, et al. Detection of active human herpesvirus-6 infection in the brain: correlation with polymerase chain reaction detection in cerebrospinal fluid. *J Infect Dis* 2007;**195**:450–4.
87. Seeley WW, Marty FM, Holmes TM, et al. Post-transplant acute limbic encephalitis: clinical features and relationship to HHV6. *Neurology* 2007;**69**:156–65.
88. Wainwright MS, Martin PL, Morse RP, et al. Human herpesvirus 6 limbic encephalitis after stem cell transplantation. *Ann Neurol* 2001;**50**:612–9.
89. Zerr DM. Human herpesvirus-6 and central nervous system disease in hematopoietic cell transplantation. *J Clin Virol* 2006;**37**:S52–7.
90. Yamane A, Mori T, Suzuki S, et al. Risk factors for developing human herpesvirus 6 (HHV-6) reactivation after allogeneic hematopoietic stem cell transplantation and its association with central nervous system disorders. *Biol Blood Marrow Transplant* 2007;**13**:100–6.
91. Zerr DM, Fann JR, Breiger D, et al. HHV-6 reactivation and its effect on delirium and cognitive functioning in hematopoietic cell transplantation recipients. *Blood* 2012;**117**:5243–9.
92. Ogata M, Satou T, Kawano R, et al. Correlations of HHV-6 viral load and plasma IL-6 concentration with HHV-6 encephalitis in allogeneic stem cell transplant recipients. *Bone Marrow Transplant* 2010;**45**:129–36.
93. Hill JA, Koo S, Guzman Suarez BB, et al. Cord-blood hematopoietic stem-cell transplant confers an increased risk for human herpesvirus-6-associated acute limbic encephalitis: a cohort analysis. *Biol Blood Marrow Transplant* 2012;**18**:1638–48.
94. Schmidt-Hieber M, Schwender J, Heinz WJ, et al. Viral encephalitis after allogeneic stem cell transplantation: a rare complication with distinct characteristics of different causative agents. *Haematologica* 2011;**96**:142–9.
95. Ishiyama K, Katagiri T, Hoshino T, Yoshida T, Yamaguchi M, Nakao S. Preemptive therapy of human herpesvirus-6 encephalitis with foscarnet sodium for high-risk patients after hematopoietic SCT. *Bone Marrow Transplant* 2011;**46**:863–9.
96. Sakai R, Kanamori H, Motohashi K, et al. Long-term outcome of human herpesvirus-6 encephalitis after allogeneic stem cell transplantation. *Biol Blood Marrow Transplant* 2011;**17**:1389–94.

97. Scheurer ME, Pritchett JC, Amirian ES, Zemke NR, Lusso P, Ljungman P. HHV-6 encephalitis in umbilical cord blood transplantation: a systematic review and meta-analysis. *Bone Marrow Transplant* 2013;**4**:574–80.
98. Shimazu Y, Kondo T, Ishikawa T, Yamashita K, Takaori-Kondo A. Human herpesvirus-6 encephalitis during hematopoietic stem cell transplantation leads to poor prognosis. *Transpl Infect Dis* 2013;**15**:195–201.
99. Muta T, Fukuda T, Harada M. Human herpesvirus-6 encephalitis in hematopoietic SCT recipients in Japan: a retrospective multicenter study. *Bone Marrow Transplant* 2009;**43**:583–5.
100. Zerr DM, Gupta D, Huang ML, Carter R, Corey L. Effect of antivirals on human herpesvirus 6 replication in hematopoietic stem cell transplant recipients. *Clin Infect Dis* 2002;**34**:309–17.
101. Tiacci E, Luppi M, Barozzi P, et al. Fatal herpesvirus-6 encephalitis in a recipient of a T-cell-depleted peripheral blood stem cell transplant from a 3-loci mismatched related donor. *Haematologica* 2000;**85**:94–7.
102. Noguchi T, Mihara F, Yoshiura T, et al. MR imaging of human herpesvirus-6 encephalopathy after hematopoietic stem cell transplantation in adults. *AJNR Am J Neuroradiol* 2006;**27**:2191–5.
103. Provenzale JM, VanLandingham KE, Lewis DV, Mukundan Jr S, White LE. Extrahippocampal involvement in human herpesvirus 6 encephalitis depicted at MR imaging. *Radiology* 2008;**249**:955–63.
104. De Almeida Rodrigues G, Nagendra S, Lee CK, De Magalhães-Silverman M. Human herpesvirus 6 fatal encephalitis in a bone marrow recipient. *Scand J Infect Dis* 1999;**31**:313–5.
105. Drobyski WR, Knox KK, Majewski D, Carrigan DR. Brief report: fatal encephalitis due to variant B human herpesvirus-6 infection in a bone marrow-transplant recipient. *N Engl J Med* 1994;**330**:1356–60.
106. Wang FZ, Linde A, Hägglund H, Testa M, Locasciulli A, Ljungman P. Human herpesvirus 6 DNA in cerebrospinal fluid specimens from allogeneic bone marrow transplant patients: does it have clinical significance? *Clin Infect Dis* 1999;**28**:562–8.
107. Hall CB, Caserta MT, Schnabel KC, et al. Persistence of human herpesvirus 6 according to site and variant: possible greater neurotropism of variant A. *Clin Infect Dis* 1998;**26**:132–7.
108. Huang LM, Kuo PF, Lee CY, Chen JY, Liu MY, Yang CS. Detection of human herpesviru-6 DNA by polymerase chain reaction in serum or plasma. *J Med Virol* 1992;**38**:7–10.
109. Yoshikawa T, Ihira M, Suzuki K, et al. Human herpesvirus 6 infection after living related liver transplantation. *J Med Virol* 2000;**62**:52–9.
110. Betts BC, Young JA, Ustun C, Cao Q, Weisdorf DJ. Human herpesvirus 6 infection after hematopoietic cell transplantation: is routine surveillance necessary? *Biol Blood Marrow Transplant* 2011;**17**:1562–8.
111. Clark DA, Nacheva EP, Leong HN, et al. Transmission of integrated human herpesvirus 6 through stem cell transplantation: implications for laboratory diagnosis. *J Infect Dis* 2006;**193**:912–6.
112. Daibata M, Taguchi T, Nemoto Y, Taguchi H, Miyoshi I. Inheritance of chromosomally integrated human herpesvirus 6 DNA. *Blood* 1999;**94**:1545–9.
113. Ward KN, Leong HN, Nacheva EP, et al. Human herpesvirus 6 chromosomal integration in immunocompetent patients results in high levels of viral DNA in blood, sera, and hair follicles. *J Clin Microbiol* 2006;**44**:1571–4.
114. Kamble RT, Clark DA, Leong HN, Heslop HE, Brenner MK, Carrum G. Transmission of integrated human herpesvirus-6 in allogeneic hematopoietic stem cell transplantation. *Bone Marrow Transplant* 2007;**40**:563–6.

115. De Bolle L, Naesens L, de Clercq E. Update on human herpesvirus 6 biology, clinical features, and therapy. *Clin Microbiol Rev* 2005;**18**:217–45.
116. Dewhurst S. Human herpesvirus type 6 and human herpesvirus type 7 infections of the central nervous system. *HERPES* 2004;**11**(Suppl. 2):105A–11A.
117. Tomblyn M, Chiller T, Einsele H, et al. Guidelines for preventing infectious complications among hematopoietic cell transplant recipients: a global perspective. *Bone Marrow Transplant* 2009;**44**:453–5.
118. Fukuda K, Straus SE, Hickie I, et al. The chronic fatigue syndrome: a comprehensive approach to its definition and study. *Ann Intern Med* 1994;**121**:953–9.
119. Komaroff AL, Cho TA. Role of infection and neurologic dysfunction in chronic fatigue syndrome. *Semin Neurol* 2011;**31**:325–37.
120. Johnson SK, DeLuca J, Fiedler N, Natelson BH. Cognitive functioning of patients with chronic fatigue syndrome. *Clin Infect Dis* 1994;**18**(Suppl. 1):S84–5.
121. Krupp LB, Sliwinski M, Masur DM, Friedberg F, Coyle PK. Cognitive functioning and depression in patients with chronic fatigue syndrome and multiple sclerosis. *Arch Neurol* 1994;**51**:705–10.
122. Marcel B, Komaroff AL, Fagioli LR, Kornish RJ, Albert MS. Cognitive deficits in patients with chronic fatigue syndrome. *Biol Psychiatry* 1996;**40**:535–41.
123. Tiersky LA, Johnson SK, Lange G, Natelson BH, DeLuca J. Neuropsychology of chronic fatigue syndrome: a critical review. *J Clin Exp Neuropsychol* 1997;**19**:560–86.
124. Daly E, Komaroff AL, Bloomingdale K, Wilson S, Albert MS. Neuropsychological functioning in patients with chronic fatigue syndrome, multiple sclerosis and depression. *App Neuropsychol* 2001;**8**:12–22.
125. DeLuca J, Christodoulou C, Diamond BJ, Rosenstein ED, Kramer N, Natelson BH. Working memory deficits in chronic fatigue syndrome: differentiating between speed and accuracy of information processing. *J Int Neuropsychol Soc* 2004;**10**:101–9.
126. Naschitz J, Rosner I, Rozenbaum M, et al. Successful treatment of chronic fatigue syndrome with midodrine: a pilot study. *Clin Exp Rheumatol* 2004;**21**:416–7.
127. Sandman CA, Barron JL, Nackoul K, Goldstein J, Fidler F. Memory deficits associated with chronic fatigue immune dysfunction syndrome. *Biol Psychiatry* 1993;**33**:618–23.
128. McDonald E, Cope H, David A. Cognitive impairment in patients with chronic fatigue: a preliminary study. *J Neurol Neurosurg Psychiatry* 1993;**56**:812–5.
129. DeLuca J, Johnson SK, Natelson BH. Information processing efficiency in chronic fatigue syndrome and multiple sclerosis. *Arch Neurol* 1993;**50**:301–4.
130. Schmaling KB, DiClementi JD, Cullum CM, Jones JF. Cognitive functioning in chronic fatigue syndrome and depression: a preliminary comparison. *Psychosom Med* 1994;**56**:383–8.
131. DeLuca J, Johnson SK, Beldowicz D, Natelson BH. Neuropsychological impairments in chronic fatigue syndrome, multiple sclerosis, and depression. *J Neurol Neurosurg Psychiatry* 1995;**58**:38–43.
132. DeLuca J, Johnson SK, Ellis SP, Natelson BH. Cognitive functioning is impaired in patients with chronic fatigue syndrome devoid of psychiatric disease. *J Neurol Neurosurg Psychiatry* 1997;**62**:151–5.
133. Michiels V, Cluydts R, Fischler B. Attention and verbal learning in patients with chronic fatigue syndrome. *J Int Neuropsychol Soc* 1998;**4**:456–66.
134. Cockshell SJ, Mathias JL. Cognitive functioning in chronic fatigue syndrome: a meta-analysis. *Psychol Med* 2010;**40**:1253–67.
135. Constant EL, Adam S, Gillain B, Lambert M, Masquelier E, Seron X. Cognitive deficits in patients with chronic fatigue syndrome compared to those with major depressive disorder and healthy controls. *Clin Neurol Neurosurg* 2011;**113**:295–302.

136. Buchwald D, Cheney PR, Peterson DL, et al. A chronic illness characterized by fatigue, neurologic and immunologic disorders, and active human herpesvirus type 6 infection. *Ann Intern Med* 1992;**116**:103–13.
137. Josephs SF, Henry B, Balachandran N, et al. HHV-6 reactivation in chronic fatigue syndrome. *Lancet* 1991;**337**:1346–7.
138. Secchiero P, Carrigan DR, Asano Y, et al. Detection of human herpesvirus 6 in plasma of children with primary infection and immunosuppressed patients by polymerase chain reaction. *J Infect Dis* 1995;**171**:273–80.
139. Zorzenon M, Rukh G, Botta GA, Colle R, Barsanti LA, Ceccherini-Nelli L. Active HHV-6 infection in chronic fatigue syndrome patients from Italy: new data. *J Chron Fatigue Syndrome* 1996;**2**:3–12.
140. Patnaik M, Komaroff AL, Conley E, Ojo-Amaize EA, Peter JB. Prevalence of IgM antibodies to human herpesvirus 6 early antigen (p41/38) in patients with chronic fatigue syndrome. *J Infect Dis* 1995;**172**:1364–7.
141. Wagner M, Krueger GRF, Ablashi DV, Whitman JE. Chronic fatigue syndrome (CFS): a critical evaluation of testing for active human herpesvirus-6 (HHV-6) infection: review of data of 107 cases. *J Chron Fatigue Syndrome* 1996;**2**:3–16.
142. Ablashi DV, Eastman HB, Owen CB, et al. Frequent HHV-6 reactivation in multiple sclerosis (MS) and chronic fatigue syndrome (CFS) patients. *J Clin Virol* 2000;**16**:179–91.
143. Nicolson GL, Gan R, Haier J. Multiple co-infections (*Mycoplasma*, *Chlamydia*, human herpes virus-6) in blood of chronic fatigue syndrome patients: association with signs and symptoms. *APMIS* 2003;**111**:557–66.

Chapter 7

HHV-6 and Multiple Sclerosis

Bridgette Jeanne Billioux[a], Roberto Alvarez Lafuente[b], and Steven Jacobson[c]

[a]The Johns Hopkins University, Baltimore, Maryland, [b]Hospital Clínico San Carlos, Madrid, Spain, [c]National Institute of Neurological Disorders and Stroke, National Institute of Health, Bethesda, Maryland

INTRODUCTION

Multiple sclerosis (MS) is an inflammatory demyelinating disease of the central nervous system (CNS). Although the cause of MS is unknown, it is thought to be an autoimmune condition mediated by autoreactive lymphocytes. The pathogenesis of MS seems to be multifactorial in nature, and it is thought to be triggered by multiple environmental factors in genetically susceptible individuals.[1] Viruses have long been suspected to play a role in the development of MS, and there seems to be increasing evidence of a link between MS and several different viruses. Although viruses may not be causal to the development of MS, there is speculation that they may act as environmental triggers in susceptible individuals. While multiple viruses have been linked with MS, including Epstein–Barr virus (EBV), human endogenous retroviruses (HERVs), varicella zoster virus (VZV), and measles, to name a few, some of the most compelling evidence for a viral etiology of MS has been found in studies of HHV-6.[2] This chapter will review the current evidence linking viruses and, more specifically, HHV-6 with MS.

MULTIPLE SCLEROSIS

Multiple sclerosis is the most common inflammatory demyelinating disease of the CNS and is estimated to affect between 2 to 2.5 million people worldwide. It is a heterogeneous disease with variable clinical presentations and pathological findings. MS is a chronic degenerative disease that involves CNS inflammation and demyelination in acute presentations, as well as axonal damage in the long term. It is classically characterized by white matter lesions disseminated in space and time.[3] MS is also characterized by gray matter atrophy, which is less well defined but may be related to neuronal degeneration caused by white matter lesions and axonal loss.[4] In the most common form of MS, relapsing-remitting MS (RRMS), the inflammation and demyelination occur episodically, correlating clinically with neurologic dysfunction.[3]

Multiple sclerosis is thought to be autoimmune in nature, although direct proof of an autoimmune cause is lacking. In this disease, it is thought that autoreactive lymphocytes cross the blood–brain barrier and infiltrate the CNS, leading to inflammation and damage to myelin and axons. Although some remyelination does occur, it is typically not durable.[5,6] There is usually recovery from the dysfunction after the inflammation resolves; however, this recovery may be incomplete. Over time, MS patients accumulate gliosis and axonal degeneration, which correlates with progression of the disease and further disability.[6] Although T cells are well recognized in the pathogenesis in MS, greater than 90% of MS patients are found to have oligoclonal bands in the cerebrospinal fluid (CSF) and an elevated immunoglobulin G (IgG) index, indicating a role for B cells in the immune response to MS, as well.[5,6]

Multiple sclerosis tends to present at a relatively early age, with a peak age of onset at 30.[3] It affects more females than males, with a ratio approaching 3:1.[7] Although most patients have a relapsing-remitting course, 65% of these patients will eventually enter a secondary progressive phase. In about 20% of patients with MS, the course is a primary progressive course from the beginning. Presenting symptoms commonly include visual disturbances such as optic neuritis and extraoccular movement abnormalities, motor and sensory symptoms, and ataxia. Other common symptoms in MS patients include spasticity, urinary dysfunction, vertigo, and fatigue. Relapsing and remitting MS patients are treated with a variety of disease-modifying agents, including interferon-γ, glatiramer acetate, natalizumab, and fingolimod; these agents reduce the frequency of relapses but do not reverse the acquired deficits, and they have unclear effects on disease progression. They have not been found to be effective in primary progressive and secondary progressive MS.[6]

Genetic Factors in MS

Multiple genetic factors have been found to play a role in the development of MS. There is a higher risk of developing MS if a family member has MS, with an odds ratio of 16.8 in siblings of MS patients compared to the general population.[8] There is also a higher concordance rate in monozygotic twins (24 to 30%) compared to dizygotic twins (3 to 5%).[9] Certain alleles of the major histocompatibility complex (MHC) have been linked to a higher risk of developing MS, particularly the HLA-DRB1 locus.[10] Epigenetic factors have also been implicated in the development of MS. There seems to be a maternal parent-of-origin effect in MS; this has been described in studies of extended family pedigrees, avuncular pairs, and half siblings.[11]

Geographical Distribution in MS

There is also an interesting geographical distribution of MS prevalence. It has long been thought that MS prevalence increases with increasing latitude away from the equator, but there are many contradictions to this notion; for example, some areas with a high frequency of MS, such as Sardinia in the Mediterranean, are relatively near the equator.[12] Another example is the Inuits, who live in a cold northern climate and have a low frequency of MS. Nevertheless, recent observations suggest that, in general, a latitudinal gradient in MS frequency does exist.[13] Genetic factors could potentially explain this apparent geographical distribution of MS. Certain races seem to be more susceptible to developing MS, including Caucasians from Scandinavia and Scotland. Conversely, among Mongolians, Japanese, Chinese, and American Indians, MS is relatively rare. MS also occurs less frequently in African blacks, Aborigines, Norwegian Lapps, and Gypsies.[3]

Immigration Studies in MS

Changes in the incidence of MS in immigration populations that cannot be explained by genetic factors have been well described. A systematic review of migrant studies of MS undertaken by Gale and Martyn[14] revealed two trends: (1) migrants moving from an area with a higher prevalence of MS to an area with a lower prevalence of MS have a decreased rate of disease, and (2) migrants who move from an area of lower to higher prevalence retain the lower risk of developing the disease, although offspring of migrants who move from an area of lower to higher prevalence will have a risk of MS approaching the host area.[14] Based on the immigration studies that took age at immigration into account, Gale and Martyn surmised that the risk of developing MS is established within the first two decades of life; however, other immigration studies found that this risk may be established within an even earlier time frame. Studies of migrants to Los Angeles County in California and King and Pierce counties in Washington State revealed that migrants to Los Angeles County had lower rates of MS than migrants to King and Pierce counties; these results seem to be even more pronounced in individuals who migrated to Los Angeles County at the age of 10 years or less when compared to individuals who migrated at older ages.[3] A study in Israel found that immigrants from Afro-Asian countries, who typically have low rates of MS, developed a risk of MS similar to that of European immigrants, who have higher rates of MS; this study noted, however, that the higher rates were only noticed in immigrants who migrated at a very early age, between infancy and preadolescence.[15] Despite these studies, other studies suggest that no particular age at immigration causes the change in risk of MS, but rather duration of the exposure to the new environment.[16,17] These immigration studies, as well as the moderate discordance in monozygotic twins, suggest a strong environmental factor in the development of MS.

Environmental Factors and MS

Many different environmental factors have been linked to MS, including vitamin D, exposure to sunlight and ultraviolet (UV) light, smoking, and viruses.

The observation of an association between MS and latitude has led to studies on the nature of vitamin D and UV exposure in relation to MS. Vitamin D levels have been found to be low in MS patients at time of diagnosis, and there also seems to be an association between low vitamin D levels and an increased risk of relapse. Similarly, studies showed that an increase in vitamin D levels in MS patients correlated with a lower risk of relapse.[18] UV exposure has also been studied; higher UV exposure at a young age seems to be associated with a decreased risk of MS.[19] Several different studies indicate a higher risk of MS in smokers; it has also been observed that parental smoking increases the risk of MS in children.[1]

VIRUSES AND MULTIPLE SCLEROSIS

Another environmental factor associated with MS is viruses. The role of viruses in MS is a somewhat controversial issue that has been of interest in one form or another for over a century. In the late 19th century, when Jean-Martin Charcot and his pupil Pierre Marie were describing MS, they postulated an infectious etiology for this disease.[2] Marie in particular asserted his hypothesis for an infectious etiology of MS when in 1884 he noted the following: "I was struck by the coincidental occurrence of *sclérose en plaques* with infectious illnesses, and by the close relationship that, from a theoretical point of view, unites these diseases. Therefore, I made an effort to renew my idea that *sclérose en plaques* often starts as an infectious process."[20] In the late 19th and early 20th century, multiple infectious agents were implicated, including bacterial, spirochetal, and viral infections; however, to date, no infectious agent has been proven to cause MS.[21] Several clusters of MS reported in the 20th century were suspected to be linked to an infectious cause. One such cluster occurred in 1947 when four of seven scientists developed MS; this group was working on a disease in lambs called *swayback*.[22] Another cluster of MS occurred in the Faroe Islands after British troops were stationed there during World War II; it was thought that British troops introduced canine distemper virus to the islands.[13] However, these and other clusters of MS are controversial and disputed by many.[3]

Although the infectious theory in MS has been a controversial topic, more interest is being placed on the role of viruses in MS, not necessarily as the sole cause of the disease but as a potential trigger or risk factor in the development of this complex disease in genetically susceptible individuals. Numerous viruses have been implicated in the development of MS, including measles, mumps, rubella, EBV, VZV, HERVs, and more recently HHV-6. Some of the most convincing data for an association between viruses and MS is found in EBV and HHV-6, two different herpes viruses.

EBV and MS

Epstein–Barr virus (also known as human herpesvirus 4) is the causative agent of infectious mononucleosis. EBV has been of interest as a potential cause or trigger of MS since the 1970s, but the association with MS has been strengthened recently in light of growing research.[23] EBV seropositivity has been found to be more common in patients with MS compared to controls, with a seroprevalence of 99% in MS patients and 89% in controls.[2] A meta-analysis studying infectious mononucleosis and MS risk found that individuals who are not infected with EBV have a risk of MS close to zero, and MS risk increases after infection with EBV; this risk was further increased if individuals developed the infection in adolescence or adulthood.[24]

The Hygiene Hypothesis

The observation that a delayed exposure to EBV leads to a higher risk of MS lends support to the hygiene hypothesis, which postulates that in more developed, and hence "hygienic," areas individuals are less likely to be exposed to infections early in life and may have abnormal immunological responses later in life. These abnormal responses may be produced when encountering these infectious agents as an adolescent or young adult.[6,25] This could at least in part explain the geographic distribution of MS and other autoimmune diseases,[3] and the hypothesis could be partly explained by the idea that infections early in life may help regulate the immune system. Another potential mechanism for this hypothesis could be related to maternal exposure to virus and transfer of antiviral antibodies. A decreased exposure to viral antigens in women prior to pregnancy leads to a reduction in the degree of protection conferred to newborns via maternal antiviral antibodies; later exposure of the child to the virus could provoke an immune response, possibly leading to autoimmune disease in the child.[26]

Animal Models of Virally Induced Demyelination

Numerous observations have led to the implication of viruses in MS. For one, there are several animal models of virally induced demyelination. One of the most well-known associations between a virus and demyelination in an animal model is seen in Theiler's murine encephalomyelitis virus (TMEV). TMEV, a picornavirus, was first isolated in 1934. TMEV causes a biphasic illness in laboratory mice, the first phase of which is a polioencephalitis causing hindlimb paralysis that resolves in a few weeks. The second phase involves a relapse of the hindlimb paralysis, occurring within three weeks. During the first phase, the virus affects the gray matter of the brain and spinal cord and is subsequently cleared over a few weeks, corresponding to clinical improvement in the mice. However, the virus persists in the white matter, and inflammatory demyelination characterizes the second phase of the illness.[2] Another example of viral-induced demyelination is seen in BALB/c mice infected with JHM virus, a coronavirus. This virus infects oligodendrocytes with resultant demyelination; however, the demyelination is not associated with inflammation or immune-mediated mechanisms, indicating that direct viral-induced damage of oligodendrocytes leads to demyelination.[27]

Murine hepatitis coronavirus also causes demyelination in mice. It causes a biphasic disease, with an acute encephalitis followed by disease relapse after a few weeks. Although the virus is cleared in a matter of weeks, viral antigen and RNA persist in the brain, inducing a chronic inflammatory response. This leads to inflammatory demyelinating lesions in the spinal cord, similar to those seen in MS. Of note, these demyelinating lesions do not occur in immune-deficient mice infected with the virus.[28] Additionally, CNS demyelination is seen in dogs with canine distemper virus, sheep with visna virus, goats with caprine arthritis-encephalitis virus, and mice with Semliki Forest virus.[2]

Recently, Japanese macaque encephalomyelitis has been described as a spontaneous disease in the Oregon National Primate Research Center's colony of Japanese macaques. Clinically, affected monkeys develop paralysis, ataxia, and ocular motor paresis. Initial symptoms are severe, and animals with the disease often do not recover; in those that do recover, relapses often occur. Pathologically, multifocal demyelination is found throughout the CNS, along with infiltrates of lymphocytes and macrophages and variable areas of axonal loss. In studying the diseased tissue, a previously unknown herpesvirus was identified, called JM rhadinovirus (JMRV); this virus was not identified in healthy macaques or in healthy-appearing tissue of the diseased macaques.[29]

Virally Induced Demyelination in Humans

In addition to the animal models mentioned above, there are several well-known examples of viral-induced demyelination occurring in humans, as well. Measles virus, a paramyxovirus, is known to cause demyelination through two separate clinical entities: acute postinfectious measles encephalomyelitis (APME) and subacute sclerosing panencephalitis (SSPE). APME, also known as measles-induced acute disseminated encephalomyelitis (ADEM), typically occurs during the resolution phase of systemic measles, or even weeks to months afterward. It is an immune-mediated process that does not involve direct viral infection of the CNS.[30] It occurs in 1 in 1000 cases of measles and very rarely following vaccination with a measles-containing vaccine.[31] This disorder is characterized by demyelination, with clinical symptoms including motor and sensory deficits, ataxia, and altered mental status. With treatment, which may include intravenous corticosteroids and intravenous immunoglobulin (IVIG), some patients may make a full recovery; however, others are left with permanent neurological deficits.[30] SSPE is related to a persistent infection with a defective measles virus. It more commonly occurs in children who contract measles before the age of two. Most patients who develop SSPE remain symptom free for 6 to 15 years after the initial infection but then later develop behavioral problems, intellectual disability, motor dysfunction, myoclonic jerks and other movement disorders, and ocular abnormalities before progressing to death within 1 to 3 years of symptom onset. Pathologically, cellular inclusion bodies, neuronal loss, and demyelination are seen.[30,32] It is known that there is a strong immune response in SSPE; despite this response in patients without immunological deficits, the virus persists in the CNS. The precise mechanism of demyelination in SSPE is unknown.[32]

Another example of virally induced demyelination occurring in humans is JC virus infection leading to progressive multifocal leukoencephalopathy (PML) in immunocompromised individuals. JC virus is a polyomavirus that is acquired in childhood or young adulthood. It is a very common virus, with at least 50% of the adult population being JCV seropositive. The exact pathogenesis of CNS infection by JC virus is unclear; it is debated whether the virus is latent in the CNS, or if immunosuppression leads to dissemination of the infection to the brain.[33] PML is characterized by JCV infection and destruction of oligodendrocytes, leading to focal demyelination.[32] This disease entity occurs in immunocompromised patients from, for example, AIDS, lymphoreticular malignancies, or immunosuppression related to organ transplantation, MS, or rheumatologic diseases. Clinically, patients may develop focal or multifocal neurologic deficits, depending on what area of the brain is affected; potential symptoms include dementia, weakness, visual disturbances, aphasia, apraxia, or ataxia. Magnetic resonance imaging (MRI) shows characteristic abnormalities localized to the subcortical white matter at the gray–white junction, typically with little to no post-contrast enhancement. No treatments have been proven effective for PML, although improvement in immune status may improve the overall clinical picture.[33]

Human T-cell leukemia virus type I (HTLV-I) as the causative agent of HTLV-I-associated myelopathy/tropical spastic paraparesis (HAM/TSP) is an additional example of a virus causing demyelination in human subjects. HTLV-I is a human retrovirus that is found predominantly in Africa, South America, the Caribbean, and southeast Japan. Of those infected with HTLV-I, fewer than 5% develop HAM/TSP.[34] This disease is characterized by a chronic progressive spastic paraparesis with sphincter disturbance and mild or no sensory loss. The disease typically occurs after a long incubation period, with an average age of onset in the fifth decade. Pathologically, an inflammatory demyelinating process affecting the thoracic spinal cord is found.[35] The pathogenesis of HAM/TSP is likely related to both viral and immunological factors. HTLV-I infection causes a proinflammatory state; additionally, it infects and causes dysfunction of regulatory T cells, leading to impaired modulation of lymphocyte activity. As a result, high levels of proinflammatory cytokines and mediators are secreted, causing demyelination and spinal cord damage.[34]

Oligoclonal Bands and IgG Index in MS

Oligoclonal bands and IgG elevation in the cerebrospinal fluid (CSF) provide further evidence to support a viral etiology in MS. The presence of oligoclonal bands and elevated IgG index in the CSF have long been used to aid in the diagnosis of MS. IgG index is obtained by calculating the ratio between CSF and serum IgG after correcting for albumin concentrations in the CSF and serum. Oligoclonal bands are detected by isoelectric focusing methods that separate proteins in the CSF and serum, including IgG. Immunoblotting is then used to detect separated IgG molecules. CSF and serum samples are compared; although there are multiple patterns of CSF oligoclonal bands, a significant pattern is one in which two or more bands are present in the CSF but absent in the serum.[36] While oligoclonal bands are detected in the majority of MS patients, the presence of oligoclonal bands is not specific to MS. Oligoclonal bands are found in a variety of inflammatory and infectious CNS disorders, including paraneoplastic disorders, CNS lupus, neurosarcoidosis, and Behçet's disease; however, the majority of cases in which oligoclonal bands are seen are infectious.[36,37] Moreover, in many infectious disorders with the presence of oligoclonal bands, including neurosyphilis, tuberculous meningitis, fungal meningitis, HAM/TSP, and SSPE, the oligoclonal bands are specific for the causative agent.[23,37] Given these findings, it is possible that the oligoclonal bands identified in MS may be an antibody response directed against a viral or other infectious agent. Oligoclonal bands in MS patients have been studied, with the goal of identifying an infectious agent against which these bands are directed; oligoclonal bands directed against *Chlamydia pneumoniae*, EBV, and HHV-6 have all been identified in MS patients.[38–40] Oligoclonal bands in MS could also be immunopathologic in nature and possibly indicative of cell-mediated immunopathology induced by viral infections.[37]

Viruses Leading to MS Exacerbation

Another piece of evidence to support a viral link to MS is the fact that viral infections are known to trigger MS relapses. Several studies have found that upper respiratory infections in particular seem to trigger exacerbations.[41–43] Moreover, it has been suggested that clinical exacerbations preceded by viral illness are often more prolonged than usual clinical relapses and may even contribute to the long-term decline of MS patients.[44]

Antiviral Effects of Interferon-β and MS

Additional data to support a potential viral etiology is the fact that some of the therapies used in MS, primarily interferon-β (IFN-β), are also active against viruses. IFN-β is a natural cytokine in humans that is expressed in response to viral infections.[45] Although the exact mechanism of action of IFN-β in MS is incompletely understood, its efficacy may be related to its antiviral properties.[46] The effects of IFN-β on HHV-6 in patients with MS are detailed later in the chapter.

HHV-6 AND MS

Although numerous viruses have been implicated in MS, HHV-6 has some of the most compelling evidence associated with it. HHV-6 is a betaherpesvirus discovered in 1986 when it was isolated from immunocompromised patients with human immunodeficiency virus (HIV) and lymphoproliferative disorders.[47,48] HHV-6 is a ubiquitous virus, with an estimated seroprevalence of greater than 95% in the adult population.[47] HHV-6 was originally classified into subtypes A and B, or HHV-6A and HHV-6B; however, HHV-6A and HHV6-B have recently been reclassified as separate viruses.[49] HHV-6B is acquired early in life, with infection usually occurring before the ages of 2 to 3 years. This primary infection can either be asymptomatic or manifest as exanthema subitum, also known as roseola infantum. Afterward, the virus becomes latent, generally found in the peripheral blood mononuclear cells (PBMCs). Less is known about the acquisition and seroprevalence of HHV-6A, partly due to a lack of appropriate serologic assays for detection.[23] HHV-6A is thought to be more neurotropic, given that it is detected more commonly in the CSF than in PBMCs.[50] HHV-6 is known to infect a variety of cells, both *in vivo* and *in vitro*, including brain tissue *in vivo* and glial cells *in vitro*. HHV-6 reactivates in immunocompromised states, such as in bone marrow transplantation, and can act as an opportunistic infection, leading in some cases to encephalitis.[47] HHV-6 has also been implicated in a variety of other neurological disorders, including mesial temporal lobe epilepsy, encephalitis in immunocompetent patients, and chronic fatigue syndrome, as well as MS.[51]

History of HHV-6 Link with MS

Human herpesvirus 6 was first implicated in MS in the early 1990s. In 1993, Sola et al.[52] found significantly higher HHV-6 serum antibody titers by immunofluorescence analysis in MS patients compared to controls; however, analysis of viral DNA in PBMCs indicated that HHV-6 DNA was rarely found in the PBMCs of either MS patients or controls. It was surmised that the higher titers seen in MS patients were more likely related to immunological impairment rather than reactivation. Shortly after, Challoner et al.[53] used representation differential analysis (RDA) in MS and control brain tissue to provide some of the first direct evidence implicating HHV-6 in the pathogenesis of MS. RDA is an unbiased search method allowing for enrichment of nonhuman DNA sequences by successive rounds of PCR amplification. Through RDA of MS and control brain tissue, the major DNA binding protein (MDBP) gene of HHV-6B was found in MS brains. When PCR analysis was performed on the MS brains and control brains, however, HHV-6 DNA was found to be comparable in both groups; this was thought to be evidence that HHV-6 is a commensal virus of the brain. Immunocytochemistry directed against HHV-6 proteins was also performed and revealed protein expression in MS brains but not in control cases. Moreover, the expression was more precisely localized to the oligodendrocytes, further suggesting an association between MS and HHV-6.

Tissue Evidence of HHV-6 in MS

Since then, more studies have additionally supported a link between HHV-6 and MS. HHV-6 DNA has been found frequently in CNS tissue. In 2000, Blumberg et al.[54] used a sensitive two-step *in situ* PCR to search for HHV-6 DNA in formalin-fixed, paraffin-embedded tissue that was archived from patients with MS. High gene expression for both HHV-6 p41 and p101 was consistently found in the white matter of MS patients, particularly in oligodendrocytes as well as neurons. Cermelli and Jacobson[55] explored the frequency of HHV-6 DNA by PCR in MS plaques compared to normal-appearing white matter in MS patients as well as controls through the use of laser microdissection. In this study, it was found that HHV-6 DNA was significantly more frequent in MS plaques compared to normal-appearing white matter in MS patients or controls. Goodman et al.[56] looked for the presence of HHV-6 DNA via *in situ* PCR in acute, untreated MS lesions. In this study, biopsy specimens were evaluated from patients who presented clinically as patients with cerebral tumors but were subsequently found to have MS based on pathology and clinical course. In all of the specimens, numerous oligodendrocytes, lymphocytes, and microglia were positive for HHV-6 DNA, although no clear HHV-6 antigens were identified in these cells. The fact that these immunomodulation-naïve specimens exhibited HHV-6 DNA indicated that HHV-6 may be associated with MS outside of potential reactivation due to the immunosuppressive therapies associated with MS. A later study by Opsahl and Kennedy[57] used fluorescent *in situ* hybridization (FISH) to study early and late viral gene expression in both lesions and normal-appearing white matter from MS patients as well as normal brain tissue. It was found that both the lesions as well as the normal-appearing tissue in MS patients had significantly higher levels of HHV-6 expression when compared to the normal tissue. However, the lesions expressed the highest levels, while the normal-appearing MS tissues exhibited intermediary levels of HHV-6. In addition, active translation of HHV-6 mRNA was found in oligodendrocytes in MS brain tissue.[57] Other studies have shown a relative lack of viral transcripts in MS brain tissues for other closely related herpesviruses (EBV, HHV-7, and HHV-8), further strengthening the association between HHV-6 and MS.[58,59]

DNA Evidence for HHV-6 in MS Outside of the CNS

Human herpesvirus 6 DNA has also been studied in fluids outside of the CNS in MS patients and controls. Akhyani et al.[60] investigated the presence of HHV-6 DNA in saliva, urine, sera, and PBMCs in a cohort of MS patients and healthy controls. HHV-6 DNA was found in the saliva and PBMCs of both groups; however, it was found in the sera and urine of 23% of MS patients and in none of the controls. Subtype analysis of the PCR products further revealed a predominance of HHV-6A variant in the MS patient samples. A larger study by Alvarez-Lafuente et al.[61] showed similar results: HHV-6 DNA was found in the sera of 14.6% of MS patients and none in healthy controls. Although HHV-6B was commonly found in the PBMCs in both controls and MS patients (30.4% and 53.4%, respectively), HHV-6A was seen more often in the PBMCs of MS patients (20.4% of patients) compared to controls (4.4% of controls). Furthermore, the HHV-6 DNA found in the sera of MS patients, but not controls, was predominantly HHV-6 variant A. HHV-6 is typically a cell-associated virus with viral particle shedding only occurring during active viral replication;[51] hence, the fact that HHV-6 DNA was found in extracellular compartments (e.g., sera and urine) in MS patients is suggestive of active HHV-6 viral replication occurring more commonly in MS patients. Similarly, Berti et al.[62] devised a longitudinal study following a cohort of 59 MS patients over 5 months; multiple serum samples were taken throughout various points of the study and were tested for HHV-6 DNA by PCR. While HHV-6 DNA was detected in the patients during both relapses and remissions, it was detected significantly more often during clinical relapses, suggesting a possible association between active HHV-6 replication and clinical MS exacerbations.

DNA Evidence for HHV-6 in MS in CSF

Numerous studies have also investigated the presence of HHV-6 DNA in CSF in MS patients compared to controls. The results vary, with multiple positive studies showing an increase in HHV-6 DNA detection in MS patients compared to controls,[61,63–67] as well as a number of negative studies showing no difference between the two groups.[68–74]

Serological Evidence for HHV-6 in MS

There is also serological evidence for an association between HHV-6 and MS. As mentioned previously, Sola et al.[52] found higher serum antibody titers in MS patients compared to controls. In a subsequent study, Soldan et al.[75] found higher IgM serum antibody response to HHV-6 early antigen (p41/38) in patients with RRMS, compared to patients with chronic progressive MS, other neurological diseases, other autoimmune diseases, and healthy controls. IgG levels were not significantly different among the different groups, given the ubiquity of HHV-6. However, elevated IgM levels in the RRMS group indicate that recent exposure or reactivation of HHV-6 may be associated with RRMS. A later study found that serum IgM and IgG antibody levels to HHV-6 were higher in patients with early MS (particularly early RRMS and clinically isolated syndromes) in comparison to SPMS patients and healthy controls, indicating a potential role of HHV-6 as a possible trigger for MS.[76] Some studies have confirmed the elevated serological titers to HHV-6 in MS,[64,65,73,74,77,78] while others have found less convincing data.[69,72,79] The differences in these studies may be attributed to differences in patient and/or control populations, or the different serological assays.

Effects of IFN-β on HHV-6 in MS

Another piece of evidence linking HHV-6 to MS is the effect of IFN-β on HHV-6 in MS patients. In a study by Hong et al.,[46] serum HHV-6 IgM antibodies and HHV-6 DNA were measured in MS patients treated with IFN-β, untreated MS patients, and healthy controls. Findings from this study suggest that treatment with IFN-β significantly decreased HHV-6 replication, given that the cell-free DNA was decreased in the treated MS group. A study by Garcia-Montojo et al.[45] also found a decrease in prevalence of HHV-6 serum DNA in MS patients after treatment with IFN-β; however, it was also found in this study that MS patients with continuous presence of HHV-6 DNA detected in the blood generally fared more poorly and experienced more frequent and severe relapses than MS patients with undetectable serum HHV-6 DNA.

Evidence for HHV-6A in MS

While the findings of Challoner et al.[53] in 1995 suggested a possible role of HHV-6 B as a trigger for MS, more recent studies seem to indicate that HHV-6A plays a greater role in the association with MS. HHV-6A DNA detection in MS patient sera and increased serum antibody to HHV-6A p31/48 protein in MS patients, as mentioned previously, suggest an association between the HHV-6A variant in MS.[60,75] In addition, Soldan et al.[80] observed an increased lymphoproliferative response to HHV-6A in MS patients that was not seen in healthy controls. In this study, lymphoproliferative responses to HHV-6A, HHV-6B, and HHV-7 cell lysates were compared in healthy controls and MS patients. Although both groups showed lymphoproliferation in response to HHV-6B lysates, the MS group showed a significantly increased response to HHV-6A lysates compared to controls (67% of MS patients compared to 33% of controls).

POTENTIAL MECHANISMS FOR HHV-6 INDUCED AUTOIMMUNITY IN MS

Molecular Mimicry

Although it is difficult to definitively establish HHV-6 as a causative agent in MS, there is an abundance of evidence associating the virus with MS. If there is a viral etiology or trigger in MS, HHV-6 would be a very likely candidate, given its ubiquity, neurotropism, and latency; its characteristically early period of infection would also fit with the idea that the risk of developing MS occurs early in life. Moreover, there are multiple potential mechanisms that could link HHV-6 as a trigger to autoimmunity in MS. Molecular mimicry has been suggested as one possible mechanism. Molecular mimicry arises when there is cross-reactivity between self epitopes and viral epitopes, possibly due to homologous amino acid sequences, leading to activation of autoreactive T cells. When this occurs, the immune system may then recognize the cross-reactive self epitopes as nonself; subsequently, an immune response will be directed against the cross-reactive self epitope, even if the virus is no longer present.[81] The U24 gene of HHV-6 has been found to share a homologous sequence (residues 4–10) with myelin basic protein (MBP) (residues 96–102). In a study by Tejada-Simon et al.,[82] it was shown that a significant percentage of T cells recognizing MBP_{93-105} cross-reacted with a synthetic peptide corresponding to HHV-6 $U24_{4-10}$ in MS patients. It was also found that T cells with specificity for both peptides were significantly increased in MS patients compared to controls.

Bystander Activation

Bystander activation is another possible mechanism by which HHV-6 could lead to an autoimmune response in MS. Bystander activation can occur when a viral infection causing direct inflammation or necrosis of a target tissue leads to nonspecific activation of autoreactive T cells.[27] Additionally, virus-specific T cells could also lead to bystander activation. In a viral infection, virus-specific T cells migrate to the area of active viral infection and encounter virally infected cells. These infected cells present viral antigens via the MHC-I molecules and are recognized by the virus-specific T cells. CD8+ T cells then release cytotoxic granules, killing the virally infected cells. In this context, the dying cells, CD8+ cells, and other inflammatory cells release inflammatory cytokines, leading to bystander damage of the uninfected surrounding cells.[83] This tissue damage and subsequent release of sequestered antigen can lead to further lymphocyte recruitment to the damaged tissue. Lymphocytes may then become reactive to self antigens (such as MBP) in this inflammatory setting, potentially leading to autoimmunity.[27,84]

Epitope Spreading

Epitope spreading is an additional mechanism by which a virus could lead to autoimmunity. This phenomenon occurs when an immune response is directed against several different epitopes, although initially the immune response was directed against a single epitope.[27] This can be seen when B cells act as antigen presenting cells (APCs) in response to a viral infection. A B cell will bind a particular epitope in an antigen, which is then internalized and processed for antigen presentation. The antigen, however, may contain other epitopes in addition to the one initially recognized by the B cell. These epitopes may fit into the binding grooves of the B cell's MHC-II molecule, leading to presentation of these additional epitopes. In this way, self antigens that were not the initial immune targets can later become targeted antigens, leading to autoimmunity.[85]

ADDITIONAL MECHANISMS BY WHICH HHV-6 MAY AFFECT MS

HHV-6 Leading to Apoptosis

In addition to mechanisms leading to autoimmunity, HHV-6 may affect the pathogenesis and course of MS through a variety of other mechanisms. HHV-6 infection, particularly with HHV-6A, could lead to cell death of neurons and oligodendrocytes. Gardell et al.[86] showed that *in vitro* exposure to HHV-6A led to apoptosis in neurons, astrocytes, and oligodendrocytes, while exposure to HHV-6B did not. Death of oligodendrocytes could lead to demyelination, while neuronal death could lead to the axonal loss seen later in the course of MS.

HHV-6 Causing Inflammation

An HHV-6 infection also leads to inflammation, which may be related to the pathogenesis of MS. HHV-6 has been reported to induce a type 1 (also known as Th-1), or proinflammatory, immune response in T cells. When T cells are infected with HHV-6A or HHV-6B, proinflammatory genes are increased and anti-inflammatory genes are decreased at the mRNA and protein levels. This leads to an increase in inflammatory cytokines such as IL-2, IL-18, and TNF-α and downregulation of anti-inflammatory cytokines such as IL-10 and IL-14.[87] A type 1 or Th-1 immune response in MS is well described to be related to worsened symptoms and disease progression.[88] Inflammatory Th-1 cytokines are typically increased during MS relapses, while anti-inflammatory cytokines are associated with MS remission.[89] Levels of inflammatory cytokines such as TNF-α in the CSF have also been shown to correlate with levels of disability and rate of progression in MS patients.[90]

HHV-6 May Impair Remyelination

Infection with HHV-6 may also interfere with remyelination in MS patients. Efficient repair of CNS demyelination depends on the ability of oligodendrocyte precursor cells to fully mature into oligodendrocytes. Dietrich et al.[91] found that infection of glial precursor cells with HHV-6 disrupts glial cell differentiation and proliferation. In the case of MS, this observed disruption could lead to fewer precursor cells being recruited to an area of demyelination and an inability of these glial precursor cells to maturate effectively into oligodendrocytes for proper remyelination.

HHV-6 and Glutamate Dysregulation

Human herpesvirus 6 could also potentially have an effect on MS pathogenesis through glutamate dysregulation. The dysregulation of glutamate has been suggested to play a role in the pathogenesis of MS, particularly through excitotoxicity.[92] It has been demonstrated that cells with persistent HHV-6 infection exhibit dysregulated glutamate uptake. This could lead to glutamate-related excitotoxicity and subsequent neurologic disease.[93]

HHV-6 and Impaired Phosphorylation of MBP

Impairment in phosphorylation has also been suggested as a potential mechanism for a role of HHV-6 in the pathogenesis of MS. It has been discovered that certain parts of myelin basic protein are phosphorylated less in MS patients, which may lead to impairment in the integrity of the myelin sheath and possibly decreased nerve conduction.[94] Tait and Straus[95] suggested that HHV-6 could lead to impaired phosphorylation of MBP. In particular, the homologous area of HHV-6 U24 may compete with MBP for phosphorylation and potentially confound signaling in which phosphorylated MBP might normally participate.

HHV-6 and HERVs

Additionally, HHV-6 may participate in the activation of human endogenous retroviruses (HERVs), which have been linked in the pathogenesis of MS. HERVs are retroviruses that entered into the human genome millions of years ago. HERVs have been found to have effects on host gene transcription and to even have effects on other viruses.[96] HERVs were first implicated in MS in 1989 when what eventually became to be known as the multiple sclerosis-associated retrovirus (MSRV) was identified in the supernatants of cell cultures from patients with MS.[97] MSRV has been identified as a new family of HERVs, HERV-W. HERV-W has been found to be more prevalent in MS patients compared to controls and has been associated with a poorer clinical outcome.[98] In particular, the *env* gene encoded by HERV-W has been implicated in the pathogenesis of MS through inflammation and potential oligodendrocyte damage. This gene encodes the protein syncytin, which has been reported to have indirect oligodendrotoxic effects by promoting the release of cytokines and reactive oxygen viruses.[99] Several different herpesviruses, including HSV-1, EBV, VZV, and HHV-6, have been found to cause reactivation of HERV-W;[96] hence, HHV-6 could play a role in MS through the reactivation of HERV-W, leading to inflammation and oligodendrocyte damage.

CONCLUSION

This chapter has detailed the existing evidence for a potential viral etiology in multiple sclerosis, with a particular focus on HHV-6. Although there is substantial evidence suggesting a viral link in MS, no direct evidence exists for a viral etiology in MS; however, there is compelling evidence for an association between MS and HHV-6. With HHV-6, particular difficulties in proving causation exist, given the fact that the virus is so ubiquitous. However, the isolation of this virus from diseased CNS tissues in MS patients, along with compelling serologic evidence, is highly suggestive of a role for this virus in MS, either as a trigger or in relation to the ongoing course of the disease. Of course, the presence of HHV-6 in these tissues in MS may be reflective of MS leading to reactivation of HHV-6; nevertheless, there is abundant evidence to suggest that HHV-6 could play a role in the inflammation, demyelination, and cell damage seen in MS. More information is needed to prove either a cause or a role for HHV-6 in the course of MS. If indeed HHV-6 is a trigger for MS, further research could examine the prevention of HHV-6 infection through vaccination. Given the high likelihood that HHV-6 has some effect on the overall course of MS, further research on the effects of antivirals active against HHV-6 in MS patients may also yield exciting new information. There is still much to be learned about this virus; it is hoped that a new nonhuman primate model of HHV-6 infection will elicit additional information with which to further establish a more causal relationship between HHV-6 and MS.[100]

REFERENCES

1. Disanto G, Morahan J, Ramagopalan S. Multiple sclerosis: risk factors and their interactions. *CNS Neurol Disord Drug Targets* 2012;**11**:545–55.
2. Tselis A. Evidence for viral etiology of multiple sclerosis. *Semin Neurol* 2011;**31**(3):307–16.
3. Milo R, Kahana E. Multiple sclerosis: geoepidemiology, genetics, and the environment. *Autoimmun Rev* 2010;**9**(5):A387–94.
4. Mühlau M, Buck D, Förschler A, et al. White-matter lesions drive deep gray-matter atrophy in early multiple sclerosis: support from structural MRI. *Mult Scler* 2013;**19**(11):1485–92.
5. Frohman EM, Racke MK, Raine CS. Multiple sclerosis—the plaque and its pathogenesis. *N Engl J Med* 2006;**354**(9):942.
6. Compston A, Coles A. Multiple sclerosis. *Lancet* 2008;**372**:1502–17.
7. Sellner J, Kraus J, Awad A, et al. The increasing incidence and prevalence of female multiple sclerosis: a critical analysis of potential environmental factors. *Autoimmun Rev* 2011;**10**(8):495–502.
8. Lin R, Charlesworth J, van der Mei I, Taylor BV. The genetics of multiple sclerosis. *Pract Neurol* 2012;**12**(5):279–88.
9. Sadovnick AD, Yee IM, Guimond C, et al. Age of onset in concordant twins and other relative pairs with multiple sclerosis. *Am J Epidemiol* 2009;**170**(3):289–96.
10. Sawcer S. Genetic risk and a primary role for cell-mediated immune mechanisms in multiple sclerosis. *Nature* 2011;**476**(7359):214–9.
11. Ramagopalan S, Knight J, Ebers G. Multiple sclerosis and the major histocompatibility complex. *Curr Opin Neurol* 2009;**22**(3):219–25.
12. Granieri E, Casetta I, Govoni V, et al. The increasing incidence and prevalence of MS in a Sardinian province. *Neurology* 2000;**55**:842–8.
13. Kurtzke JF. Epidemiology and etiology of multiple sclerosis. *Phys Med Rehabil Clin North Am* 2005;**16**:327–49.

14. Gale CR, Martyn CN. Migrant studies in multiple sclerosis. *Prog Neurobiol* 1995;**47**:425–48.
15. Alter M, Kahana E, Lowenson R. Migration and risk of multiple sclerosis. *Neurology* 1978;**28**:1089–93.
16. Hammond SR, English DR, McLeod JG. The age-range of risk of developing multiple sclerosis: evidence from a migrant population in Australia. *Brain* 2000;**123**(Pt 5):968–74.
17. Kahana E, Alter M, Zilber N, The Israeli MS Study Group Environmental factors determine multiple sclerosis risk in migrant to Israel. *Mult Scler* 2008;**14**(Suppl. 1):S69–70.
18. Weinstock-Guttman B, Mehta BK, Ramanathan M, et al. Vitamin D and multiple sclerosis. *Neurologist* 2012;**18**(4):179–83.
19. Van der Mei IA, Ponsonby AL, Dwyer T, et al. Past exposure to sun, skin phenotype, and risk of multiple sclerosis: case-control study. *BMJ* 2003;**327**(7410):316.
20. Marie P. Sclérose en plaques et maladies infectieuses. *Prog Med* 1884;**12**:287–9.
21. Murray TJ. The history of multiple sclerosis: the changing frame of the disease over the centuries. *J Neurol Sci* 2009;**277**(Suppl. 1):S3–S8.
22. Acheson ED. Epidemiology of multiple sclerosis. *Br Med Bull* 1977;**33**(1):9–14.
23. Virtanen JO, Jacobson S. Viruses and multiple sclerosis. *CNS Neurol Disord Drug Targets* 2012;**11**(5):1–17.
24. Thacker Evan L. Infectious mononucleosis and risk for multiple sclerosis: a meta-analysis. *Ann Neurol* 2006;**59**(3):499–503.
25. Rook G. Hygiene hypothesis and autoimmune diseases. *Clin Rev Allerg Immunol* 2012;**42**:5–15.
26. Bach JF. The effect of infections on susceptibility to autoimmunity and allergic diseases. *N Engl J Med* 2002;**347**(12):911–20.
27. Grigoriadis N, Hadjigeorgiou GM. Virus-mediated autoimmunity in multiple sclerosis. *J Autoimmune Dis* 2006;**3**:1.
28. Matthews AE, Weiss SR, Paterson Y. Murine hepatitis virus: a model for virus-induced CNS demyelination. *J Neurovirol* 2002;**8**(2):76–85.
29. Axthelm MK, Bourdette DN, Marracci GH, et al. Japanese macaque encephalomyelitis: a spontaneous multiple sclerosis-like disease in a nonhuman primate. *Ann Neurol* 2011;**70**(3):362–73.
30. Buchanan R, Bonthius DJ. Measles virus and associated central nervous system sequelae. *Semin Pediatr Neurol* 2012;**19**:107–14.
31. Chowdhary J, Ashraf SM, Khajuria K. Measles with acute disseminated encephalomyelitis (ADEM). *Indian Pediatr* 2009;**46**(1):72–4.
32. Fazakerley JK, Walker R. Virus demyelination. *J Neurovirol* 2003;**9**:148–64.
33. Aksamit AJ. Progressive multifocal leukoencephalopathy. *Continuum (Minneap Minn)* 2012;**18**(6):1374–91.
34. Souza A, Tanajura A, Toledo-Cornell C, et al. Immunopathogenesis and neurological manifestations associated to HTLV-1 infection. *Rev Soc Bras Med Trop* 2012;**45**(5):545–52.
35. Casseb J, Penalva-de-Oliveira AC. The pathogenesis of tropical spastic paraparesis/human T-cell leukemia type I-associated myelopathy. *Braz J Med Biol Res* 2000;**33**:1395–401.
36. Link H, Huang Y-M. Oligoclonal bands in multiple sclerosis cerebrospinal fluid: an update on methodology and clinical usefulness. *J Neuroimmunol* 2006;**180**:17–28.
37. Owens GP, Gilden D, Burgoon MP, et al. Viruses and multiple sclerosis. *Neuroscientist* 2011;**17**(6):659–76.
38. Yao SY, Stratton CW, Mitchell WM, et al. CSF oligoclonal bands in MS include antibodies against *Chlamydophila* antigens. *Neurology* 2001;**56**(9):1168–76.
39. Cepok S, Zhou D, Srivastava R, et al. Identification of Epstein–Barr virus proteins as putative targets of the immune response in multiple sclerosis. *J Clin Invest* 2005;**115**(5):1352–60.

40. Virtanen JO, Pietiläinen-Nicklén J, Uotila L, et al. Intrathecal human herpesvirus 6 antibodies in multiple sclerosis and other demyelinating diseases presenting as oligoclonal bands in cerebrospinal fluid. *J Neuroimmunol* 2011;**237**(1–2):93–7.
41. Kriesel JD, Sibley WA. The case for rhinoviruses in the pathogenesis of multiple sclerosis. *Mult Scler* 2005;**11**(1):1–4.
42. Panitch HS. Influence of infection on exacerbations of multiple sclerosis. *Ann Neurol* 1994;**36**(Suppl):S25–8.
43. Edwards S, Zvartau M, Clarke H, et al. Clinical relapses and disease activity on magnetic resonance imaging associated with viral upper respiratory tract infections in multiple sclerosis. *J Neurol Neurosurg Psychiatry* 1998;**64**(6):736–41.
44. Buljevac D, Flach HZ, Hop WC, et al. Prospective study on the relationship between infections and multiple sclerosis exacerbations. *Brain* 2002;**125**(Pt 5):952–60.
45. Garcia-Montojo M, De Las Heras V, Dominguez-Mozo M, et al. HHV-6 and Multiple Sclerosis Study Group. Human herpesvirus 6 and effectiveness of interferon β1b in multiple sclerosis patients. *Eur J Neurol* 2011;**18**(8):1027–35.
46. Hong J, Tejada-Simon MV, Rivera VM, et al. Anti-viral properties of interferon beta treatment in patients with multiple sclerosis. *Mult Scler* 2002;**8**(3):237–42.
47. De Bolle L, Naesens L, De Clercq E. Update on human herpesvirus 6 biology, clinical features, and therapy. *Clin Microbiol Rev* 2005;**18**(1):217–45.
48. Salahuddin SZ, Ablashi DV, Markham PD, et al. Isolation of a new virus, HBLV, in patients with lymphoproliferative disorders. *Science* 1986;**234**(4776):596–601.
49. Blashi D, Agut H, Alvarez-Lafuente R, et al. Classification of HHV-6A and HHV-6B as distinct viruses. *Arch Virol* 2013 6. [Epub ahead of print] PMID: 24193951.
50. Hall CB, Caserta MT, Schnabel KC, et al. Persistence of human herpesvirus 6 according to site and variant: possible greater neurotropism of variant A. *Clin Infect Dis* 1998;**26**(1):132–7.
51. Yao K, Crawford JR, Komaroff AL, et al. Review part 2: human herpesvirus-6 in central nervous system diseases. *J Med Virol* 2010;**82**(10):1669–78.
52. Sola P, Merelli E, Marasca R, et al. Human herpesvirus 6 and multiple sclerosis: survey of anti-HHV-6 antibodies by immunofluorescence analysis and of viral sequences by polymerase chain reaction. *J Neurol Neurosurg Psychiatry* 1993;**56**(8):917–9.
53. Challoner PB, Smith KT, Parker JD, et al. Plaque-associated expression of human herpesvirus 6 in multiple sclerosis. *Proc Natl Acad Sci USA* 1995;**92**(16):7440–4.
54. Blumberg BM, Mock DJ, Powers JM, et al. The HHV6 paradox: ubiquitous commensal or insidious pathogen? A two-step *in situ* PCR approach. *J Clin Virol* 2000;**16**(3):159–78.
55. Cermelli C, Berti R, Soldan SS, et al. High frequency of human herpesvirus 6 DNA in multiple sclerosis plaques isolated by laser microdissection. *J Infect Dis* 2003;**187**(9):1377–87.
56. Goodman AD, Mock DJ, Powers JM, et al. Human herpesvirus 6 genome and antigen in acute multiple sclerosis lesions. *J Infect Dis* 2003;**187**(9):1365–76.
57. Opsahl ML, Kennedy PG. Early and late HHV-6 gene transcripts in multiple sclerosis lesions and normal appearing white matter. *Brain* 2005;**128**(Pt 3):516–27.
58. Opsahl ML, Kennedy PG. Investigating the presence of human herpesvirus 7 and 8 in multiple sclerosis and normal control brain tissue. *J Neurol Sci* 2006;**240**(1–2):37–44.
59. Opsahl ML, Kennedy PG. An attempt to investigate the presence of Epstein–Barr virus in multiple sclerosis and normal control brain tissue. *J Neurol* 2007;**254**(4):425–30.
60. Akhyani N, Berti R, Brennan MB, et al. Tissue distribution and variant characterization of human herpesvirus (HHV)-6: increased prevalence of HHV-6A in patients with multiple sclerosis. *J Infect Dis* 2000;**182**(5):1321–5.

61. Alvarez-Lafuente R, Martín-Estefanía C, de Las Heras V, et al. Active human herpesvirus 6 infection in patients with multiple sclerosis. *Arch Neurol* 2002;**59**(6):929–33.
62. Berti R, Brennan MB, Soldan SS, et al. Increased detection of serum HHV-6 DNA sequences during multiple sclerosis (MS) exacerbations and correlation with parameters of MS disease progression. *J Neurovirol* 2002;**8**(3):250–6.
63. Wilborn F, Schmidt CA, Brinkmann V, et al. A potential role for human herpesvirus type 6 in nervous system disease. *J Neuroimmunol* 1994;**49**(1–2):213–4.
64. Liedtke W, Malessa R, Faustmann PM, Eis-Hübinger AM. Human herpesvirus 6 polymerase chain reaction findings in human immunodeficiency virus associated neurological disease and multiple sclerosis. *J Neurovirol* 1995;**1**(3–4):253–8.
65. Ablashi DV, Lapps W, Kaplan M, et al. Human herpesvirus-6 (HHV-6) infection in multiple sclerosis: a preliminary report. *Mult Scler* 1998;**4**(6):490–6.
66. Fillet AM, Lozeron P, Agut H, et al. HHV-6 and multiple sclerosis. *Nat Med* 1998;**4**(5):537. author reply, 538.
67. Tejada-Simon MV, Zang YC, Hong J, et al. Detection of viral DNA and immune responses to the human herpesvirus 6 101-kilodalton virion protein in patients with multiple sclerosis and in controls. *J Virol* 2002;**76**(12):6147–54.
68. Martin C, Enbom M, Söderström M, et al. Absence of seven human herpesviruses, including HHV-6, by polymerase chain reaction in CSF and blood from patients with multiple sclerosis and optic neuritis. *Acta Neurol Scand* 1997;**95**(5):280–3.
69. Enbom M, Wang FZ, Fredrikson S, et al. Similar humoral and cellular immunological reactivities to human herpesvirus 6 in patients with multiple sclerosis and controls. *Clin Diagn Lab Immunol* 1999;**6**(4):545–9.
70. Goldberg SH, Albright AV, Lisak RP, González-Scarano F. Polymerase chain reaction analysis of human herpesvirus-6 sequences in the sera and cerebrospinal fluid of patients with multiple sclerosis. *J Neurovirol* 1999;**5**(2):134–9.
71. Mirandola P, Stefan A, Brambilla E, et al. Absence of human herpesvirus 6 and 7 from spinal fluid and serum of multiple sclerosis patients. *Neurology* 1999;**53**(6):1367–8.
72. Taus C, Pucci E, Cartechini E, et al. Absence of HHV-6 and HHV-7 in cerebrospinal fluid in relapsing-remitting multiple sclerosis. *Acta Neurol Scand* 2000;**101**(4):224–8.
73. Virtanen JO, Färkkilä M, Multanen J, et al. Evidence for human herpesvirus 6 variant A antibodies in multiple sclerosis: diagnostic and therapeutic implications. *J Neurovirol* 2007;**13**(4):347–52.
74. Kuusisto H, Hyöty H, Kares S, et al. Human herpes virus 6 and multiple sclerosis: a Finnish twin study. *Mult Scler* 2008;**14**(1):54–8.
75. Soldan SS, Berti R, Salem N, et al. Association of human herpes virus 6 (HHV-6) with multiple sclerosis: increased IgM response to HHV-6 early antigen and detection of serum HHV-6 DNA. *Nat Med* 1997;**3**(12):1394–7.
76. Villoslada P, Juste C, Tintore M, et al. The immune response against herpesvirus is more prominent in the early stages of MS. *Neurology* 2003;**60**(12):1944–8.
77. Ablashi DV, Eastman HB, Owen CB, et al. Frequent HHV-6 reactivation in multiple sclerosis (MS) and chronic fatigue syndrome (CFS) patients. *J Clin Virol* 2000;**16**(3):179–91.
78. Friedman JE, Lyons MJ, Cu G, et al. The association of the human herpesvirus-6 and MS. *Mult Scler* 1999;**5**(5):355–62.
79. Riverol M, Sepulcre J, Fernandez-Diez B, et al. Antibodies against Epstein–Barr virus and herpesvirus type 6 are associated with the early phases of multiple sclerosis. *J Neuroimmunol* 2007;**192**(1–2):184–5.
80. Soldan SS, Leist TP, Juhng KN, et al. Increased lymphoproliferative response to human herpesvirus type 6A variant in multiple sclerosis patients. *Ann Neurol* 2000;**47**(3):306–13.

81. Fujinami RS, Oldstone MB. Amino acid homology between the encephalitogenic site of myelin basic protein and virus: mechanism for autoimmunity. *Science* 1985;**230**(4729):1043–5.
82. Tejada-Simon MV, Zang YC, Hong J, et al. Cross-reactivity with myelin basic protein and human herpesvirus-6 in multiple sclerosis. *Ann Neurol* 2003;**53**(2):189–97.
83. Fujinami RS, von Herrath MG, Christen U, et al. Molecular mimicry, bystander activation, or viral persistence: infections and autoimmune disease. *Clin Microbiol Rev* 2006;**19**(1):80–94.
84. Horwitz MS, Sarvetnick N. Viruses, host responses, and autoimmunity. *Immunol Rev* 1999;**169**:241–53.
85. Salinas GF, Braza F, Brouard S, et al. The role of B lymphocytes in the progression from autoimmunity to autoimmune disease. *Clin Immunol* 2013;**146**(1):34–45.
86. Gardell JL, Dazin P, Islar J, et al. Apoptotic effects of human herpesvirus-6A on glia and neurons as potential triggers for central nervous system autoimmunity. *J Clin Virol* 2006;**37**(Suppl 1):S11–6.
87. Mayne M, Cheadle C, Soldan SS, et al. Gene expression profile of herpesvirus-infected T cells obtained using immunomicroarrays: induction of proinflammatory mechanisms. *J Virol* 2001;**75**(23):11641–50.
88. Oreja-Guevara C, Ramos-Cejudo J, Aroeira LS, et al. TH1/TH2 cytokine profile in relapsing-remitting multiple sclerosis patients treated with glatiramer acetate or natalizumab. *BMC Neurol* 2012;**12**(1):95.
89. Amedei A, Prisco D, D'Elios MM. Multiple sclerosis: the role of cytokines in pathogenesis and in therapies. *Int J Mol Sci* 2012;**13**(10):13438–60.
90. Sharief MK, Hentges R. Association between tumor necrosis factor-alpha and disease progression in patients with multiple sclerosis. *N Engl J Med* 1991;**325**:467–72.
91. Dietrich J, Blumberg BM, Roshal M, et al. Infection with an endemic human herpesvirus disrupts critical glial precursor cell properties. *J Neurosci* 2004;**24**(20):4875–83.
92. Bolton C, Paul C. Glutamate receptors in neuroinflammatory demyelinating disease. *Mediators Inflamm* 2006;**2**:93684.
93. Fotheringham J, Williams EL, Akhyani N, Jacobson S. Human herpesvirus 6 (HHV-6) induces dysregulation of glutamate uptake and transporter expression in astrocytes. *J Neuroimmune Pharmacol* 2008;**3**(2):105–16.
94. Kim JK, Mastronardi FG, Wood DD, et al. Multiple sclerosis: an important role for post-translational modifications of myelin basic protein in pathogenesis. *Mol Cell Proteomics* 2003;**2**(7):453–62.
95. Tait AR, Straus SK. Phosphorylation of U24 from human herpes virus type 6 (HHV-6) and its potential role in mimicking myelin basic protein (MBP) in multiple sclerosis. *FEBS Lett* 2008;**582**(18):2685–8.
96. Perron H, Bernard C, Bertrand JB, et al. Endogenous retroviral genes, herpesviruses and gender in multiple sclerosis. *J Neurol Sci* 2009;**286**(1–2):65–72.
97. Perron H, Geny C, Laurent A, et al. Leptomeningeal cell line from multiple sclerosis with reverse transcriptase activity and viral particles. *Res Virol* 1989;**140**(6):551–61.
98. Sotgiu S, Mameli G, Serra C, et al. Multiple sclerosis-associated retrovirus and progressive disability of multiple sclerosis. *Mult Scler* 2010;**16**(10):1248–51.
99. Ruprecht K, Obojes K, Wengel V, et al. Regulation of human endogenous retrovirus W protein expression by herpes simplex virus type 1: implications for multiple sclerosis. *J Neurovirol* 2006;**12**(1):65–71.
100. Leibovitch E, Wohler JE, Cummings Macri SM, et al. Novel marmoset (*Callithrix jacchus*) model of human herpesvirus 6A and 6B infections: immunologic, virologic and radiologic characterization. *PLoS Pathog* 2013;**9**(1):e1003138.

Chapter 8

Human Herpes Virus 6B in Mesial Temporal Lobe Epilepsy

Pitt Niehusmann[a], Jin-Mei Li[b], and Dong Zhou[b]
[a]*University of Bonn Medical Center, Bonn, Germany,* [b]*West China Hospital of Sichuan University, Sichuan, China*

INTRODUCTION

Temporal lobe epilepsy is a frequent form of localization-related epilepsy,[1] and mesial temporal lobe epilepsy (MTLE) with hippocampal sclerosis (HS) is often associated with resistance to pharmacotherapy but shows excellent results on resective epilepsy surgery. Hence, patients suffering from MTLE with HS account for the most frequent subgroup in surgical cohorts in studies from tertiary care centers.[2] HS is characterized by pyramidal cell loss within the hippocampal subfields CA1 to CA4 and associated reactive astrogliosis (see Figure 8.1).[3]

Based on frequently reported "initial precipitating insults," it has been suggested that MTLE-associated HS is a progressive disorder affecting the organization of the mesial temporal lobe.[4] Three main periods in the development of the disorder were assumed: (1) an early childhood insult, (2) a latency period, and (3) manifestation of pharmacoresistent epilepsy accompanied by the appearance of neuropathological changes. The early childhood insults include birth trauma, head injury, and (meningo-) encephalitis. Complex febrile seizures are observed in 34 to 70% of MTLE patients with associated HS.[4]

As discussed in detail in Chapter 4 of this book, some data point to an association between HHV-6B and febrile seizures. Recent data from the FEBSTAT study show that HHV-6B and HHV-7 infection together account for one third of febrile status epilepticus.[5] Although findings from retrospective studies argue for an association between febrile seizures and the development of MTLE, others have failed to confirm a causative correlation.[6–8]

The role of HHV-6 in acute encephalitis is also described earlier in this book (see Chapter 5). Due to its close connection with the limbic system, the olfactory pathway has been proposed as the means for HHV-6 entry into the central nervous system (CNS),[9] similar to other neurotropic viruses.

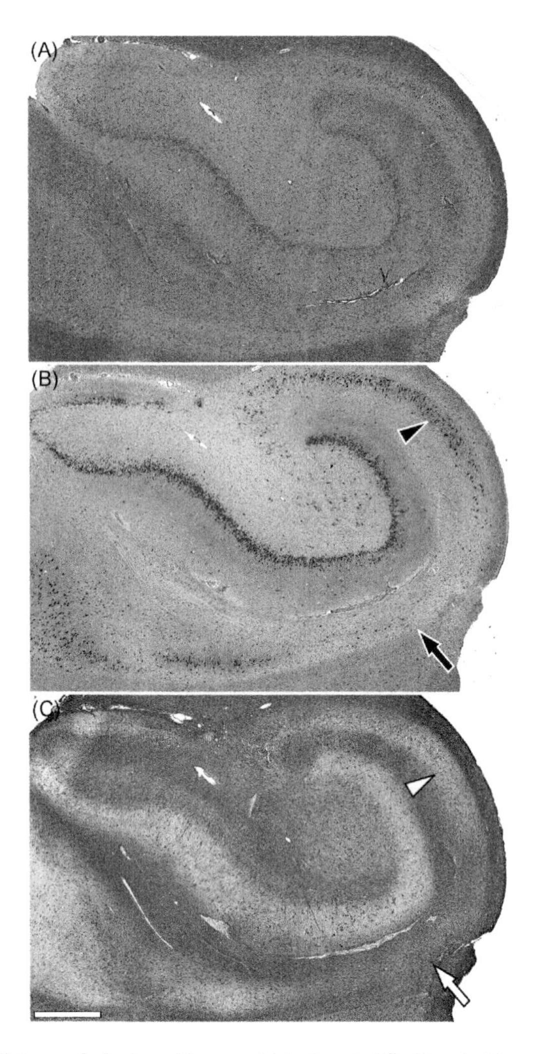

FIGURE 8.1 Histomorphologic and immunohistochemical findings in hippocampal sclerosis. (A) Hematoxylin and eosin (H&E) staining and (B) immunohistochemical analysis with antibodies against NeuN show a typical pattern of hippocampal sclerosis with distinctive neuronal cell loss in the CA1 (black arrow) and CA3/4 regions. The CA2 subfield (black arrowhead) is less severely affected in the illustrated case. (C) Immunohistochemistry with antibodies against glial fibrillary acidic protein (GFAP) reveals strong reactive astrogliosis in areas with severe neuronal cell loss (white arrow), but less gliosis in the CA2 region (white arrowhead). Scale bar = 1000 μm.

The almost ubiquitous HHV-6 infection in childhood,[10] the known neurotropism of the virus, its association with febrile status epilepticus, and its capacity to establish lifelong latent infection make it reasonable to discuss the potential role of HHV-6 in MTLE. For obvious reasons, no CNS tissue-based analyses from the silent latency period between early childhood insult (i.e., febrile seizures) and subsequent MTLE manifestation exist. However, as summarized in Table 8.1,

TABLE 8.1 Review of HHV-6 PCR Studies Based on Epilepsy-Surgical Removed Tissue

Publication	No. of HHV-6 Tested TLE Patients (HHV-6+)	Analyzed Tissue and Method	History of Febrile Seizures in TLE Patients (HHV-6+)	History of Encephalitis/ Meningitis in TLE Patients (HHV-6+)[a]	Other Epilepsy Patients and Autopsy Samples (HHV-6+)
Uesugi et al.[11]	15 (6)	PCR for HHV-6 and HSV from fresh frozen tissue	2 (0)[b]	4 (3)[b]	0 (0)
Donati et al.[13]	8 (4)	Quantitative real-time PCR from immediately processed tissue and immunohistochemistry	3 (1)	0 (0)	7 (0)
Fotheringham et al.[14]	16 (11)	Quantitative real-time PCR from immediately processed tissue, culture of astrocytes and immunohistochemistry	5 (3)	1 (1)[c]	7 (0)
Eeg-Olofsson et al.[12]	36 (7)	Nested PCR for six different herpes viruses from fresh frozen tissue	11 (2)	9 (3)[d]	12 (3)
Karatas et al.[15]	33 (3)	Real-time PCR from formalin-fixed and paraffin-embedded tissue	3 (3)	1 (0)	7 (0)
Niehusmann et al.[16]	38 (8)	Nested PCR from formalin-fixed and paraffin-embedded tissue and immunohistochemistry	6 (0)	12 (8)[e]	10 (0)
Li et al.[17]	32 (9)	Nested PCR done of minced and nitrogen frozen or 10% formalin-fixed tissues and immunohistochemistry	11 (7)	0 (0)	12 (1)
Overall	178 (48)	—	41 (16)	27 (15)	55 (4)

[a] *According to reported data.*
[b] *History of febrile seizures not explicitly mentioned in publication but concluded from given data of "exanthema subitum status" at the ages of 6 and 18 months.*
[c] *A putative history of encephalitis is not noted in the patient data; however, one patient was diagnosed with chronic encephalitis.*
[d] *Including three patients with diagnosis of Rasmussen's encephalitis.*
[e] *Including two patients with diagnosis of limbic encephalitis and autoantibodies to glutamic acid decarboxylase (GAD-Ab) and one patient with subacute HSV encephalitis.*
Note: HHV-6+, positive for HHV-6 DNA.

various groups have sought to determine the presence of HHV-6 DNA in surgically removed tissue from patients with pharmacoresistent MTLE and have reported varying results.

IMMUNOHISTOCHEMICAL AND PCR STUDIES ON EPILEPSY BIOPSY SAMPLES

A Japanese group reported in 2000 on their polymerase chain reaction (PCR) findings in surgically removed tissue from patients with temporal lobe epilepsy who underwent anterior temporal lobectomy.[11] Fresh frozen samples from 15 patients were screened for HHV-6 and herpex simplex virus (HSV) DNA. In two additional patients PCR for measles virus was performed. The authors did not provide information regarding mesial or neocortical focus of the temporal lobe epilepsy nor did they differentiate between HHV-6A and HHV-6B. However, hippocampal and lateral temporal lobe tissues were analyzed, and HHV-6 DNA was detected in six patients (hippocampal, $n=3$; lateral, $n=2$; both localizations, $n=1$). Two patients had a history of status epilepticus during exanthema subitum, and one patient suffered from a status epilepticus of unknown origin, but in none of the corresponding samples was HHV-6 DNA observed. In contrast, three other patients had a history of either measles encephalitis or not further classified (sub-) coma. HHV-6 DNA was found in samples from all three of these patients, including in temporolateral and temporomesial tissue from the patient with a diagnosis of measles encephalitis. The remaining positive results were from patients without a history of meningitis or encephalitis.

Eeg-Olofsson and colleagues used a nested PCR approach to determine the presence of the DNA of six different herpes viruses in biopsy samples from a heterogenic group of 36 patients with temporal lobe epilepsy and 12 control samples.[12] Specific PCR for differentiation between HHV-6A and HHV-6B was not performed. The specimens from epilepsy patients were obtained from the temporal cortex and kept frozen until DNA extraction. Control samples were obtained from neurosurgery or autopsy and consisted of mainly temporal but also frontal and parietal cortex. These samples were analyzed within 24 hours after surgery or autopsy without previous freezing. The patient collective included patients with cortical dysplasia, ganglioglioma, Rasmussen's encephalitis, vascular malformation, and porencephaly. Gliosis was the most frequently diagnosed pathology, whereas HS was not reported. Overall, nine patients had a history of meningitis or encephalitis, including three patients with Rasmussen's encephalitis, and febrile seizures were documented in 11 cases. HHV-6 DNA was detected in samples from seven patients. Two patients were also diagnosed with Rasmussen's encephalitis, and one had a history of meningitis. Two patients with febrile seizures were also positive for HHV-6 DNA.

In two subsequent studies with biopsy specimens from 38 patients with either temporal lobe epilepsy or other focal epilepsy syndromes, HHV-6 was demonstrated exclusively in patients with temporal seizure focus.[13,14] In the first study, Donati and colleagues[13] reported their findings in surgically resected tissue from eight patients with MTLE and seven patients with neocortical epilepsy. None of the patients had a history of encephalitis, but previous febrile seizures were reported in three patients. In several cases, when immediately processed hippocampal tissue as well as lateral temporal lobe tissue were analyzed for HHV-6B DNA, a higher amount of HHV-6B DNA was found in the temporal mesial biopsy samples. HHV-6B DNA was detected in four cases with MTLE, including one case with a history of febrile seizures, but in none of the samples from patients with neocortical epilepsy. Western blot analysis with a monoclonal antibody specific for the HHV-6 p41 antigen was positive using material from the hippocampus of the patient with the highest values of HHV-6 DNA in the PCR experiments. However, western blot analysis using lateral temporal lobe tissue from the same patient, which was also positive for HHV-6 DNA on PCR, showed no specific HHV-6 antigen. Immunohistochemical staining (HHV-6gp116/54/64 antibody) confirmed the presence of HHV-6 antigen in the three samples with the highest amounts of HHV-6 DNA. Double immunofluorescence analysis revealed co-localization of glial fibrillary acidic protein (GFAP) with HHV-6 antigen, suggesting that astrocytes are the preferentially infected cells.

In the second study, Fotheringham et al.[14] analyzed specimens from 15 patients with MTLE and seven patients with epilepsy other than MTLE.[14] They also described an additional patient who presented with typical temporal lobe epilepsy but then developed a diffuse epileptic syndrome. After a mesial temporal lobectomy and two additional focal resections were performed without a satisfactory seizure outcome, a hemispherectomy was performed. Upon neuropathological examination chronic encephalitis was diagnosed; however, the authors did not report if any patient had a history of CNS infection. Five patients with MTLE suffered from febrile seizures. A quantitative virus-specific real-time PCR technique was used to detect HHV-6B DNA in samples from 10 patients with MTLE and in tissue from each brain resection from the additional patient with chronic encephalitis. Once again, the highest levels of HHV-6 were found in hippocampal tissue. A high percentage of patients with detection of HHV-6 DNA suffered from febrile seizures, but no statistically significant association was observed.

Using real-time PCR, Karatas et al.[15] investigated the presence of viral DNA in formalin-fixed and paraffin-embedded specimens obtained from 33 patients with pharmacologically intractable MTLE and seven control samples. Analysis included PCR for HSV-1, HSV-2, cytomegalovirus (CMV), HHV-6, and HHV-8, but HHV-6A/6B-specific PCR was not performed. A review of the clinical data revealed febrile seizures in 75% of MTLE patients, but only in one patient with a history of meningitis. HHV-6 was detected in three patients, HSV-1 in two patients, and HHV-8 in one patient. All patients with HHV-6 had a history of febrile seizures. In a further study,

Niehusmann et al.[16] stratified 35 patients with temporal lobe epilepsy in four cliniconeuropathological predefined groups. Nested PCR experiments for HHV-6 after DNA extraction from formalin-fixed and paraffin-embedded hippocampal tissue showed 55% positivity in MTLE patients with a history of encephalitis. In contrast, no HHV-6 DNA was detected in patients with a well-documented history of complex febrile seizures, lesion-associated temporal lobe epilepsy (e.g., ganglioglioma, vascular malformation), or HS without any history of precipitating injury. Notably, 10 autopsy control samples were negative, whereas samples from additional patients suffering from limbic encephalitis with autoantibodies to glutamergic acid decarboxylase (GAD, $n=2$) or subacute HSV encephalitis ($n=1$) were also positive for HHV-6 DNA. Sequence analysis revealed HHV-6B DNA but no HHV-6A DNA in the positive samples. Double immunofluorescence analysis with a HHV-6 gp116/54/64 antibody and a GFAP antibody confirmed astrocytic localization of HHV-6 antigens.

A recent study from China reported positive nested PCR for HHV-6B in 9 of 32 hippocampal samples from patients with MTLE.[17] One of 12 control samples was also positive for HHV-6B. Nested PCR analysis for HHV-6A, HSV-1, and HSV-2 was negative for both MTLE patients and controls. Immunoreactivity with an anti-HHV-6B antibody was observed in all PCR-positive samples with the exception of the controls. Patients with a history of encephalitis were excluded from the study. Febrile seizures were reported in 11 cases and were significantly associated with HHV-6B detection. Real-time PCR, western blot analysis, and immunohistochemistry revealed upregulated NF-κB expression in HHV-6B-positive samples. The authors argued that this might be a sign of inflammatory responses.

A major problem is the limited comparability between the findings from the different studies, which has several causes:

- The analyzed patient populations vary substantially. Whereas some studies included patients with a history of encephalitis, other excluded such patients *a priori* from analysis or reported the clinical data only marginally. In particular, the diagnostic criteria for febrile seizures or differentiation among simple, complex, or prolonged febrile seizures are rarely given, although criteria exist.[18] In addition, it is not always stated if the patients suffered from *mesial* temporal lobe epilepsy or if cases with *neocortical* temporal lobe epilepsy were also included.
- There are important differences in the handling of the analyzed tissue samples. Long formalin fixation decreases PCR sensitivity and may denature proteins; therefore, virus may have been missed in studies basing on formalin-fixed tissues.
- Study designs ranged from nested PCRs to quantitative real-time PCRs. Substantial variations in the sensitivity of different HHV-6 PCR assays have been reported.[19]

Most authors of the studies on HHV-6 detection in surgically removed tissue recognized conflicting results compared to previous publications and referred to the above-mentioned differences in patients and methods. The existence of variations in specific HHV-6B prevalence in geographical regions has also been suggested.[17]

DISCUSSION

Despite the limitations on comparing the seven studies shown in Table 8.1 that analyzed surgically removed tissue, the summarized data show some interesting results. Virus detection was reported in 27.0% of all patients with temporal lobe epilepsy. Higher percentages were observed in patients with a history of febrile seizures (39.0%) or a history of encephalitis (55.6%). Control samples were analyzed in six studies, but only 4 of 55 control samples (7.3%) were positive for HHV-6 DNA. Whenever performed, HHV-6 virus differentiation revealed exclusively HHV-6B. An association between a neuropathological diagnosis of HS and the presence of HHV-6 DNA was not observed.

The low presence of HHV-6 DNA in controls is in contrast to previous data. Using a nested PCR approach, Chan et al.[20] reported detecting HHV-6A/B DNA in fresh autopsy frontal cortical brain tissue from 42.9% of 84 subjects. Analyzing the distribution of HHV-6A/B DNA in adult brain from postmortem cases, the same group found positive results in 34 of 40 patients (85.0%) and in 97 of 400 samples (24.3%).[21] Temporal lobe specimens were positive for HHV-6B DNA in 20.0%; however, the mean age of patients was approximately 65 years in both studies, in contrast to a mean age of approximately 33 years in the controls of some MTLE studies.[13,17] The presence of HHV-6 antigens was not determined in the autopsy studies. Interestingly, the Chinese MTLE study observed no HHV-6B antigen by immunohistochemistry in the one control patient with a positive HHV-6B PCR result.[17] This finding was interpreted as latent viral persistence without actual pathological relevance. HHV-6 DNA detection in autoptic brain tissue but an absence of HHV-6 protein have also been reported.[22]

Emerging evidence points to chronic inflammatory processes as important features in the development of MTLE. Animal models and clinical observations in humans indicate that the upregulation of proinflammatory molecules in epileptogenesis and pharmacological targeting of proinflammatory pathways show antiepileptic effects.[23] Interestingly, viral infections serve as a useful tool in animal models of epilepsy, such as Theiler's murine encephalomyelitis virus in mice.

The reported upregulation of NF-κB expression in MTLE patients with evidence of hippocampal HHV-6B antigen expression may also indicate inflammation.[17] In addition, increased proinflammatory cytokine expression was observed in HHV-6-infected astrocytoma cell lines.[24] Nevertheless, inflammatory infiltrates in MTLE patients with HHV-6B detection were more the

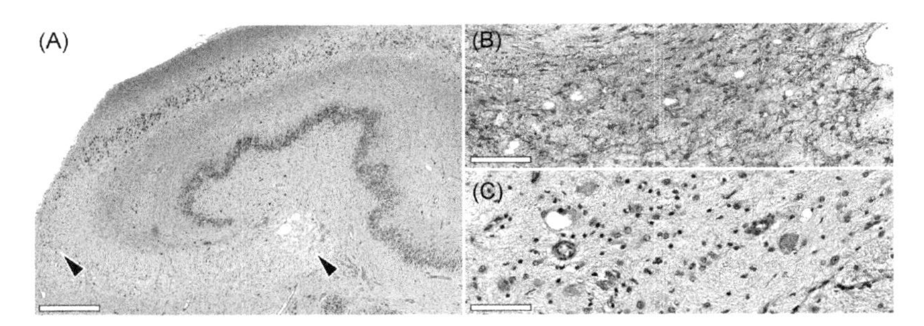

FIGURE 8.2 Hippocampal sclerosis, reactive astrogliosis, and inflammatory infiltrates in a MTLE patient with limbic encephalitis. Images are of a patient (ID 37) from the series published by Niehusmann and co-workers.[16] This patient suffered from limbic encephalitis with detection of autoantibodies to glutamergic acid decarboxylase (GAD). Nested PCR and immunofluorescence analysis were positive for HHV-6 (not shown). (A) Immunohistochemistry with antibodies against NeuN showed a so-called "end folium sclerosis" with severe neuronal cell loss in the hilar region (CA4/3, black arrowheads). In the same areas, (B) pronounced cellular astrogliosis (GFAP staining) and (C) T-lymphocytic infiltrates (CD3 staining) were observed. Scale bar: A = 1000 μm, B = 200 μm, C = 100 μm.

exception than the rule. Only in a patient diagnosed with chronic encephalitis and two patients with antibody-related nonparaneoplastic limbic encephalitis were increased lymphocytes in the hippocampus reported (Figure 8.2).[14,16]

Cerebrospinal fluid (CSF) pleocytosis is rare in patients with HS, even if associated with antibody-related limbic encephalitis. Recent analysis of preoperative CSF samples from surgically treated MTLE patients showed negative HHV-6 PCR results, although HHV-6 was detected in brain tissue (unpublished data, P.N.). That negative HHV-6 PCR in CSF does not exclude viral CNS invasion has been reported before.[25]

Hippocampus samples with severe sclerosis reflect the final stage of a probably progressive disorder, and we can only speculate about the pathology during the latency period. As an unspecific reaction in chronic processes, astrogliosis is an accompanying feature of neuronal cell loss in HS (see Figures 8.1C and 8.2B) but has also been found in HHV-6-positive MTLE samples without HS.[12] Because it is known that astrocytes contribute to the regulation of extracellular neurotransmitter levels,[26] the repeatedly demonstrated astrocytic localization of HHV-6 infection is of special interest (Figure 8.3).[13,14,16]

Indeed, Fotheringham et al.[14] showed in a nice series of experiments evidence for an altered function of HHV-6B-infected astrocytes. Cultured astrocytes from several MTLE patients were first analyzed for expression of the HHV-6 surface protein gp116/54/64 by immunofluorescence. Specific immunoreactivity for viral antigen was observed only in samples with previous HHV-6B DNA detection by PCR. Interestingly, decreased expression of the glutamate transporter EAAT2 was found in cultured primary astrocytes isolated from HHV-6 positive samples and after *in vitro* infection of primary

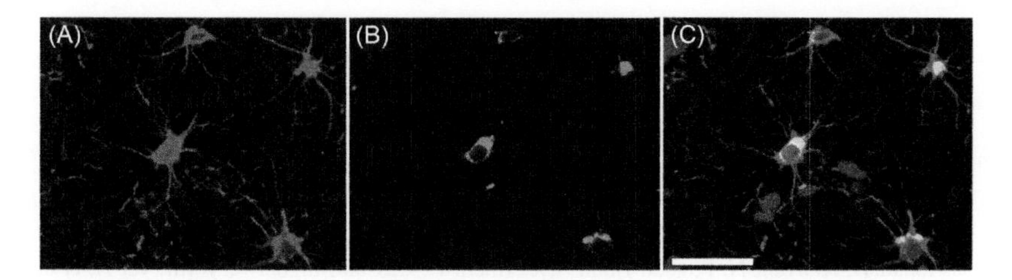

FIGURE 8.3 HHV-6 immunohistochemistry. Similar to what has been reported by several groups (see text for more details), double immunofluorescence analysis with (A) GFAP antibodies (red) and (B) HHV-6 gp116/54/64-antibodies (green) shows astrocytic localization of HHV-6 antigens in a patient with detection of HHV-6B DNA by real-time PCR. (C) Co-localization can be seen in the combined images. Diamidino-2-phenylindole (DAPI) was used for counterstaining of nuclei (blue). Scale bar = 50 μm. *(unpublished data).*

astrocytes. The authors hypothesized that decreased EAAT2 levels may be involved in glutamatergic excitotoxicity.

Due to its far-reaching therapeutical consequences, it is important to further elucidate if HHV-6B infection contributes to epileptogenesis in MTLE. Standardized virological investigations of larger and clinically well-characterized patient cohorts are necessary to address this point. The use of transgenic mice could help to establish an animal model of HHV-6B infection in MTLE.

REFERENCES

1. Semah F, Picot MC, Adam C, Broglin D, Arzimanoglou A, Bazin B, et al. Is the underlying cause of epilepsy a major prognostic factor for recurrence? *Neurology* 1998;**51**(5):1256–62.
2. Bien CG, Raabe AL, Schramm J, Becker A, Urbach H, Elger CE. Trends in presurgical evaluation and surgical treatment of epilepsy at one centre from 1988-2009. *J Neurol Neurosurg Psychiatry* 2013;**84**(1):54–61.
3. Blumcke I, Pauli E, Clusmann H, Schramm J, Becker A, Elger C, et al. A new clinico-pathological classification system for mesial temporal sclerosis. *Acta Neuropathol* 2007;**113**(3):235–44.
4. Blümcke I, Thom M, Wiestler OD. Ammon's horn sclerosis: a maldevelopmental disorder associated with temporal lobe epilepsy. *Brain Pathol* 2002;**12**(2):199–211.
5. Epstein LG, Shinnar S, Hesdorffer DC, Nordli DR, Hamidullah A, Benn EK, et al. Human herpesvirus 6 and 7 in febrile status epilepticus: the FEBSTAT study. *Epilepsia* 2012;**53**(9):1481–8.
6. Abou-Khalil B, Andermann E, Andermann F, Olivier A, Quesney LF. Temporal lobe epilepsy after prolonged febrile convulsions: excellent outcome after surgical treatment. *Epilepsia* 1993;**34**(5):878–83.
7. Cendes F, Andermann F, Dubeau F, Gloor P, Evans A, Jones-Gotman M, et al. Early childhood prolonged febrile convulsions, atrophy and sclerosis of mesial structures, and temporal lobe epilepsy: an MRI volumetric study. *Neurology* 1993;**43**(6):1083–7.
8. Shinnar S. Febrile seizures and mesial temporal sclerosis. *Epilepsy Curr* 2003;**3**(4):115–8.
9. Harberts E, Yao K, Wohler JE, Maric D, Ohayon J, Henkin R, et al. Human herpesvirus-6 entry into the central nervous system through the olfactory pathway. *Proc Natl Acad Sci USA* 2011;**108**(33):13734–9.

10. Caserta MT, Mock DJ, Dewhurst S. Human herpesvirus 6. *Clin Infect Dis* 2001;**33**(6):829–33.
11. Uesugi H, Shimizu H, Maehara T, Arai N, Nakayama H. Presence of human herpesvirus 6 and herpes simplex virus detected by polymerase chain reaction in surgical tissue from temporal lobe epileptic patients. *Psychiatry Clin Neurosci* 2000;**54**(5):589–93.
12. Eeg-Olofsson O, Bergstrom T, Andermann F, Andermann E, Olivier A, Rydenhag B. Herpesviral DNA in brain tissue from patients with temporal lobe epilepsy. *Acta Neurol Scand* 2004;**109**(3):169–74.
13. Donati D, Akhyani N, Fogdell-Hahn A, Cermelli C, Cassiani-Ingoni R, Vortmeyer A, et al. Detection of human herpesvirus-6 in mesial temporal lobe epilepsy surgical brain resections. *Neurology* 2003;**61**(10):1405–11.
14. Fotheringham J, Donati D, Akhyani N, Fogdell-Hahn A, Vortmeyer A, Heiss JD, et al. Association of human herpesvirus-6B with mesial temporal lobe epilepsy. *PLoS Med* 2007;**4**(5):e180.
15. Karatas H, Gurer G, Pinar A, Soylemezoglu F, Tezel GG, Hascelik G, et al. Investigation of HSV-1, HSV-2, CMV, HHV-6 and HHV-8 DNA by real-time PCR in surgical resection materials of epilepsy patients with mesial temporal lobe sclerosis. *J Neurol Sci* 2008;**264**(1–2):151–6.
16. Niehusmann P, Mittelstaedt T, Bien CG, Drexler JF, Grote A, Schoch S, et al. Presence of human herpes virus 6 DNA exclusively in temporal lobe epilepsy brain tissue of patients with history of encephalitis. *Epilepsia* 2010;**51**(12):2478–83.
17. Li JM, Lei D, Peng F, Zeng YJ, Li L, Xia ZL, et al. Detection of human herpes virus 6B in patients with mesial temporal lobe epilepsy in West China and the possible association with elevated NF-κB expression. *Epilepsy Res* 2011;**94**(1–2):1–9.
18. Berg AT, Shinnar S. Complex febrile seizures. *Epilepsia* 1996;**37**(2):126–33.
19. Flamand L, Gravel A, Boutolleau D, Alvarez-Lafuente R, Jacobson S, Malnati MS, et al. Multicenter comparison of PCR assays for detection of human herpesvirus 6 DNA in serum. *J Clin Microbiol* 2008;**46**(8):2700–6.
20. Chan PK, Ng HK, Hui M, Ip M, Cheung JL, Cheng AF. Presence of human herpesviruses 6, 7, and 8 DNA sequences in normal brain tissue. *J Med Virol* 1999;**59**(4):491–5.
21. Chan PK, Ng HK, Hui M, Cheng AF. Prevalence and distribution of human herpesvirus 6 variants A and B in adult human brain. *J Med Virol* 2001;**64**(1):42–6.
22. Cuomo L, Trivedi P, Cardillo MR, Gagliardi FM, Vecchione A, Caruso R, et al. Human herpesvirus 6 infection in neoplastic and normal brain tissue. *J Med Virol* 2001;**63**(1):45–51.
23. Ravizza T, Balosso S, Vezzani A. Inflammation and prevention of epileptogenesis. *Neurosci Lett* 2011;**497**(3):223–30.
24. Yoshikawa T, Asano Y, Akimoto S, Ozaki T, Iwasaki T, Kurata T, et al. Latent infection of human herpesvirus 6 in astrocytoma cell line and alteration of cytokine synthesis. *J Med Virol* 2002;**66**(4):497–505.
25. Mannonen L, Herrgard E, Valmari P, Rautiainen P, Uotila K, Aine MR, et al. Primary human herpesvirus-6 infection in the central nervous system can cause severe disease. *Pediatr Neurol* 2007;**37**(3):186–91.
26. Danbolt NC. Glutamate uptake. *Prog Neurobiol* 2001;**65**(1):1–105.

Chapter 9

HHV-6B and HHV-7 in Exanthema Subitum and Related Skin Diseases

Tetsushi Yoshikawa
Fujita Health University School of Medicine, Toyoake, Aichi, Japan

INTRODUCTION

Human herpesvirus 6 (HHV-6) is now classified as two distinct viruses: HHV-6A and HHV-6B.[1,2] HHV-6B is ubiquitous throughout Europe, the United States, and Japan,[3–6] whereas HHV-6A is endemic in Sub-Saharan Africa.[7] It is well known that primary HHV-6B infection can cause exanthem subitum,[8] which is a common febrile exanthematous disease in infants and young children; however, the clinical features of primary HHV-6A infection remain unclear. Although it has been demonstrated that HHV-7 is also associated with exanthem subitum,[9] the precise incidence of HHV-7-associated exanthem subitum is uncertain.

The typical clinical course of exanthem subitum is the appearance of skin eruptions that coincide with waning fever on the third or fourth day. The pathogenesis of the skin eruptions in the disease remains unclear because histological analysis of skin tissues from affected patients has been limited. Moreover, the elucidation of the pathogenesis of skin eruptions has been hampered by the lack of suitable animal models of primary HHV-6B infection. Both primary HHV-6B infection and its reactivation are associated with skin diseases. HHV-6B infection is associated with fever and skin rash resembling acute graft-versus-host disease (GVHD)[10–12] and may play an important role in the development of acute GVHD in hematopoietic stem cell transplant (HSCT) recipients.[13–15] Furthermore, several reports have documented HHV-6B reactivation in patients with drug-induced hypersensitivity syndrome (DIHS)/drug reaction with eosinophilia and systemic symptom (DRES).[16–18] Thus, not only primary infection but also reactivation of these two roseoloviruses can cause skin eruptions. This chapter focuses on exanthem subitum and other skin diseases linked to HHV-6 and HHV-7 infections.

Human Herpesviruses HHV-6A, HHV-6B & HHV-7.

EXANTHEM SUBITUM AND HHV-6B

As previously described, most infants and young children receive primary HHV-6B infection by 2 years after birth.[19–22] Horizontal transmission via salivary HHV-6B shedding from seropositive individuals is considered to be the main route of viral infection.[23,24] Although vertical transmission of the virus was considered to be rare,[25,26] congenital HHV-6B infection by reactivated virus from chromosomally integrated HHV-6B has been recently demonstrated.[27,28]

Primary infection with HHV-6B causes exanthem subitum, a common febrile disease in infants and young children.[8] In Japan, 70 to 80% of children with primary HHV-6B infection that visited hospitals with high fever also exhibited a typical clinical course of exanthem subitum.[29,30] The other patients with primary HHV-6B infection demonstrated a subclinical infection such as fever without rash[31] or rash without fever,[32] or no clinical symptoms. In a previous study, we determined that the skin rashes were primarily papular (rubella-like; 54%), with some occurrences of macular (measles-like; 40%) and maculopapular (6%) rashes (Figure 9.1A). Rashes initially appeared on the face, trunk, or both, and subsequently spread to other locations.[30] The frequency of other clinical features such as diarrhea, edematous eyelids, bulging fontanelle, and Nagayama's spot (Figure 9.1B) was also determined in a retrospective study (Table 9.1).[30]

Meanwhile, in the United States, only 9 to 17% of children with primary HHV-6B infections who visited an emergency room developed exanthem subitum; the majority developed undefined febrile illnesses.[21,33] A prospective population-based study was conducted outside of the acute-care setting to identify the clinical feature of primary HHV-6B infection.[22] A cohort of 277 children was followed up from birth through the first 2 years of life using polymerase chain reaction (PCR) to monitor salivary shedding of HHV-6B, and parents maintained a daily log of signs and symptoms of illnesses in

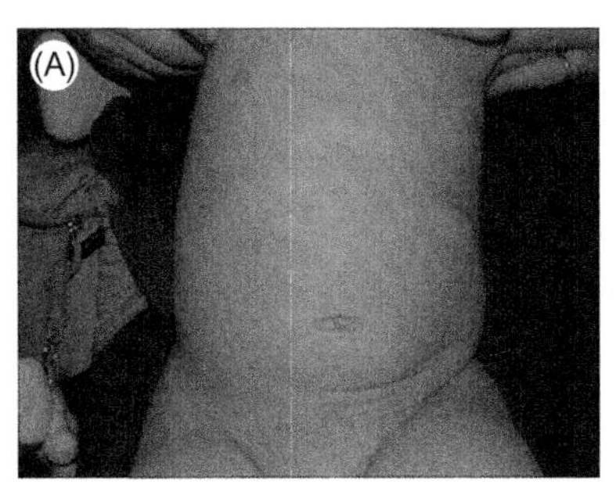

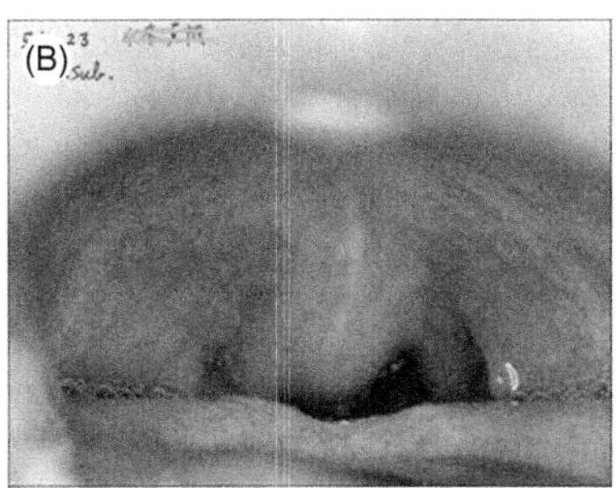

FIGURE 9.1 (A) Typical skin rash was observed on the trunk of a patient with primary HHV-6B infection. (B) Nagayama's spot was demonstrated on the palate of a patient with primary HHV-6B infection.

their children. Compared to children with unrelated illnesses, those with primary HHV-6B infection were more likely to have rash ($p = 0.003$) and roseola ($p = 0.002$). The actual incidences of these rash (31%) and roseola (24%) symptoms were low. Taken together, these findings suggest that the clinical course of primary HHV-6B infection differs between North America and Japan; however, the reasons underlying this difference remain unclear.

Although primary HHV-6B infection is generally benign and self-limiting, it can be associated with several complications, including the fatal outcomes listed in Table 9.2.[34–44] Among these complications, neurological complications are the most important because they occur more frequently than any other complication. Neurovirulence of HHV-6B has been demonstrated by both *in vivo*[45] and *in vitro*[46,47] studies, which are described in detail elsewhere in this book. In some cases, primary HHV-6B infection is associated with febrile seizure[48–52] and encephalitis.[37,53] According to our recent survey, the annual incidence of exanthem subitum-associated encephalitis was estimated to be 60 cases per year in Japan, with two fatal cases; almost half of the patients had severe neurological sequelae, including quadriplegia and mental retardation.[54] Several *in vivo* studies[53,55] suggested that host immune responses, such as the release of

TABLE 9.1 Frequency of Signs and Symptoms in Infants with Virologically Confirmed Exanthem Subitum

Category of Findings	No. of Patients Evaluated	No. (%) with Positive Findings
Prodromal symptoms[a]	63	9 (14)
Fever[b]	176	173 (98)
Rash	176	172 (98)
Diarrhea	171	116 (68)
Edematous eyelids	60	18 (30)
Nagayama's spots[c]	49	32 (65)
Cough	173	90 (50)
Cervical lymph node swelling	61	19 (31)
Bulging fontanelle	152	39 (26)
Convulsion	173	13 (8)

[a]*Nonspecific symptoms such as listlessness or irritability.*
[b]*Equal to or greater than 37.5°C.*
[c]*Erythematous papules on the mucosa of the soft palate and the base of uvula.*
Source: *Asano Y, et al. Pediatrics 1994;* **93***: 104–8.*

TABLE 9.2 Complications Caused by Primary HHV-6B Infection

Category of Complications	Incidence	Fatal Case
Febrile seizure	Common	No
Encephalitis/encephalopathy	Rare	Yes
Facial nerve paralysis	Rare	No
Guillain–Barré syndrome	Rare	No
Myelitis	Rare	No
Hemophagocytic syndrome	Rare	Yes
Idiopathic thrombocytopenic purpura	Rare	No
Hepatitis	Common	Yes
Intussusceptions	Rare	No
Rhabdomyolysis	Rare	No
Myocarditis	Rare	Yes

cytokines, and not direct invasion of the virus may play an important role in the pathogenesis of HHV-6B encephalitis at the time of primary infection. In addition to HHV-6B encephalitis,[36] other fatal outcomes have been documented, including virus-associated hemophagocytic syndrome, fulminant hepatitis,[34] fatal disseminated infection,[56] and myocarditis.[40]

Previous studies have demonstrated the therapeutic efficacy of ganciclovir and foscarnet as antivirals against HHV-6. Although acyclovir is less toxic than both ganciclovir and foscarnet, *in vitro* analysis determined that acyclovir had no antiviral activity against HHV-6.[57,58] In most patients with primary HHV-6B infection, antiviral treatment is not necessary because the resulting disease is benign and self-limiting. The efficacy of antiviral treatments, such as ganciclovir and foscarnet, has been demonstrated in patients with post-transplant HHV-6B encephalitis caused by viral reactivation.[59,60] Antiviral treatments may be beneficial for patients who have a primary HHV-6B infection with severe complications as shown in Table 9.2. However, as HHV-6B viremia occurs during the febrile period of exanthem subitum,[61] the administration of an antiviral drug during the febrile period may be effective. Thus, a rapid diagnostic method for active HHV-6B infection such as quantitative PCR from plasma or serum is necessary to determine the indication for antiviral treatments during febrile period of exanthem subitum.

EXANTHEM SUBITUM AND HHV-7

Similar to HHV-6B, HHV-7 is ubiquitous in Japan.[5,62] Based on sero-epidemiological analyses[5,62] and large cohort analyses to elucidate acquisition of the two viruses,[63] the timing of primary HHV-7 infection was generally later than primary HHV-6B infection. Both of the viruses persistently infect the salivary glands after primary viral infection. Although HHV-7 is frequently isolated from saliva collected from seropositive individuals, only viral DNA (not infectious virus) is detected in saliva collected from HHV-6B seropositive individuals.[64,65] Despite the high frequency of HHV-7 shedding into saliva, the underlying reasons for the timing of HHV-7 acquisition after HHV-6B infection remains unclear.

Primary HHV-7 infection was shown to cause exanthem subitum that had clinical features similar to that of primary HHV-6B infection.[66,67] We identified 15 patients with primary HHV-7 infection during a 19-month observation period, and 14 out of the 15 patients demonstrated typical clinical features of exanthem subitum.[66] In a large cohort study designed to elucidate the acquisition of HHV-7 infection in the United States, 30 cases of HHV-7 viremia were detected: 23 (77%) patients with primary and 7 (23%) patients with reactivated HHV-7 infection. Clinical manifestations such as fever, upper respiratory illness, and gastroenteritis were similar in patients with primary and reactivated HHV-7 infections, except that seizures occurred more frequently in patients with reactivated infections.[63] Additionally, other investigators also demonstrated that HHV-7 was associated with febrile seizures and encephalitis.[68,69]

A significant increase of anti-HHV-6 antibody levels was observed in patients with prior HHV-6 infection at the time of primary HHV-7 infection.[70] Subsequent analysis identified a cross-reaction between HHV-6 and HHV-7 antibodies. Additionally, HHV-7 infection can induce reactivation of latent HHV-6 in peripheral blood mononuclear cells;[70] therefore, these factors should be considered for the accurate diagnosis of primary HHV-7 infection.

HHV-6B REACTIVATION IN SEVERE FORMS OF DRUG ALLERGY

It is well known that viral infections can be involved in the development of some drug reactions, such as ampicillin-induced eruption in mononucleosis. HHV-6B infection is involved in the pathogenesis of DIHS, which is characterized by the clinical triad of fever, rash, and internal organ involvement due to drug exposure. An association between HHV-6B and drug allergy is described elsewhere in this book. The typical clinical course of the disease is fever, maculopapular eruptions, lymphadenopathy, hepatitis, and leukocytosis at 2 to 6 weeks following the initiation of drug administration. Several studies have demonstrated that HHV-6B reactivation occurs about 2 weeks after the onset of DIHS.[16–18,71] In addition to HHV-6B, recent studies have demonstrated that several other herpesviruses including HHV-7 are also reactivated in these patients.[72,73] HHV-6B reactivation may account for a prolonged course, slow resolution, and recurrence of signs and symptoms of DIHS.[71] Various drugs, including sulfasalazine, allopurinol, ibuprofen, phenobarbital, carbamazepine, zonisamide, phenytoin, and sodium valproate, have been proposed as drugs that can induce HHV-6B infection associated DIHS.[74] Neurological complications have been reported in DIHS patients with HHV-6B reactivation.[75,76] In addition to the encephalitis, several other complications, including liver damage, acute renal failure, and myocarditis, have been linked to patient mortality. In addition to these complications, recently fulminant type 1 diabetes mellitus was reported in DIHS patients with HHV-6 reactivation.[77]

Recently, in patients with clinical symptoms similar to DIHS, we found HHV-6B reactivation after occupational exposure to trichloroethylene in China.[78] Fifty-nine patients and 59 healthy exposed workers from the patients' factories were selected on an age-matched basis and enrolled in a study to elucidate an association between disease and viral infection. Marked increases in anti-HHV-6 IgG titer (≥256) were observed in 14 (25%) patients, while no abnormal increases were detected in the controls ($p < 0.001$).[78]

Many mechanisms have been proposed to explain the association between HHV-6 infection and DIHS. Within 10 days after the onset of DIHS, increased numbers of monomyeloid precursors were detected in the peripheral blood of the patients. HHV-6 antigen was detected in the monomyeloid precursor cells of the patients with viral reactivation.[79] These findings suggest that monomyeloid precursors appear at the early stage of DIHS and may play an important role in HHV-6 reactivation in DIHS patients. Furthermore, the induction of cytokines during the allergic reaction may also be involved in the reactivation of HHV-6B in DIHS patients. Elevated TNF-alpha and IL-6 levels were reported to precede HHV-6B reactivation in DIHS patients. On the other hand, in DIHS patients IL-6 levels are significantly decreased to undetectable levels during viral infection and, subsequently, serum IL-6 levels increase after viral infection. Thus, these inflammatory cytokines may play an important role in the induction of HHV-6B reactivation, which in turn may be associated with the enhanced expression of these cytokines.[80]

ACUTE GRAFT-VERSUS-HOST DISEASE AND HHV-6B

Several clinical conditions, such as skin rash resembling acute GVHD, bone marrow suppression, interstitial pneumonitis, and encephalitis, may be related to HHV-6B infection after HSCT. Moreover, an association between HHV-6B infection and acute GVHD is suggested in HSCT recipients. Although the exact frequency of HHV-6B reactivation is difficult to determine and varies depending on the method used for monitoring viral infection (namely, PCR versus viral isolation), approximately 40 to 50% of HSCT recipients develop HHV-6B infection around 2 to 4 weeks after HSCT. Many studies have been carried out to elucidate the clinical features of "HHV-6B diseases" in HSCT recipients. Collectively, these studies have suggested that HHV-6B plays an important role in the development of acute GVHD[14,81] or the development of skin rash after allogeneic HSCT.[13,82,83]

Our previous report demonstrated the isolation of HHV-6B from the blood of the three HSCT recipients 15 days after transplant. Two of the three patients, subsequently treated for acute GVHD, had fever and macular rash at the time of viremia.[10] To further explore the possible association between HHV-6B infection and acute GVHD, two large cohort studies were conducted. Twenty-five HSCT recipients were enrolled in the initial study to monitor HHV-6 infection using virus isolation and serological analysis.[11] HHV-6B infection was confirmed in 12 of 25 (48%) recipients. Four of these 12 recipients developed skin rashes, and three of those four recipients also experienced a febrile episode during the period of viral isolation. None of the remaining 13 patients developed any of these symptoms. In the subsequent larger cohort study, we found HHV-6 viremia in 9 of 15 (60%) patients who developed a skin rash within 1 month after HSCT compared with no cases (0%) with a skin rash at more than 1 month after transplantation ($p = 0.008$).[12] A similar finding was reported by Volin et al.[84] based on an HHV-6 antigenemia assay, which is a good tool for the evaluation of active viral infection.

To clarify the association between HHV-6B and skin rash in transplant recipients, a study was undertaken to detect the virus genome within skin tissues of recipients.[85] Additionally, several cytokines were implicated in the pathogenesis of HHV-6B-associated skin rash in HSCT recipients.[86,87] However, it remains unclear whether HHV-6B causes acute GVHD or whether the virus causes an erythematous illness similar to acute GVHD. In addition, GVHD may induce HHV-6B reactivation, and the virus may not be associated with the observed skin eruptions. The appropriate strategy for patient management differs depending on the particular circumstances of HHV-6B infection. To determine a precise association between HHV-6B reactivation and acute GVHD or skin rash and fever, further studies are needed.

PATHOGENESIS OF SKIN MANIFESTATIONS DUE TO HHV-6B INFECTION

The pathogenesis of HHV-6B-related skin eruptions remains unclear, largely due to the difficulty in obtaining skin tissues for histological analysis from patients with exanthem subitum as well as the absence of suitable animal models of primary HHV-6 infection. The absence of rash after the resolution of fever was demonstrated in immunocompromised infants during primary HHV-6B infection.[88,89] Thus, the host immune response against the virus appears to play an important pathogenic role in the skin rash in patients with

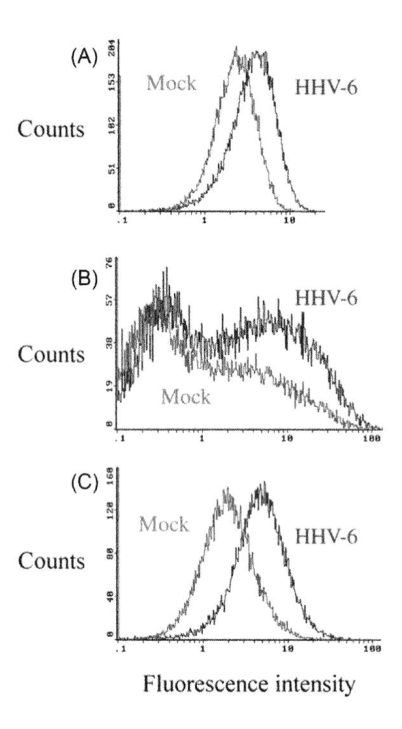

FIGURE 9.2 Flow cytometric profiles of the expression of (A) HLA-ABC, (B) HLA-DR, and (C) intercellular adhesion molecule 1 (ICAM-1) on the surface of HHV-6- or mock-infected A431 cells. Nonspecific binding of each antibody was excluded, using control isotype antibody. Although low levels of vascular adhesion molecule 1 (VCAM-1) and endothelial leukocyte adhesion molecule 1 (ELAM-1) expression were observed on uninfected A431 cells, no change in the levels of the surface molecules was demonstrated between HHV-6- and mock-infected A431 cells (data not shown). Culture media from both HHV-6- and mock-infected cells did not alter the expression of these three surface molecules (data not shown).[90]

exanthem subitum. An *in vitro* study demonstrated that HHV-6 can infect an epidermal cell line, and human leukocyte antigen (HLA)-A, -B, and -C, HLA-DR, and intercellular adhesion molecule 1 (ICAM-1) expression was upregulated in HHV-6-infected cells (Figure 9.2).[90] Moreover, human umbilical endothelial cells were also susceptible to HHV-6 infection.[91,92] HHV-6B infection of epidermal cells or vascular endothelial cells via virus-infected mononuclear cells may play an important role in the pathogenesis of skin eruptions in exanthem subitum. Furthermore, the upregulation of cell surface molecules on the HHV-6B-infected cells could recruit virus-infected mononuclear cells in the skin tissues. In addition to HHV-6B infection in local skin tissues, the host immune response against HHV-6B may also play an important role in the development of skin rashes in patients with exanthem subitum. It is likely that the mechanisms underlying the development of skin rashes in HSCT recipients would be different and more complicated than those of exanthem subitum patients. Further in-depth studies based on virological and immunological analyses are necessary to elucidate the full spectrum of the pathogenesis of skin rashes caused by HHV-6B infection.

REFERENCES

1. Ablashi DV, Balachandran N, Josephs SF, Hung CL, Krueger GR, Kramarsky B, et al. Genomic polymorphism, growth properties, and immunologic variations in human herpesvirus-6 isolates. *Virology* 1991;**184**:545–52.
2. Schirmer EC, Wyatt LS, Yamanishi K, Rodriguez WJ, Frenkel N. Differentiation between two distinct classes of viruses now classified as human herpesvirus 6. *Proc Natl Acad Sci USA* 1991;**88**:5922–6.
3. Dewhurst S, McIntyre K, Schnabel K, Hall CB. Human herpesvirus 6 (HHV-6) variant B accounts for the majority of symptomatic primary HHV-6 infections in a population of U.S. infants. *J Clin Microbiol* 1993;**31**:416–8.
4. Hall CB, Caserta MT, Schnabel KC, Long C, Epstein LG, Insel RA, et al. Persistence of human herpesvirus 6 according to site and variant: possible greater neurotropism of variant A. *Clin Infect Dis* 1998;**26**:132–7.
5. Tanaka-Taya K, Kondo T, Mukai T, Miyoshi H, Yamamoto Y, Okada S, et al. Seroepidemiological study of human herpesvirus-6 and -7 in children of different ages and detection of these two viruses in throat swabs by polymerase chain reaction. *J Med Virol* 1996;**48**:88–94.
6. Aberle SW, Mandl CW, Kunz C, Popow-Kraupp T. Presence of human herpesvirus 6 variants A and B in saliva and peripheral blood mononuclear cells of healthy adults. *J Clin Microbiol* 1996;**34**:3223–5.
7. Bates M, Monze M, Bima H, Kapambwe M, Clark D, Kasolo FC, et al. Predominant human herpesvirus 6 variant A infant infections in an HIV-1 endemic region of Sub-Saharan Africa. *J Med Virol* 2009;**81**:779–89.
8. Yamanishi K, Okuno T, Shiraki K, Takahashi M, Kondo T, Asano Y, et al. Identification of human herpesvirus-6 as a causal agent for exanthem subitum. *Lancet* 1988;**1**(8594):1065–7.
9. Tanaka K, Kondo T, Torigoe S, Okada S, Mukai T, Yamanishi K. Human herpesvirus 7: another causal agent for roseola (exanthem subitum). *J Pediatr* 1994;**125**:1–5.
10. Asano Y, Yoshikawa T, Suga S, Nakashima T, Yazaki T, Fukuda M, et al. Reactivation of herpesvirus type 6 in children receiving bone marrow transplants for leukemia. *N Engl J Med* 1991;**324**:634–5.
11. Yoshikawa T, Suga S, Asano Y, Nakashima T, Yazaki T, Sobue R, et al. Human herpesvirus-6 infection in bone marrow transplantation. *Blood* 1991;**78**:1381–4.
12. Yoshikawa T, Asano Y, Ihira M, Suzuki K, Ohashi M, Suga S, et al. Human herpesvirus 6 viremia in bone marrow transplant recipients: clinical features and risk factors. *J Infect Dis* 2002;**185**:847–53.
13. Appleton AL, Peiris JS, Taylor CE, Sviland L, Cant AJ. Human herpesvirus 6 DNA in skin biopsy tissue from marrow graft recipients with severe combined immunodeficiency. *Lancet* 1994;**344**(8933):1361–2.
14. Wilborn F, Brinkmann V, Schmidt CA, Neipel F, Gelderblom H, Siegert W. Herpesvirus type 6 in patients undergoing bone marrow transplantation: serologic features and detection by polymerase chain reaction. *Blood* 1994;**83**:3052–8.

15. Zerr DM, Corey L, Kim HW, Huang ML, Nguy L, Boeckh M. Clinical outcomes of human herpesvirus 6 reactivation after hematopoietic stem cell transplantation. *Clin Infect Dis* 2005;**40**:932–40.
16. Tohyama M, Yahata Y, Yasukawa M, Inagi R, Urano Y, Yamanishi K, et al. Severe hypersensitivity syndrome due to sulfasalazine associated with reactivation of human herpesvirus 6. *Arch Dermatol* 1998;**134**:1113–7.
17. Descamps V, Bouscarat F, Laglenne S, Aslangul E, Veber B, Descamps D, et al. Human herpesvirus 6 infection associated with anticonvulsant hypersensitivity syndrome and reactive haemophagocytic syndrome. *Br J Dermatol* 1997;**137**:605–8.
18. Descamps V, Valance A, Edlinger C, Fillet AM, Grossin M, Lebrun-Vignes B, et al. Association of human herpesvirus 6 infection with drug reaction with eosinophilia and systemic symptoms. *Arch Dermatol* 2001;**137**:301–4.
19. Yoshikawa T, Suga S, Asano Y, Yazaki T, Kodama H, Ozaki T. Distribution of antibodies to a causative agent of exanthem subitum (human herpesvirus-6) in healthy individuals. *Pediatrics* 1989;**84**:675–7.
20. Okuno T, Takahashi K, Balachandra K, Shiraki K, Yamanishi K, Takahashi M, et al. Seroepidemiology of human herpesvirus 6 infection in normal children and adults. *J Clin Microbiol* 1989;**27**:651–3.
21. Hall CB, Long CE, Schnabel KC, Caserta MT, McIntyre KM, Costanzo MA, et al. Human herpesvirus-6 infection in children. A prospective study of complications and reactivation. *N Engl J Med* 1994;**331**:432–8.
22. Zerr DM, Meier AS, Selke SS, Frenkel LM, Huang ML, Wald A, et al. A population-based study of primary human herpesvirus 6 infection. *N Engl J Med* 2005;**352**:768–76.
23. Okuno T, Mukai T, Baba K, Ohsumi Y, Takahashi M, Yamanishi K. Outbreak of exanthem subitum in an orphanage. *J Pediatr* 1991;**119**:759–61.
24. Mukai T, Yamamoto T, Kondo T, Kondo K, Okuno T, Kosuge H, et al. Molecular epidemiological studies of human herpesvirus 6 in families. *J Med Virol* 1994;**42**:224–7.
25. Okuno T, Oishi H, Hayashi K, Nonogaki M, Tanaka K, Yamanishi K. Human herpesviruses 6 and 7 in cervixes of pregnant women. *J Clin Microbiol* 1995;**33**:1968–70.
26. Ohashi M, Yoshikawa T, Ihira M, Suzuki K, Suga S, Tada S, et al. Reactivation of human herpesvirus 6 and 7 in pregnant women. *J Med Virol* 2002;**67**:354–8.
27. Hall CB, Caserta MT, Schnabel K, Shelley LM, Marino AS, Carnahan JA, et al. Chromosomal integration of human herpesvirus 6 is the major mode of congenital human herpesvirus 6 infection. *Pediatrics* 2008;**122**:513–20.
28. Hall CB, Caserta MT, Schnabel KC, Shelley LM, Carnahan JA, Marino AS, et al. Transplacental congenital human herpesvirus 6 infection caused by maternal chromosomally integrated virus. *J Infect Dis* 2010;**201**:505–7.
29. Asano Y, Nakashima T, Yoshikawa T, Suga S, Yazaki T. Severity of human herpesvirus-6 viremia and clinical findings in infants with exanthem subitum. *J Pediatr* 1991;**118**:891–5.
30. Asano Y, Yoshikawa T, Suga S, Kobayashi I, Nakashima T, Yazaki T, et al. Clinical features of infants with primary human herpesvirus 6 infection (exanthem subitum, roseola infantum). *Pediatrics* 1994;**93**:104–8.
31. Suga S, Yoshikawa T, Asano Y, Yazaki T, Hirata S. Human herpesvirus-6 infection (exanthem subitum) without rash. *Pediatrics* 1989;**83**:1003–6.
32. Asano Y, Suga S, Yoshikawa T, Urisu A, Yazaki T. Human herpesvirus type 6 infection (exanthem subitum) without fever. *J Pediatr* 1989;**115**:264–5.
33. Pruksananonda P, Hall CB, Insel RA, McIntyre K, Pellett PE, Long CE, et al. Primary human herpesvirus 6 infection in young children. *N Engl J Med* 1992;**326**:1445–50.

34. Asano Y, Yoshikawa T, Suga S, Yazaki T, Kondo K, Yamanishi K. Fatal fulminant hepatitis in an infant with human herpesvirus-6 infection. *Lancet* 1990;**335**(8693):862–3.
35. Asano Y, Yoshikawa T, Suga S, Hata T, Yamazaki T, Yazaki T. Simultaneous occurrence of human herpesvirus 6 infection and intussusception in three infants. *Pediatr Infect Dis J* 1991;**10**:335–7.
36. Asano Y, Yoshikawa T, Kajita Y, Ogura R, Suga S, Yazaki T, et al. Fatal encephalitis/encephalopathy in primary human herpesvirus-6 infection. *Arch Dis Child* 1992;**67**:1484–5.
37. Suga S, Yoshikawa T, Asano Y, Kozawa T, Nakashima T, Kobayashi I, et al. Clinical and virological analyses of 21 infants with exanthem subitum (roseola infantum) and central nervous system complications. *Ann Neurol* 1993;**33**:597–603.
38. Tajiri H, Tanaka-Taya K, Ozaki Y, Okada S, Mushiake S, Yamanishi K. Chronic hepatitis in an infant, in association with human herpesvirus-6 infection. *J Pediatr* 1997;**131**:473–5.
39. Yoshikawa T, Morooka M, Suga S, Niinomi Y, Kaneko T, Shinoda K, et al. Five cases of thrombocytopenia induced by primary human herpesvirus 6 infection. *Acta Paediatr Jpn* 1998;**40**:278–81.
40. Yoshikawa T, Ihira M, Suzuki K, Suga S, Kito H, Iwasaki T, et al. Fatal acute myocarditis in an infant with human herpesvirus 6 infection. *J Clin Pathol* 2001;**54**:792–5.
41. Miyake F, Yoshikawa T, Suzuki K, Ohashi M, Suga S, Asano Y. Guillain–Barré syndrome after exanthem subitum. *Pediatr Infect Dis J* 2002;**21**:569–70.
42. Pitkaranta A, Lahdenne P, Piiparinen H. Facial nerve palsy after human herpesvirus 6 infection. *Pediatr Infect Dis J* 2004;**23**:688–9.
43. Ohsaka M, Houkin K, Takigami M, Koyanagi I. Acute necrotizing encephalopathy associated with human herpesvirus-6 infection. *Pediatr Neurol* 2006;**34**:160–3.
44. Fujino M, Ohashi M, Tanaka K, Kato T, Asano Y, Yoshikawa T. Rhabdomyolysis in an infant with primary human herpesvirus 6 infection. *Pediatr Infect Dis J* 2012;**31**:1202–3.
45. Caserta MT, Hall CB, Schnabel K, McIntyre K, Long C, Costanzo M, et al. Neuroinvasion and persistence of human herpesvirus 6 in children. *J Infect Dis* 1994;**170**:1586–9.
46. He J, McCarthy M, Zhou Y, Chandran B, Wood C. Infection of primary human fetal astrocytes by human herpesvirus 6. *J Virol* 1996;**70**:1296–300.
47. Donati D, Martinelli E, Cassiani-Ingoni R, Ahlqvist J, Hou J, Major EO, et al. Variant-specific tropism of human herpesvirus 6 in human astrocytes. *J Virol* 2005;**79**:9439–48.
48. Jee SH, Long CE, Schnabel KC, Sehgal N, Epstein LG, Hall CB. Risk of recurrent seizures after a primary human herpesvirus 6-induced febrile seizure. *Pediatr Infect Dis J* 1998;**17**:43–8.
49. Kondo K, Nagafuji H, Hata A, Tomomori C, Yamanishi K. Association of human herpesvirus 6 infection of the central nervous system with recurrence of febrile convulsions. *J Infect Dis* 1993;**167**:1197–200.
50. Barone SR, Kaplan MH, Krilov LR. Human herpesvirus-6 infection in children with first febrile seizures. *J Pediatr* 1995;**127**:95–7.
51. Suga S, Suzuki K, Ihira M, Yoshikawa T, Kajita Y, Ozaki T, et al. Clinical characteristics of febrile convulsions during primary HHV-6 infection. *Arch Dis Child* 2000;**82**:62–6.
52. Epstein LG, Shinnar S, Hesdorffer DC, Nordli DR, Hamidullah A, Benn EK, et al. Human herpesvirus 6 and 7 in febrile status epilepticus: the FEBSTAT study. *Epilepsia* 2012;**53**:1481–8.
53. Ichiyama T, Ito Y, Kubota M, Yamazaki T, Nakamura K, Furukawa S. Serum and cerebrospinal fluid levels of cytokines in acute encephalopathy associated with human herpesvirus-6 infection. *Brain Dev* 2009;**31**:731–8.
54. Yoshikawa T, Ohashi M, Miyake F, Fujita A, Usui C, Sugata K, et al. Exanthem subitum-associated encephalitis: nationwide survey in Japan. *Pediatr Neurol* 2009;**41**:353–8.

55. Kawamura Y, Sugata K, Ihira M, Mihara T, Mutoh T, Asano Y, et al. Different characteristics of human herpesvirus 6 encephalitis between primary infection and viral reactivation. *J Clin Virol* 2011;**51**:12–19.
56. Prezioso PJ, Cangiarella J, Lee M, Nuovo GJ, Borkowsky W, Orlow SJ, et al. Fatal disseminated infection with human herpesvirus-6. *J Pediatr* 1992;**120**:921–3.
57. Agut H, Collandre H, Aubin JT, Guetard D, Favier V, Ingrand D, et al. *In vitro* sensitivity of human herpesvirus-6 to antiviral drugs. *Res Virol* 1989;**140**:219–28.
58. Burns WH, Sandford GR. Susceptibility of human herpesvirus 6 to antivirals *in vitro*. *J Infect Dis* 1990;**162**:634–7.
59. Mookerjee BP, Vogelsang G. Human herpes virus-6 encephalitis after bone marrow transplantation: successful treatment with ganciclovir. *Bone Marrow Transplant* 1997;**20**:905–6.
60. Tokimasa S, Hara J, Osugi Y, Ohta H, Matsuda Y, Fujisaki H, et al. Ganciclovir is effective for prophylaxis and treatment of human herpesvirus-6 in allogeneic stem cell transplantation. *Bone Marrow Transplant* 2002;**29**:595–8.
61. Asano Y, Yoshikawa T, Suga S, Yazaki T, Hata T, Nagai T, et al. Viremia and neutralizing antibody response in infants with exanthem subitum. *J Pediatr* 1989;**114**:535–9.
62. Yoshikawa T, Asano Y, Kobayashi I, Nakashima T, Yazaki T, Suga S, et al. Seroepidemiology of human herpesvirus 7 in healthy children and adults in Japan. *J Med Virol* 1993;**41**:319–23.
63. Hall CB, Caserta MT, Schnabel KC, McDermott MP, Lofthus GK, Carnahan JA, et al. Characteristics and acquisition of human herpesvirus (HHV) 7 infections in relation to infection with HHV-6. *J Infect Dis* 2006;**193**:1063–9.
64. Ihira M, Yoshikawa T, Ohashi M, Enomono Y, Akimoto S, Suga S, et al. Variation of human herpesvirus 7 shedding in saliva. *J Infect Dis* 2003;**188**:1352–4.
65. Yoshikawa T, Ihira M, Taguchi H, Yoshida S, Asano Y. Analysis of shedding of 3 beta-herpesviruses in saliva from patients with connective tissue diseases. *J Infect Dis* 2005;**192**:1530–6.
66. Suga S, Yoshikawa T, Nagai T, Asano Y. Clinical features and virological findings in children with primary human herpesvirus 7 infection. *Pediatrics* 1997;**99**:E4.
67. Torigoe S, Kumamoto T, Koide W, Taya K, Yamanishi K. Clinical manifestations associated with human herpesvirus 7 infection. *Arch Dis Child* 1995;**72**:518–9.
68. Torigoe S, Koide W, Yamada M, Miyashiro E, Tanaka-Taya K, Yamanishi K. Human herpesvirus 7 infection associated with central nervous system manifestations. *J Pediatr* 1996;**129**:301–5.
69. Caserta MT, Hall CB, Schnabel K, Long CE, D'Heron N. Primary human herpesvirus 7 infection: a comparison of human herpesvirus 7 and human herpesvirus 6 infections in children. *J Pediatr* 1998;**133**:386–9.
70. Tanaka-Taya K, Kondo T, Nakagawa N, Inagi R, Miyoshi H, Sunagawa T, et al. Reactivation of human herpesvirus 6 by infection of human herpesvirus 7. *J Med Virol* 2000;**60**:284–9.
71. Tohyama M, Hashimoto K, Yasukawa M, Kimura H, Horikawa T, Nakajima K, et al. Association of human herpesvirus 6 reactivation with the flaring and severity of drug-induced hypersensitivity syndrome. *Br J Dermatol* 2007;**157**:934–40.
72. Yagami A, Yoshikawa T, Asano Y, Koie S, Shiohara T, Matsunaga K. Drug-induced hypersensitivity syndrome due to mexiletine hydrochloride associated with reactivation of human herpesvirus 7. *Dermatology* 2006;**213**:341–4.
73. Kano Y, Hiraharas K, Sakuma K, Shiohara T. Several herpesviruses can reactivate in a severe drug-induced multiorgan reaction in the same sequential order as in graft-versus-host disease. *Br J Dermatol* 2006;**155**:301–6.

74. Tohyama M, Hashimoto K. New aspects of drug-induced hypersensitivity syndrome. *J Dermatol* 2011;**38**:222–8.
75. Fujino Y, Nakajima M, Inoue H, Kusuhara T, Yamada T. Human herpesvirus 6 encephalitis associated with hypersensitivity syndrome. *Ann Neurol* 2002;**51**:771–4.
76. Descamps V, Collot S, Houhou N, Ranger-Rogez S. Human herpesvirus-6 encephalitis associated with hypersensitivity syndrome. *Ann Neurol* 2003;**53**:280.
77. Chiou CC, Chung WH, Hung SI, Yang LC, Hong HS. Fulminant type 1 diabetes mellitus caused by drug hypersensitivity syndrome with human herpesvirus 6 infection. *J Am Acad Dermatol* 2006;**54**:S14–7.
78. Huang H, Kamijima M, Wang H, Li S, Yoshikawa T, Lai G, et al. Human herpesvirus 6 reactivation in trichloroethylene-exposed workers suffering from generalized skin disorders accompanied by hepatic dysfunction. *J Occup Health* 2006;**48**:417–23.
79. Hashizume H, Aoshima M, Ito T, Seo N, Takigawa M, Yagi H. Emergence of circulating monomyeloid precursors predicts reactivation of human herpesvirus-6 in drug-induced hypersensitivity syndrome. *Br J Dermatol* 2009;**161**:486–8.
80. Yoshikawa T, Fujita A, Yagami A, Suzuki K, Matsunaga K, Ihira M, et al. Human herpesvirus 6 reactivation and inflammatory cytokine production in patients with drug-induced hypersensitivity syndrome. *J Clin Virol* 2006;**37**:S92–6.
81. Boutolleau D, Fernandez C, Andre E, Imbert-Marcille BM, Milpied N, Agut H, et al. Human herpesvirus (HHV)-6 and HHV-7: two closely related viruses with different infection profiles in stem cell transplantation recipients. *J Infect Dis* 2003;**187**:179–86.
82. Appleton AL, Sviland L, Peiris JS, Taylor CE, Wilkes J, Green MA, et al. Human herpes virus-6 infection in marrow graft recipients: role in pathogenesis of graft-versus-host disease. Newcastle upon Tyne Bone Marrow Transport Group. *Bone Marrow Transplant* 1995;**16**:777–82.
83. Cone RW, Huang ML, Corey L, Zeh J, Ashley R, Bowden R. Human herpesvirus 6 infections after bone marrow transplantation: clinical and virologic manifestations. *J Infect Dis* 1999;**179**:311–8.
84. Volin L, Lautenschlager I, Juvonen E, Nihtinen A, Anttila VJ, Ruutu T. Human herpesvirus 6 antigenaemia in allogeneic stem cell transplant recipients: impact on clinical course and association with other beta-herpesviruses. *Br J Haematol* 2004;**126**:690–6.
85. Yoshikawa T, Ihira M, Ohashi M, Suga S, Asano Y, Miyazaki H, et al. Correlation between HHV-6 infection and skin rash after allogeneic bone marrow transplantation. *Bone Marrow Transplant* 2001;**28**:77–81.
86. Fujita A, Ihira M, Suzuki R, Enomoto Y, Sugiyama H, Sugata K, et al. Elevated serum cytokine levels are associated with human herpesvirus 6 reactivation in hematopoietic stem cell transplantation recipients. *J Infect* 2008;**57**:241–8.
87. Kitamura K, Asada H, Iida H, Fukumoto T, Kobayashi N, Niizeki H, et al. Relationship among human herpesvirus 6 reactivation, serum interleukin 10 levels, and rash/graft-versus-host disease after allogeneic stem cell transplantation. *J Am Acad Dermatol* 2008;**58**:802–9.
88. Yoshikawa T, Kobayashi I, Asano Y, Nakashima T, Yazaki T, Kojima S, et al. Clinical features of primary human herpesvirus-6 infection in an infant with acute lymphoblastic leukemia. *Am J Pediatr Hematol Oncol* 1993;**15**:424–6.
89. Yoshikawa T, Ihira M, Suzuki K, Suga S, Asano Y, Asonuma K, et al. Primary human herpesvirus 6 infection in liver transplant recipients. *J Pediatr* 2001;**138**:921–5.

90. Yoshikawa T, Goshima F, Akimoto S, Ozaki T, Iwasaki T, Kurata T, et al. Human herpesvirus 6 infection of human epidermal cell line: pathogenesis of skin manifestations. *J Med Virol* 2003;**71**:62–8.
91. Caruso A, Favilli F, Rotola A, Comar M, Horejsh D, Alessandri G, et al. Human herpesvirus-6 modulates RANTES production in primary human endothelial cell cultures. *J Med Virol* 2003;**70**:451–8.
92. Wu CA, Shanley JD. Chronic infection of human umbilical vein endothelial cells by human herpesvirus-6. *J Gen Virol* 1998;**79**:1247–56.

Chapter 10

HHV-6A and HHV-6B in Autoimmune Disease

Francesco Broccolo
University of Milano–Bicocca, Monza, Italy

INTRODUCTION

In recent years, several reports have provided important information linking HHV-6 to autoimmune diseases (ADs), including multiple sclerosis,[1–7] autoimmune connective tissue diseases,[8–10] and Hashimoto's thyroiditis.[11] In addition, it has been suggested that HHV-6 infection might be related to the onset of autoimmune disorders, including Sjögren's syndrome,[12] purpura fulminans, severe autoimmune acquired protein S deficiency,[13] severe and acute autoimmune hepatitis,[14,15] and autoimmune hemolytic anemia/neutropenia.[16]

This chapter discusses the autoimmune diseases mentioned above that are associated with HHV-6 and the mechanisms proposed for HHV-6-induced autoimmunity. HHV-6 might trigger autoimmunity through lysis of infected cells by exposing high amounts of normally sequestered cell antigens. The virus could also induce aberrant expression of histocompatibility molecules, thereby promoting the presentation of autoantigens. Another potential trigger is represented by molecular mimicry, with the synthesis of viral proteins that resemble cellular molecules as a mechanism of immune escape. Based on the similarity in peptide sequences between viral proteins and self-proteins, it has been postulated that viral infections could activate cross-reactive T cells that are able to recognize both viral and self-antigens, which could then trigger an autoimmune response and cause tissue damage.[17–19]

HHV-6 AND MULTIPLE SCLEROSIS

Human herpesvirus 6 (HHV-6) has long been cited as a potential candidate virus for the etiology of multiple sclerosis (MS), an inflammatory, demyelinating disease of the central nervous system believed to be initiated and mediated by autoreactive T cells directed against myelin antigens. It develops in young adults with a complex genetic predisposition and is thought to require an inciting environmental insult such as a viral infection to trigger the disease.[20] HHV-6 infects and can establish latency in the central nervous system (CNS). HHV-6A and HHV-6B have virus-specific tropisms in human glial cells, suggesting that the two HHV-6 viruses might be responsible for distinct disease outcomes due to the infection pattern in these cells *in vivo*.[21] HHV-6A (which has more often been associated with MS) established a productive infection in astrocytes with cytopathic effect and high virus DNA loads, while the HHV-6B-infected astrocytes showed no morphological changes but maintained low levels of intracellular viral DNA without detectable RNA. A relationship between variant A of HHV-6 and MS was first suggested by immunohistochemical demonstration of viral antigen in oligodendrocytes of MS white matter lesions but not in control brains.[1] Since this initial report, several studies have supported an association of HHV-6 and MS by the demonstration of elevated antibody titers to HHV-6 antigens compared to controls and by amplification of HHV-6 DNA from the serum, cerebrospinal fluid (CSF), and brain tissue of MS patients.[1–7,20–28] Exacerbation of relapsing-remitting MS has been linked to higher viral loads in serum and in peripheral blood mononuclear cells (PBMCs), suggesting an association between HHV-6 reactivation and disease relapses.[2,25,31]

Abundant clinical studies have highlighted a correlation between MS and several parameters assessing for HHV-6 infection. For example, levels of HHV-6DNA in serum, characteristic of ongoing infection, are significantly increased in MS patients when compared to healthy donors or with patients with other diseases.[29] HHV-6DNA was also detected at higher frequencies in the CSF and in the peripheral blood mononuclear cells of MS patients.[2,6,25,29] Moreover, the levels of HHV-6-specific immunoglobulin G (IgG) and IgM in the serum and in the CSF were reported to be higher in MS patients in several studies,[2] although this phenomenon does not appear to be specific for HHV-6. Soldan et al.[2] also showed that lymphoproliferative responses against HHV-6 antigens were increased in MS patients. Their analysis of brain biopsies and postmortem tissues indicated that HHV-6DNA was present more frequently in the brain of MS patients than in control brains, and that it was also more frequent in MS lesions than in normal areas of the same brains. Immunohistochemistry analyses confirmed the presence of viral proteins in oligodendrocytes and astrocytes in the brain from MS patients, with a higher frequency in demyelinating plaques.[3,23,28,30] Most interestingly, viral loads were detected more frequently, and levels of HHV-6-specific IgG were increased in MS patients experiencing disease exacerbation,[20,30,32] thus suggesting a correlation between HHV-6 infection and MS relapses.

Because recognition that HHV-6A and HHV-6B are two distinct viruses has only recently occurred, many of the initial studies do not discriminate between the two viruses. However, based on few reports, it appears that HHV-6A is found more frequently than HHV-6B in the serum of MS patients.[22] Especially in the case of active infection, Alvarez-Lafuente et al.[30] found only HHV-6A. In contrast, in another study, intrathecal HHV-6B IgG levels were more abundant than HHV-6A IgG in MS patients, and only HHV-6B-specific IgM levels were found.[32] The potential association between HHV-6A or HHV-6B infection and MS has often been discussed and remains controversial. Some studies have provided contradictory results,[33,34] raising methodological and technical questions, especially concerning the choice of control groups and the immunological state of the included patients, who often receive immunosuppressive treatments that may provoke latent herpesvirus reactivation by itself.

Pathogenic Hypotheses for HHV-6-Induced Multiple Sclerosis

New studies investigating the biology of HHV-6 have provided insight toward understanding how HHV-6 may play a role in MS pathology. By inducing molecular mimicry or excessive complement activation, HHV-6 reactivation may have the potential to trigger autoimmunity and tissue damage associated with MS lesion development. Reports have suggested that the constitutive presence of active HHV-6 infection in glial cells in inflamed CNS tissue could result in virus-triggered immunopathologies in MS.[30] A mechanism of molecular mimicry involving HHV-6 has been proposed as one mechanism by which the autoimmune process could be triggered. One study showed that certain residues on the HHV-6 genome are identical to residues of myelin basic protein. Importantly, both T cells and antibody responses to this peptide sequence were found elevated in MS patients.[35] Moreover, *in vitro* infection of glial precursor cells was found to impair cell replication and increase the expression of oligodendrocyte markers, suggesting that HHV-6 infection of the CNS may influence the neural repair mechanism; lymphoproliferative responses to HHV-6A antigens have also been demonstrated to be greater in MS patients than in controls.[36] Yet, whether HHV-6 infection is the etiologic cause, a factor for disease progression, or a consequence of MS remains unclear and requires further investigation. Taken together, epidemiological data and the presence of active HHV-6 infection in some MS brain samples suggest a possible role for HHV-6 in perpetuating tissue damage in MS.

Several studies suggest that such a mechanism could be involved in HHV-6-induced neuroinflammation. One study reported that 15 to 25% of HHV-6-specific T-cell clones obtained from healthy donors or MS patients were cross-reactive to myelin basic protein (MBP), one of the autoantigens implicated in MS pathology.[37] In fact, MBP and the U24 protein from HHV-6 were later shown to share an identical amino acid sequence of seven residues. Moreover, T cells directed against an MBP peptide also recognized an HHV-6 peptide, with both peptides containing the identical sequence. Interestingly, cross-reactive cells were more frequent in MS patients than in controls.[35] These data were further confirmed by a more recent study in which the presence of cross-reactive CD8+ cytotoxic T cells was detected.[38] Altogether, these studies suggest that HHV-6 infection can activate T-cell responses, which can simultaneously be directed against myelin sheaths, thus strongly supporting the potential role for HHV-6 in autoimmune diseases affecting the CNS.

Moreover, the fact that HHV-6 is a ubiquitous virus that infects the vast majority of humans raises a relevant question about why only a minority of individuals are affected by MS. In this regard, a complex interaction between these pathogens and individual genetic backgrounds represents the most reasonable explanation. Indeed, beside the well-known risk-conferring genes belonging to the HLA-DR locus,[33] another study has suggested that some killer immunoglobulin-like receptor (KIR) genes are strongly underrepresented among MS patients and there is an increased risk of disease susceptibility among carriers of the natural HLA-C ligands (HLA-C1).[34] KIRs are major histocompatibility complex (MHC) class I-specific regulatory receptors utilized by human natural killer (NK) cells and CD8 T cells.[35] Several lines of evidences link differences in KIR expression to differential responses to invading pathogens and autoimmune disorders.[36,37] Because we have demonstrated that NK recognition and specific killing of HHV-6-infected lymphocytes are tightly regulated by KIRs and their MHC ligands,[38] it is important to establish the role of KIRs in MS with a particular attention on HHV-6 viremic MS patients. However, current evidence implicates roles for both environmental and genetic factors in the onset and progression of the disease.[39–43]

Finally, CD46–HHV-6 interaction is another recently proposed mechanism: HHV-6 could participate in neuroinflammation in the context of MS by promoting inflammatory processes through CD46 binding. CD46, or membrane cofactor protein (MCP), is a member of the complement regulatory protein family, and it is also known to be a receptor for different viruses and bacteria. In 1999, CD46 was identified as the principal receptor molecule for both HHV-6A and HHV-6B on the surface of human cells.[44] The transmembrane protein CD46 is the only known entry receptor for both HHV-6A and HHV-6B entry. Santoro et al.[44] demonstrated that CD46 was selectively and progressively downregulated from the target cell surface during the course of HHV-6 infection; the downmodulation of this complement regulatory protein, secondary to HHV-6 infection, might induce complement activation and, therefore, an increase in the vulnerability of the cell to complement lysis.[45] Two forms of CD46 have been described—membrane and soluble (sCD46)—but no conclusive evidence is documented in the literature indicating whether posttranslational events or alternative splicing produces the soluble form,[46] whose levels have been shown to be increased in the serum of patients with autoimmune disorders including Sjögren's syndrome,[47] systemic lupus erythematosus,[48] and MS.[49] A physical association between the HHV-6 virion and sCD46 was found in the serum of patients with MS with HHV-6 DNA, but not in the serum of controls, suggesting that the presence of CD46–HHV-6 complexes might contribute to the increased levels of CD46 found in the serum of patients with MS.[50]

In a study performed by Hammarstedt et al.[51] with the objective of identifying possible host proteins associated with HHV-6, western blot analyses showed that the cellular complement protein CD46 was associated with the purified and infectious virions; the authors suggested that the relevance of the association in disease and especially in autoimmunity should be further investigated. As it has been demonstrated that CD46 is selectively and progressively downregulated from the target cell surface during the course of HHV-6 infection,[44] the increased levels of CD46 expression reported in this study in patients with MS with HHV-6 infection could be related to the increased levels of the soluble form of CD46 described in patients with MS;[49,50] therefore, this could constitute an immunopathogenic factor that should be investigated in MS. However, there are many theories that could be explored to explain the different ways in which CD46–HHV-6 interaction can play a role in MS pathogenesis. One of them is related to the persistence of the virus in the serum. It has been reported that some patients with MS had HHV-6 genomes in their sera in consecutive samplings performed over 1-year follow-up.[52] If HHV-6 is attached to the receptor in its soluble form, then different epitopes of the virus that share homology with human proteins and also different host cell proteins that could be incorporated into the envelope of the virus[51] would be more exposed to the host immunological system, thus increasing the possibility of their playing a role in autoimmunity. Furthermore, CD46 is highly expressed at the blood–brain barrier (BBB),[53] which is composed of tight junctions that prevent the entry of large proteins into the CNS. Crossing this barrier is precisely regulated and crucial for the immune surveillance of the brain, but CD46 could be mediating access of HHV-6 to the brain by promoting passage through the BBB. As we have previously mentioned, however, CD46 is also a member of the complement regulatory protein family that confers protection against activated complement-mediated lysis by inactivating C3b/C4b deposited on the membrane of autologous cells; therefore, the increased levels of CD46 demonstrated in the serum and CSF of patients with MS may be indicative of increased activation of the complement system in MS, both peripherally and intrathecally, and could lend further support to the possible contribution of complement in disease pathogenesis. Finally, CD46 has also been related to T cells; it is known that CD3/CD46 costimulation promotes T-cell proliferation with a potency comparable to that of CD28,[54] and the importance of CD46 in the regulation of the immune response through the induction of Tr1 cells and interleukin-10 (IL-10) production is well known.[55] As the role of regulatory T cells in patients with MS has been previously demonstrated by various groups,[56,57] an alteration in CD46 in patients with MS could lead to further damage and inflammation.[55]

Finally, this complement regulatory protein also plays an important role in the adaptive immune response, as it can modulate T-cell responses depending on which cytoplasmic tail is expressed,[58] and it can induce CD4+ T cells toward a Tr1 phenotype, with high IL-10 production.[59] One could then hypothesize that HHV-6A and HHV-6B, by binding to their receptor, could modulate its functions. In support of this theory, a clinical study indicated that an increase inHHV-6 viral load was correlated with enhanced CD46 expression in MS patients,[60] and several alterations in CD46 functions were described. The

CD46-induced IL-10 secretion by T cells was strongly decreased,[61] whereas CD46-dependent IL-23 production by dendritic cells (DCs) and IL-17 expression by T cells were enhanced.[62,63] This suggests that HHV-6 could participate in neuroinflammation in the context of MS by promoting inflammatory processes through CD46 binding.

HHV-6 AND AUTOIMMUNE CONNECTIVE TISSUE DISEASES

Autoimmune connective tissue diseases (ACTDs) encompass a very large group of diseases, including systemic sclerosis or scleroderma (SSc), systemic lupus erythematosus (SLE), discoid lupus erythematosus (DLE), dermatomyositis (DM), rheumatoid arthritis (RA), and other conditions causing chronic inflammation that can affect many organs and systems. Although the etiology of ACTD remains unclear, clinical, epidemiological, and laboratory findings suggest that several viral infections may be involved in these diseases.[64] Reactivation of human herpesvirus 6 (HHV-6), as suggested by the high rates of viral isolation, occurs frequently in patients with collagen vascular diseases.[8] Moreover, Hoffmann et al.[9] demonstrated active HHV-6 infection in a 37-year-old woman affected by SLE and histiocytic necrotizing lymphadenitis (Kikuchi–Fujimoto disease). More recently, frequent reactivation of HHV-6 in active ACTD (especially in lupus erythematosus) has been detected, suggesting a role in the pathogenesis of these diseases.[10]

Pathogenic Hypotheses for HHV-6-Induced Autoimmune Connective Tissue Diseases

A number of infectious agents including members of the Herpesviridae family and parvovirus B19 (B19V) have been proposed as possible triggering factors in ACTD, mainly in SSc.[17–19,64–66] Homology between viruses and autoantibody targets suggests that molecular mimicry may play a role in the initiation of antibody response in disorders characterized by diffuse vascular damage. Four pathogenic hypotheses have been proposed: molecular mimicry, endothelial cell damage, super-antigen stimulation, and microchimerism;[17–19,64–68] however, evidence for a direct association is still lacking, even though several studies have provided important information linking infectious agents to ACTD.[68] Indeed, infectious agents have the potential to initiate autoreactivity through polyclonal activation, the release of previously sequestered antigens, or molecular mimicry. Evidence from animal models indicates that molecular mimicry of host proteins by a pathogen can induce autoimmune diseases,[18] although it appears to be an infrequent occurrence in the majority of viral infections. More commonly, viruses induce autoimmunity by cell death, predominantly by increased apoptosis, resulting in the release of self antigens, and increased apoptosis has been suggested as a major pathogenetic mechanism in ACTD.[19]

HHV-6 IN HASHIMOTO'S THYROIDITIS

Hashimoto's thyroiditis (HT), or chronic lymphocytic thyroiditis, is a common autoimmune disease with unknown etiology, and its prevalence has been increasing over the past 50 years.[69,70] Together with genetic factors, environmental factors are thought to be important in triggering autoimmune thyroid diseases (AITDs), and viral infections have been suggested as possible environmental triggers,[71] although no conclusive evidence is available. Also, herpesviruses have been suggested as potential cofactors and have occasionally been detected in AITDs.[72,73] Thyroid cells infected with human cytomegalovirus were shown to act as antigen-presenting cells and therefore might be involved in autoimmunity.[74] Patients with Graves' disease display a higher frequency of Epstein–Barr virus (EBV)-infected B cells secreting antibody to thyroid-stimulating hormone receptor (TSHR),[75] and AITD patients have elevated antibody titers against EBV antigens.[76] Human herpesvirus 6 DNA has been detected in HT tissue specimens, but not in tissues from patients with Graves' disease or multinodular goiter.[73] A recent study[11] has linked HHV-6A to Hashimoto's thyroiditis, a common autoimmune thyroid disease (AITD). The study found that HHV-6 was detected significantly more frequently among thyroid fine-needle aspirates (FNAs) from HT individuals than controls (82% versus 10%, respectively), and low-grade acute infection was identified in all HHV-6-positive HT samples compared to 0% of controls. Furthermore, the presence of HHV-6 infection was found localized mainly to thyrocytes, rather than in lymphocytes infiltrating the lesion, and increased prevalence of latent HHV-6 infection was seen in PBMCs overall. In addition, the group demonstrated that thyroid cells infected with both HHV-6A and HHV-6B became susceptible to NK-mediated killing, providing evidence of a potential mechanism for HHV-6-induced autoimmunity. These findings are consistent with the possibility that the thyroid of HT patients may constitute a site of active HHV-6 infection and replication.

Pathogenic Hypotheses for HHV-6-Induced Hashimoto's Thyroiditis

Recently, Caselli et al.[11] provided evidence indicating that HHV-6 may induce a *de novo* expression of HLA class II molecules in thyrocytes, which may thus behave as functional antigen-presenting cells for CD4+ T lymphocytes. Intriguingly, in this same study the enhanced HHV-6-specific T-cell responses were observed in all HT patients, with a marked increase in the number of CD4+ T lymphocytes recognizing HHV-6 antigens, particularly the subset of polyfunctional CD4+ T cells secreting both interferon gamma (IFN-γ) and IL-2. These findings are consistent with an abnormal, probably persistent, immune response to HHV-6 antigens in HT patients, possibly favored by the local upregulation of HLA class II molecules on thyrocytes induced by HHV-6 infection. These HHV-6-specific responses are likely embedded in a context of global overactivation of the CD4+ T-cell compartment, as suggested by the increased responses of CD4+ T cells producing IL-2 or IFN-γ/IL-2 to the tetanus toxoid (TT) antigen. Nevertheless, HT patients showed significantly higher numbers of CD4+ and CD8+ T cells secreting IFN-γ only in response to the U94 antigen, suggesting a possible role of these effectors in mediating the killing of U94-expressing thyrocytes. In addition to abnormal HHV-6-specific T-cell responses, innate immunity triggered by HHV-6 may also contribute to HT development. In fact, PBMCs from HT patients showed a markedly enhanced cytotoxic activity to HHV-6-infected thyrocytes compared to control PBMCs derived from healthy donors. These findings, together with the observation that NK cells of HT patients show high levels of basal degranulation even when cultured with uninfected thyrocytes, suggest that these patients might suffer from an inherent NK cell alteration. However, further studies are required to fully elucidate this association and the mechanisms underlying the possible role of HHV-6 as a trigger of HT.

REFERENCES

1. Challoner PB, Smith KT, et al. Plaque-associated expression of human herpes virus 6 in multiple sclerosis. *Proc Natl Acad Sci USA* 1995;**92**:7440–4.
2. Soldan SS, Berti R, Salem N, et al. Association of human herpes virus 6 (HHV-6) with multiple sclerosis: increased IgM response to HHV-6 early antigen and detection of serum HHV-6 DNA. *Nat Med* 1997;**3**(2):1394–7.
3. Jacobson S, Soldan SS, Berti R. HHV-6 and multiple sclerosis. *Nat Med* 1998;**4**:538.
4. Coates AR, Bell J. HHV-6 and multiple sclerosis. *Nat Med* 1998;**4**(5):537–8.
5. Fillet AM, Lozeron P, Agut H, et al. HHV-6 and multiple sclerosis. *Nat Med* 1998;**4**(5):537–8.
6. Ablashi DV, Lapps W, Kaplan M, Whitman JE, Richert JR, Pearson GR. Human herpesvirus 6 (HHV-6) infection in multiple sclerosis: a preliminary report. *Mult Scler* 1998;**4**(6):490–6.
7. Friedman JE, Lyons MJ, Cu G, et al. The association of the human herpesvirus-6 and MS. *Mult Scler* 1999;**5**(5):355–62.
8. Krueger GR, Sander C, Hoffmann A, Barth A, Koch B, et al. Isolation of human herpesvirus-6 (HHV-6) from patients with collagen vascular diseases. *In Vivo* 1991;**5**:217–25.
9. Hoffmann A, Kirn E, Kuerten A, Sander C, Krueger GR, et al. Active human herpesvirus-6 (HHV-6) infection associated with Kikuchi–Fujimoto disease and systemic lupus erythematosus (SLE). *In Vivo* 1991;**5**:265–9.
10. Broccolo F, Drago F, Paolino S, Cassina G, Gatto F, et al. Reactivation of human herpesvirus 6 (HHV-6) infection in patients with connective tissue diseases. *J Clin Virol* 2009;**46**:43–6.
11. Caselli E, Zatelli MC, Rizzo R, Benedetti S, Martorelli D, Trasforini G, Cassai E, degli Uberti EC, Di Luca D, Dolcetti R. Virologic and immunologic evidence supporting an association between HHV-6 and Hashimoto's thyroiditis. *PLoS Pathog* 2012;**8**(10):e1002951.
12. Ranger-Rogez S, Vidal E, Liozon F, Denis F. Primary Sjögren's syndrome and antibodies to human herpesvirus type 6. *Clin Infect Dis* 1994;**19**:1159–60.
13. Boccara O, Lesage F, Regnault V, Lasne D, Dupic L, et al. Nonbacterial purpura fulminans and severe autoimmune acquired protein S deficiency associated with human herpesvirus-6 active replication. *Br J Dermatol* 2009;**161**:181–3.

14. Potenza L, Luppi M, Barozzi P, Rossi G, Cocchi S, et al. HHV-6A in syncytial giant-cell hepatitis. *N Engl J Med* 2008;**359**:593–602.
15. Grima P, Chiavaroli R, Calabrese P, Tundo P. Severe hepatitis with autoimmune features following a HHV-6: a case report. *Cases J* 2008;**1**:110.
16. Yagasaki H, Kato M, Shimizu N, Shichino H, Chin M, Mugishima H. Autoimmune hemolytic anemia and autoimmune neutropenia in a child with erythroblastopenia of childhood (TEC) caused by human herpesvirus-6 (HHV-6). *Ann Hematol* 2011;**90**:851–2.
17. McClain MT, Heinlen LD, Dennis GJ, Roebuck J, Harley JB, James JA. Early events in lupus humoral autoimmunity suggest initiation through molecular mimicry. *Nat Med* 2005;**11**:85–9.
18. Lunardi C, Bason C, Corrocher R, Puccetti A. Induction of endothelial cell damage by hCMV molecular mimicry. *Trends Immunol* 2005;**26**:19–24.
19. Mackay IR, Leskovsek NV, Rose NR. Cell damage and autoimmunity: a critical appraisal. *J Autoimmun* 2008;**30**:5–11.
20. Hafler DA. Multiple sclerosis. *J Clin Invest* 2004;**113**:788–94.
21. Donati D, Martinelli E, Cassiani-Ingoni R, Ahlqvist J, Hou J, Major EO, Jacobson S. Variant-specific tropism of human herpesvirus 6 in human astrocytes. *J Virol* 2005; **79**:9439–48.
22. Akhyani N, Berti R, Brennan MB, et al. Tissue distribution and variant characterization of human herpesvirus (HHV)-6: increased prevalence of HHV-6A in patients with multiple sclerosis. *J Infect Dis* 2000;**182**(5):1321–5.
23. Sanders VJ, Felisan S, Waddell A, Tourtellotte WW. Detection of Herpesviridae in postmortem multiple sclerosis brain tissue and controls by polymerase chain reaction. *J Neurovirol* 1996;**2**(4):249–58.
24. Kim JS, Lee KS, Park JH, Kim MY, Shin WS. Detection of human herpesvirus 6 variant A in peripheral blood mononuclear cells from multiple sclerosis patients. *Eur Neurol* 2000;**43**:170–3.
25. Chapenko S, Millers A, Nora Z, Logina I, Kukaine R, Murovska M. Correlation between HHV-6 reactivation and multiple sclerosis disease activity. *J Med Virol* 2003;**69**:111–7.
26. Opsahl ML, Kennedy PG. Early and late HHV-6 gene transcripts in multiple sclerosis lesions and normal appearing white matter. *Brain* 2005;**128**:516–27.
27. Rotola A, Merlotti I, Caniatti L, Caselli E, Granieri E, et al. Human herpesvirus 6 infects the central nervous system of multiple sclerosis patients in the early stages of the disease. *Mult Scler* 2004;**10**:348–54.
28. Goodman AD, Mock DJ, Powers JM, Baker JV, Blumberg BM. Human herpesvirus 6 genome and antigen in acute multiple sclerosis lesions. *J Infect Dis* 2003;**187**:1365–76.
29. Berti R, Brennan MB, Soldan SS, Ohayon JM, Casareto L, McFarland HF, Jacobson S. Increased detection of serum HHV-6 DNA sequences during multiple sclerosis (MS) exacerbations and correlation with parameters of MS disease progression. *J Neurovirol* 2002;**8**:250–6.
30. Alvarez-Lafuente R, De las Heras V, Bartolomé M, Picazo JJ, Arroyo R. Relapsing–remitting multiple sclerosis and human herpesvirus 6 active infection. *Arch Neurol* 2004;**61**:1523–7.
31. Wilborn F, Schmidt CA, et al. A potential role for human herpesvirus type 6 in nervous system disease. *J Neuroimmunol* 1994;**49**:213–4.
32. Simpson S, Taylor B, Dwyer DE, et al. Anti-HHV-6 IgG titer significantly predicts subsequent relapse risk in multiple sclerosis. *Mult Scler* 2012;**18**:799–806.
33. Martin C, Enbom M, Soderstrom M, Fredrikson S, Dahl H, Lycke J, Bergström T, Linde A. Absence of seven human herpesviruses including HHV-6, by polymerase chain reaction in

CSF and blood from patients with multiple sclerosis and optic neuritis. *Acta Neurol Scand* 1997;**95**:280–3.
34. Mirandola P, Stefan A, Brambilla E, Campadelli-Fiume G, Grimaldi LM. Absence of human herpesvirus 6 and 7 from spinal fluid and serum of multiple sclerosis patients. *Neurology* 1999;**53**:1367–8.
35. Tejada-Simon MV, Zang YCQ, Hong J, Rivera VM, Zhang JZ. Cross-reactivity with myelin basic protein and human herpesvirus-6 in multiple sclerosis. *Ann Neurol* 2003;**53**:189–97.
36. Dietrich J, Blumberg BM, Roshal M, Baker JV, Hurley SD, Mayer-Pröschel M, Mock DJ. Infection with an endemic human herpesvirus disrupts critical glial precursor cell properties. *J Neurosci* 2004;**19**:4875–83.
37. Cirone M, Cuomo L, Zompetta C, Ruggieri S, Frati L, et al. Human herpesvirus 6 and multiple sclerosis: a study of T cell cross-reactivity to viral and myelin basic protein antigens. *J Med Virol* 2002;**68**:268–72.
38. Cheng W, Ma Y, Gong F, et al. Cross-reactivity of autoreactive T cells with MBP and viral antigens in patients with MS. *Front Biosci* 2012;**17**:1648–58.
39. Dyment DA, Ebers GC, Sadovnick AD. Genetics of multiple sclerosis. *Lancet Neurol* 2004;**3**:104–10.
40. Oksenberg JR, Baranzini SE, Sawcer S, Hauser SL. The genetics of multiple sclerosis: SNPs to pathways to pathogenesis. *Nat Rev Genet* 2008,**9**.516–26.
41. Ebers GC. Environmental factors and multiple sclerosis. *Lancet Neurol* 2008;**7**:268–77.
42. Ramagopalan SV, Deluca GC, Degenhardt A, Ebers GC. The genetics of clinical outcome in multiple sclerosis. *J Neuroimmunol* 2008;**201**:183–99.
43. Patsopoulos NA, et al. Genome-wide meta-analysis identifies novel multiple sclerosis susceptibility loci. *Ann. Neurol* 2011;**70**:897–912.
44. Santoro F, Kennedy PE, Locatelli G, Malnati MS, Berger EA, Lusso P. CD46 is a cellular receptor for human herpesvirus 6. *Cell* 1999;**99**:817–27.
45. Assem MM, Gad WH, El-Sharkawy NM, et al. Prevalence of anti human herpes virus-6 IgG and its receptor in acute leukemia (membrane cofactor protein: MCP, CD46). *J Egypt Natl Cancer Inst* 2005;**17**(1):29–34.
46. Dhiman N, Jacobson RM, Poland GA. Measles virus receptors: SLAM and CD46. *Rev Med Virol* 2004;**14**:217–29.
47. Cuida M, Legler DW, Eidsheim M, Jonsson R. Complement regulatory proteins in the salivary glands and saliva of Sjögren's syndrome patients and healthy subjects. *Clin Exp Rheumatol* 1997;**15**:615–23.
48. Kawano M, Seya T, Koni I, Mabuchi H. Elevated serum levels of soluble membrane cofactor protein (CD46, MCP) in patients with systemic lupus erythematosus (SLE). *Clin Exp Immunol* 1999;**116**:542–6.
49. Soldan SS, Fogdell-Hahn A, Brennan MB, et al. Elevated serum and cerebrospinal fluid levels of soluble human herpesvirus type 6 cellular receptor, membrane cofactor protein, in patients with multiple sclerosis. *Ann Neurol* 2001;**50**:486–93.
50. Fogdell-Hahn A, Soldan SS, Shue S, et al. Co-purification of soluble membrane cofactor protein (CD46) and human herpesvirus 6 variant A genome in serum from multiple sclerosis patients. *Virus Res* 2005;**110**:57–63.
51. Hammarstedt M, Ahlqvist J, Jacobson S, Garoff H, Fogdell-Hahn A. Purification of infectious human herpesvirus 6A virions and association of host cell proteins. *Virol J* 2007;**4**:101.
52. Alvarez-Lafuente R, Montojo-García M. De las Heras V, Bartolomé M, Arroyo R. Clinical parameters and HHV-6 active replication in relapsing–remitting multiple sclerosis patients. *J Clin Virol* 2006;**37**:S24–6.

53. Shusta EV, Zhu C, Boado RJ, Pardridge WM. Subtractive expression cloning reveals high expression of CD46 at the blood–brain barrier. *J Neuropathol Exp Neurol* 2002;**61**:597–604.
54. Astier A, Trescol-Biemont MC, Azocar O, Lamouille B, Rabourdin-Combe C. Cutting edge: CD46, a new costimulatory molecule for T cells, that induces p120CBL and LAT phosphorylation. *J Immunol* 2000;**164**:6091–5.
55. Astier AL, Hafler DA. Abnormal Tr1 differentiation in multiple sclerosis. *J Neuroimmunol* 2007;**191**:70–8.
56. Haas J, Hug A, Viehover A, et al. Reduced suppressive effect of CD4+ CD25 high regulatory T cells on the T cell immune response against myelin oligodendrocyte glycoprotein in patients with multiple sclerosis. *Eur J Immunol* 2005;**35**:3343–52.
57. Viglietta V, Baecher-Allan C, Weiner HL, Hafler DA. Loss of functional suppression by CD4+ CD25+ regulatory T cells in patients with multiple sclerosis. *J Exp Med* 2004;**199**:971–9.
58. Marie JC, Astier AL, Rivailler P, Rabourdin-Combe C, Wild TF, Horvat B. Linking innate and acquired immunity: divergent role of CD46 cytoplasmic domains in T cell-induced inflammation. *Nat Immunol* 2002;**3**(7):659–66.
59. Kemper C, Chan AC, Green JM, Brett KA, Murphy M, Atkinson JP. Activation of human CD4+ cells with CD3 and CD46 induces a T-regulatory cell 1 phenotype. *Nature* 2003;**421**(6921):388–92.
60. Alvarez-Lafuente R, Garcia-Montojo M, De Las Heras V, Dominguez-Mozo MI, Bartolome M, Arroyo R. CD46 expression and HHV-6 infection in patients with multiple sclerosis. *Acta Neurol Scand* 2009;**120**:246–50.
61. Astier AL, Meiffren G, Freeman S, Hafler DA. Alterations in CD46-mediated Tr1 regulatory T cells in patients with multiple sclerosis. *J Clin Invest* 2006;**116**(12):3252–7.
62. Vaknin-Dembinsky A, Murugaiyan G, Hafler DA, Astier AL, Weiner HL. Increased IL-23 secretion and altered chemokine production by dendritic cells upon CD46 activation in patients with multiple sclerosis. *J Neuroimmunol* 2008;**195**:140–5.
63. Yao K, Graham J, Akahata Y, Oh U, Jacobson S. Mechanism of neuroinflammation: enhanced cytotoxicity and IL-17 production via CD46 binding. *J Hepatol* 2010;**5**:469–78.
64. Hamamdzic D, Kasman LM, LeRoy EC. The role of infectious agents in the pathogenesis of systemic sclerosis. *Curr Opin Rheumatol* 2002;**14**:694–8.
65. Sekigawa I, Nawata M, Seta N, Yamada M, Iida N, Hashimoto H. Cytomegalovirus infection in patients with systemic lupus erythematosus. *Clin Exp Rheumatol* 2002;**20**:559–64.
66. Lehmann HW, von Landenberg P, Modrow S. Parvovirus B19 infection and autoimmune disease. *Autoimmun Rev* 2003;**2**:218–23.
67. Harley JB, Harley IT, Guthridge JM, James JA. The curiously suspicious: a role for Epstein–Barr virus in lupus. *Lupus* 2006;**15**:768–77.
68. Grossman C, Dovrish Z, Shoenfeld Y, Amital H. Do infections facilitate the emergence of systemic sclerosis?. *Autoimmun Rev* 2011;**10**:244–7.
69. Hueston W.J.Treatment of hypothyroidism. *Am Fam Physician* 64: 1717–1724.
70. Staii A, Mirocha S, Todorova-Koteva K, Glinberg S, Jaume JC. Hashimoto thyroiditis is more frequent than expected when diagnosed by cytology which uncovers a pre-clinical state. *Thyroid Res* 2010;**3**:11.
71. Mori K, Yoshida K. Viral infection in induction of Hashimoto's thyroiditis: a key player or just a bystander?. *Curr Opin Endocrinol Diabetes Obes* 2010;**17**:418–24.
72. Scotet E, Peyrat MA, Saulquin X, Retiere C, Couedel C, et al. Frequent enrichment for CD8 T cells reactive against common herpes viruses in chronic inflammatory lesions: towards a reassessment of the physiopathological significance of T cell clonal expansions found in autoimmune inflammatory processes. *Eur J Immunol* 1999;**29**:973–85.

73. Thomas D, Liakos V, Michou V, Kapranos N, Kaltsas G, et al. Detection of herpes virus DNA in post-operative thyroid tissue specimens of patients with autoimmune thyroid disease. *Exp Clin Endocrinol Diabetes* 2008;**116**:35–9.
74. Khoury EL, Pereira L, Greenspan FS. Induction of HLA-DR expression on thyroid follicular cells by cytomegalovirus infection *in vitro*. Evidence for a dual mechanism of induction. *Am J Pathol* 1991;**138**:1209–23.
75. Fan JL, Desai RK, Dallas JS, Wagle NM, Seetharamaiah GS, et al. High frequency of B cells capable of producing anti-thyrotropin receptor antibodies in patients with Graves' disease. *Clin Immunol Immunopathol* 1994;**71**:69–74.
76. Vrbikova J, Janatkova I, Zamrazil V, Tomiska F, Fucikova T. Epstein–Barr virus serology in patients with autoimmune thyroiditis. *Exp Clin Endocrinol Diabetes* 1996;**104**:89–92.

Chapter 11

HHV-6A and HHV-6B in Drug-Induced Hypersensitivity Syndrome/Drug Reaction with Eosinophilia and Systemic Symptoms

Vincent Descamps[a], Mikoko Tohyama[b], Yoko Kano[c], and Tetsuo Shiohara[c]

[a]Hôpital Bichat, Université Paris-Diderot, Paris, France; [b]Ehime University Graduate School of Medicine, Shitsukawa, Toon-city, Ehime, Japan; [c]Kyorin University School of Medicine, Shinkawa Mitaka, Tokyo, Japan

INTRODUCTION

Drug-induced hypersensitivity syndrome (DIHS) and drug reaction with eosinophilia and systemic symptoms (DRESS) are severe adverse drug reactions characterized by skin rashes, fever, leukocytosis with eosinophilia and/or atypical lymphocytosis, lymph node enlargement, and liver dysfunction. Various terms have been used to refer to these syndromes based on the generic names of the culprit drugs, such as allopurinol hypersensitivity syndrome, phenytoin syndrome, and dapson syndrome. It was then demonstrated that there is an association between this drug reaction and human herpesvirus 6 (HHV-6) reactivation. It has been recently shown that not only HHV-6 but also various herpesviruses reactivate during the course of this disease. This chapter focuses on the clinical manifestations, herpesvirus reactivations, and pathophysiology of DIHS/DRESS.

DIHS/DRESS DEFINED

Drug-induced hypersensitivity syndrome (DIHS) and drug reaction with eosinophilia and systemic symptoms (DRESS) are life-threatening multiorgan system reactions caused by a limited number of drugs.[1–3] DRESS is characterized

Human Herpesviruses HHV-6A, HHV-6B & HHV-7.

TABLE 11.1 Diagnostic Criteria for DIHS Established by a Japanese Consensus Group

1. Maculopapular rash developed >3 weeks after starting with a limited number of drugs
2. Prolonged clinical symptoms after discontinuation of the causative drug
3. Fever (>38°C)
4. Liver abnormalities (ALT >100 U/L)[a]
5. Leukocyte abnormalities (at least one present)
   a. Leukocytosis ($>11 \times 10^9$/L)
   b. Atypical lymphocytosis (>5%)
   c. Eosinophilia ($>1.5 \times 10^9$/L)
6. Lymphadenopathy
7. HHV-6 reactivation

[a]*This can be replaced by other organ involvement, such as renal involvement.*
Note: The diagnosis is confirmed by the presence of the seven criteria above (typical DIHS) or of five of the seven (atypical DIHS).
Source: *Picard D et al. Sci Transl Med* 2010; **2**: 46–62. With permission.

by the association of fever, rash, and lymphadenopathies and involvement of internal organs, including hepatitis, pneumonitis, renal failure, myocarditis, encephalitis, pancreatitis, or hemophagocytic syndrome. These manifestations are associated with hematological abnormalities, including lymphocytosis with atypical lymphocytes, and eosinophilia.

The acronym DRESS was proposed by Roujeau et al. in 1996 to encompass the principal manifestations of this syndrome and its clinical presentations. In 2006, the Japanese Research Committee on Severe Cutaneous Adverse Reaction (J-SCAR) included the reactivation of HHV-6 in a set of criteria for diagnosis of this syndrome (Table 11.1), termed drug-induced hypersensitivity syndrome (DIHS).[2] The RegiSCAR group, a scientific network of people interested in severe cutaneous adverse drug reactions (SCARs), proposed another set of criteria that may be used for a retrospective diagnosis of DRESS.[3,4]

Among other cutaneous drug adverse reactions, DIHS/DRESS is clinically characterized by (1) a long delay between the first drug intake and development of the reaction (2 weeks to 3 months); (2) a limited number of culprit drugs; (3) an unexplained cross-reactivity to multiple drugs with different structures; (4) a paradoxical deterioration of clinical symptoms after withdrawal of the causative drug; and (5) a long duration of the syndrome (more than 2 weeks) with some flare-ups even though the causative drug has been stopped.[2,3] One of the major characteristics is its association with a sequential reactivation of human herpesviruses, especially HHV-6 before and during the course of this syndrome.[5,6]

EPIDEMIOLOGY

The incidence of DIHS/DRESS is estimated to be one case among 1,000 to 10,000 drug exposures for certain causative drugs.[7] This severe adverse reaction is much more frequent than Stevens–Johnson syndrome (SJS)/toxic epidermal necrolysis (TEN), which has an incidence of 1.2 to 6 cases per 1,000,000 person-years. DIHS and DRESS are life-threatening syndromes with a mortality rate of 10 to 20%. In most cases they cannot be predicted. Common drugs associated with DIHS/DRESS include allopurinol, aromatic anticonvulsants (carbamazepine, phenytoin, and phenobarbital), sodium valproate, mexiletine, minocycline, sulfones, and sulfonamides.[8]

Genetic factors have been suggested to contribute to the development of DIHS/DRESS. Genetic predisposition is supported by the association of certain human leukocyte antigen (HLA) molecules with DIHS/DRESS induced by certain drugs.[9] These associations are often observed in only some populations or ethnicities. For instance, in Japan, the allele frequency of HLA-B*1301 was much higher in patients with DIHS induced by aromatic amine anticonvulsants (15.4%) than that reported in the general Japanese population (1.3%). The allele HLA-B*5801 was associated with DIHS/DRESS in the Han Chinese (Taiwan) population. Abacavir-induced hypersensitivity was associated with the presence of HLA-B*5701, but 45% of patients with this allele do not develop the hypersensitivity syndrome. Interestingly, the sensitivity and specificity of this association have been demonstrated across ethnicities. Today, HLA-B*5701 screening before prescribing abacavir is required. Elucidation of associations between genetic background and drug-induced DIHS/DRESS, such as has occurred for abacavir, will provide the ability to predict the potential development of this drug adverse syndrome. Genetic susceptibility probably accounts for many aspects of DIHS/DRESS pathophysiology, pharmacogenomics, and genetic control for human herpesvirus reactivation and antiviral immune response.

DIHS/DRESS PATHOPHYSIOLOGY

DIHS and DRESS are considered to be models of severe drug hypersensitivity reactions. Their pathophysiology remains incompletely elucidated.[10] Today, their pathophysiology must take into account several well-described characteristics:

- Reactivations of human herpesviruses have been demonstrated to be the hallmark of DIHS/DRESS. There is an accumulating body of evidence that suggests that some herpesviruses, especially HHV-6, contribute to its pathogenesis.[11–13] These viruses reactivate sequentially in DIHS/DRESS, as has been observed in transplant recipients, but the mechanism that induces herpesvirus reactivation is not yet understood.
- A limited number of drugs are associated with DIHS/DRESS, suggesting some specific characteristics of these drugs; they do not have structural homologies but may have cross-reactivities.[2,3]

- Some drugs associated with DIHS/DRESS may increase *in vitro* replication of HHV-6 and Epstein–Barr virus (EBV), suggesting a possible direct action of these drugs in the epigenetic control of herpesviruses biology.[14]
- A limited number of patients will experience DIHS/DRESS, suggesting an immunogenetic predisposing background.
- The T-cell immune response is largely directed toward herpesvirus antigens. In a series of 40 patients, EBV-specific CD8+ were overrepresented, with up to 21% of total cytotoxic T cells (compared to <0.1% in controls). The severity of DIHS/DRESS was principally related to both the severity of the antiviral immune response and the level of production of TH1 cytokines (interferon gamma, interleukin-2, TNF-α).[6] Flare-ups have been correlated with sequential herpesvirus reactivation.
- The T-cell immune response may encompass an immune response to the drug itself. The results of patch tests and the lymphocyte transformation test (LTT) with the causative drugs indicate that drug-specific T cells play a role in this syndrome.[10]
- The drug hypersensitivity reactions have been attributed to drug interaction with HLA molecules. Recent data have demonstrated that small molecule drugs may bind within the antigen binding cleft of the HLA molecule in a manner that alters the repertoire of HLA-bound peptides.[15]
- Dramatic expansions of functional regulatory T (Treg) cells are observed in the acute phase of DIHS/DRESS, whereas a gradual loss of Treg function occurs after resolution. In addition, a profound decrease in B cells and hypogammaglobulinemia is observed at the onset of DIHS/DRESS.[16]

Numerous theories have been proposed, but we propose the following likely scenario:

1. Before the first clinical manifestations of the syndrome, DIHS/DRESS-associated drugs induce a reactivation of some herpesviruses. The mechanism of this transition from herpesvirus latency to reactivation is not clear; causative drugs may act directly on epigenetic control of herpesvirus latency and may induce a state of immunosuppression.
2. A dramatic expansion of Treg may be one consequence of T-cell infection after HHV-6 reactivation. T cells and dendritic cells are the targets of HHV-6 infection. Expanded Treg cell populations might serve to prevent the activation and expansion of antiviral T cells and to decrease the number of B cells.[15]
3. The first manifestations of DIHS/DRESS, including fever, pharyngitis, cervical lymphadenopathy, general malaise, and atypical lymphocytes, are similar to those of a severe viral infection because they are the consequence of the strong herpesvirus reactivation.[17]
4. Severe manifestations of DIHS/DRESS with internal organ involvement are the consequence of an immune response against viral antigens.[6] Discontinuation of the drugs may increase clinical manifestations by

restoring the immune response. This strong antiviral immune response may explain why, in the early phases of DIHS/DRESS, the herpesvirus genome may not be detected. This may be considered to be equivalent to a viral-associated immune reconstitution syndrome.[10]

5. This strong immune response is favored by the presence of the drug or its metabolites. As demonstrated with abacavir, drug molecules may interact within the HLA binding site and modify the human leukocyte antigen (HLA) repertoire.[16] This interaction has dramatic consequences for the recognition of some peptides by T cells, as demonstrated for abacavir and HLA-B*5701. The immune response to viral peptide and self antigens might be increased.
6. Strong antiviral reactions are also probably conditioned by an immunogenetic background (HLA genotype, cytokine polymorphisms).
7. A bystander anti drug immune response may develop, but in another scenario this "drug allergy" is responsible for the initial symptoms and herpesvirus reactivation.
8. Herpesviruses may cooperate, as observed in transplant recipients, in sequential herpesvirus reactivation.[10]
9. A decrease in the Treg suppressive effect increases susceptibility to the development of autoimmune disorders. Cross-mimicking immune responses with viral antigen and self antigens may be responsible for the development of autoimmune disorders (e.g., thyroiditis, pancreatitis, lupus, GVH-like manifestations).[10,18]

This pathophysiology explains why corticosteroids are the mainstay of DIHS/DRESS treatment because they control the immune response, but a rapid reduction in corticosteroid dose is associated with flare-up of the disease. It remains to be determined whether or not an antiviral treatment in association with corticosteroids would be beneficial from the viewpoint of long-term control of the disease outcome and sequelae.

CLINICAL FEATURES AND LABORATORY FINDINGS

Initial Findings

DIHS/DRESS typically develops 2 to 6 weeks after the initiation of drug administration.[1,2] The initial signs and symptoms are fever, general fatigue, and cutaneous eruptions. Additionally, lymphadenopathy and internal organ involvement such as hepatitis, renal dysfunction, and pulmonary disease develop. Hematological abnormalities, including leukocytosis, atypical lymphocytosis, and eosinophilia, also occur to varying degrees. The individual symptoms are not specific and may suggest other diseases, such as viral infection. High fever is the most common clinical sign, as with other severe adverse drug reactions, including Stevens–Johnson syndrome and toxic epidermal necrolysis. Despite discontinuation of the causative drugs, the elevated body temperature does not return to baseline levels.

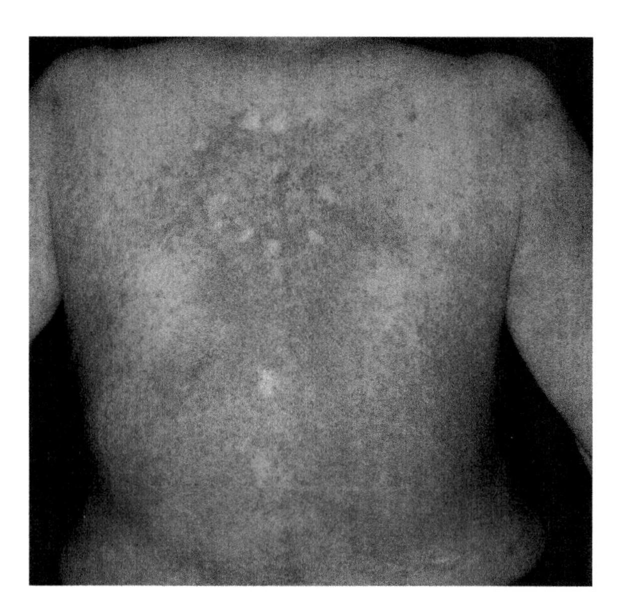

FIGURE 11.1 Generalized maculopapular rash.

The skin manifestations of DIHS/DRESS include a maculopapular rash (Figure 11.1), erythema multiforme-like eruption, pustular eruption, and erythroderma. A purpuric rash is often observed, especially in the legs (Figure 11.2).[19,20] Oral mucosal involvement may be observed to a lesser extent; however, skin rashes rarely progress to Stevens–Johnson syndrome and toxic epidermal necrolysis with mucosal involvement.[21] Thus, it is difficult to diagnose DIHS/DRESS based on the skin rash type; nevertheless, facial eruption with edema and perinasal/perioral infiltrated red papules may provide important clues to the diagnosis of DIHS/DRESS (Figure 11.3).[1,19,20,22]

Lymphadenopathy is usually observed in the cervical region and is painful. In some cases, generalized lymphadenopathy is confirmed on physical examination or by computed tomography. Lymphadenopathy is generally resolved before the detection of HHV-6 reactivation.

Liver involvement is common (>80%).[20,23] Hepatomegaly may be found on physical examination, and serum liver transaminase levels are increased due to hepatocellular damage. A marked increase in serum alanine aminotransferase and aspartate aminotransferase, with variable increases in γ-glutamyl transpeptidase and alkaline phosphatase, are detected.[10] Cholestatic hepatitis and hyperbilirubinemia are observed at a low frequency. Hepatitis sometimes precedes the development of a fever and skin rash.

Renal involvement occurs less often than hepatitis. Serum creatinine and urea are increased, and creatinine clearance is decreased. In patients with underlying renal impairment, the dysfunction sometimes progresses to acute

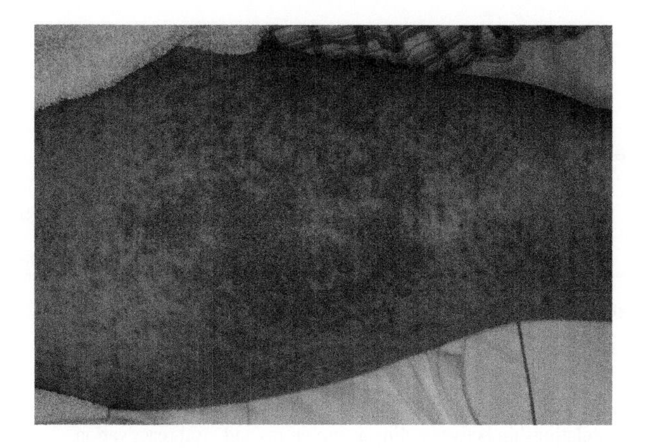

FIGURE 11.2 Purpuric rash in the thigh.

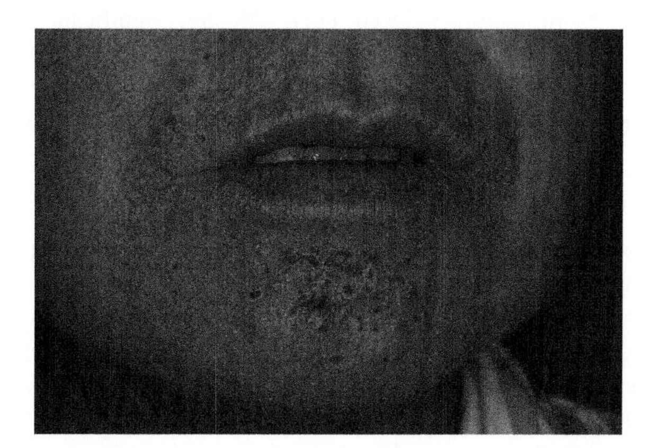

FIGURE 11.3 Perioral-infiltrated red papules and crusting.

renal failure, and dialysis may be required. Pulmonary involvement in the initial phase of DIHS/DRESS is rare, but interstitial pneumonia can occur.

Hematological abnormalities include leukocytosis, eosinophilia, and/or the appearance of atypical lymphocytes. Leukocytosis may appear after discontinuation of the causative drug and may be further aggravated for 1 to 2 weeks. Eosinophilia develops in 50 to 70% of DIHS/DRESS patients.[20,23] Although the appearance of atypical lymphocytes is commonly observed in cases of viral infection (e.g., EBV, cytomegalovirus, measles virus), viruses are not usually detected during this stage of DIHS/DRESS.[24] HHV-6 usually reactivates later.

It is conceivable that a drug allergy is responsible for these initial symptoms; therefore, clinical symptoms may vary according to the causative drug. For example, although hepatitis is common in DIHS/DRESS, renal involvement frequently occurs in allopurinol-induced DIHS/DRESS without liver

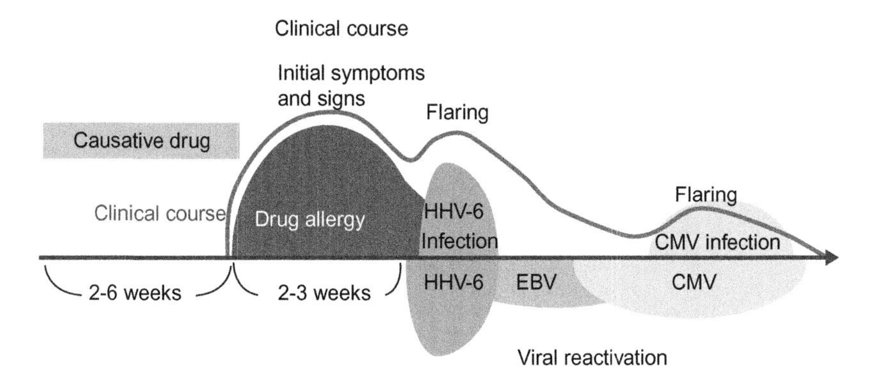

FIGURE 11.4 Relationship between clinical symptoms and viral reactivation.

involvement.[20,25] In dapson-induced DIHS/DRESS, significant hyperbilirubinemia and jaundice are observed at a high frequency, as compared to other types of drug-induced DIHS/DRESS.[26] Pulmonary involvement is often observed among patients who have minocycline-induced DIHS/DRESS.[25]

Flare-Ups and the Development of New Symptoms

In severe cases, flare-ups involving clinical signs such as fever, eruption, or hepatitis often occur after improvement of the patient's initial symptoms and more than 2 weeks after the onset of DIHS/DRESS (Figure 11.4). Recent reports have demonstrated that reactivated viruses participate in the flaring of symptoms as described in the Viral Reactivation section, below. Additionally, new symptoms may rarely develop during the course of DIHS/DRESS, such as central nervous system (CNS) disorders, myocarditis, hemophagocytic syndrome, gastrointestinal disorders, and interstitial pneumonia. Recently, it was noted that fulminant type 1 diabetes is associated with DIHS. The involvement of viruses, including HHV-6 or cytomegalovirus (CMV), has been suggested in some of these cases.

HISTOPATHOLOGY

Skin

A skin biopsy will show dense dermal infiltration of lymphocytes, as compared with an ordinary drug rash. Lymphocytes are mainly observed around dermal vessels with or without eosinophils. No epidermal change or exocytosis of a few lymphocytes with mild spongiosis is usually observed (Figure 11.5). However, apoptotic keratinocytes and vacuolar changes in basal cells are frequently observed in erythema multiforme-like eruptions.[19] When the skin manifestation shows the features of SJS/TEN, apoptotic keratinocytes and epidermal necrosis with exocytosis of lymphocytes are frequently observed (Figure 11.6).

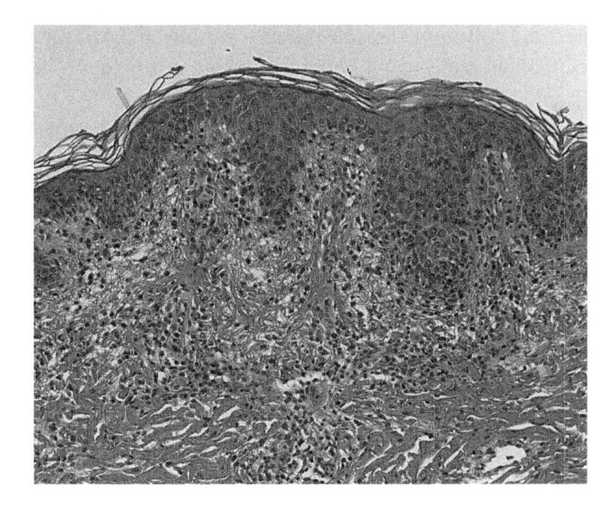

FIGURE 11.5 Histological finding of skin section from maculopapular rash.

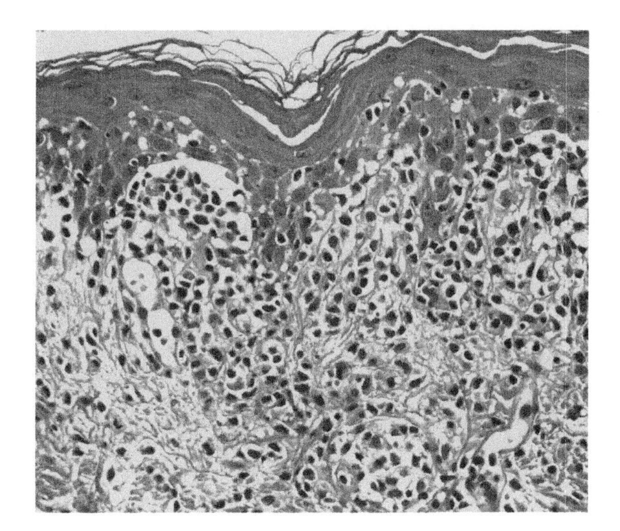

FIGURE 11.6 Histological finding of SJS-like skin lesion.

Lymph Nodes

Histopathological findings of the lymph nodes include benign lymphoid hyperplasia or features mimicking lymphoma.[1] A histological pattern resembling malignant lymphoma has been observed, mainly in cases of anticonvulsant-induced DIHS/DERSS. In these cases, the lymph nodes show hyperplasia of the reticulum cells and other elements with frequent mitoses, focal necrosis, and phagocytosis, but no Reed–Sternberg cells.[1] However, in some cases of anticonvulsant-, antibiotic-, or dapson-induced DIHS/DRESS, the histological findings of lymphadenopathy were compatible with malignant lymphoma.[27–29] It may be difficult to assess whether lymphadenopathy is the result of drug hypersensitivity by histological examination without clinical information.

Liver

Liver biopsy shows a hepatocellular pattern or mixed hepatocellular and cholestatic pattern that is commonly observed in drug-induced hepatitis. In severe hepatitis, massive necrosis or multiple disseminated necrotic foci can occur, resulting in liver failure.[1]

VIRAL REACTIVATION

HHV-6

HHV-6 is commonly reactivated 2 to 3 weeks after the onset of illness in most DIHS patients (see Figure 11.4).[24] The criteria for DIHS proposed by the Japanese Research Committee on Severe Cutaneous Adverse Reactions include the detection of HHV-6 reactivation.[2] On the other hand, in DRESS, the HHV-6 detection rate is lower than that in DIHS.[8] This difference may be explained by the differing ranges of severity between the two diseases: DIHS is regarded as a more severe form of DRESS.[2,8] A certain level of severity in patients with a drug allergy may be required for HHV-6 reactivation. HHV-6 can be detected in serum for 3 to 7 days by polymerase chain reaction (PCR) or viral culture, accompanied by viral replication in peripheral blood.[11,24] When HHV-6 DNA disappears from the serum, a significant increase in anti-HHV-6 IgG titers occurs in DIHS/DRESS (Figure 11.7).[24] A rise in anti-HHV-6 IgM is not always observed. HHV-6 DNA is detectable by PCR for a longer time in peripheral blood samples than in serum samples, even if the anti-HHV-6 IgG titer has already increased. HHV-6 reactivation sometimes

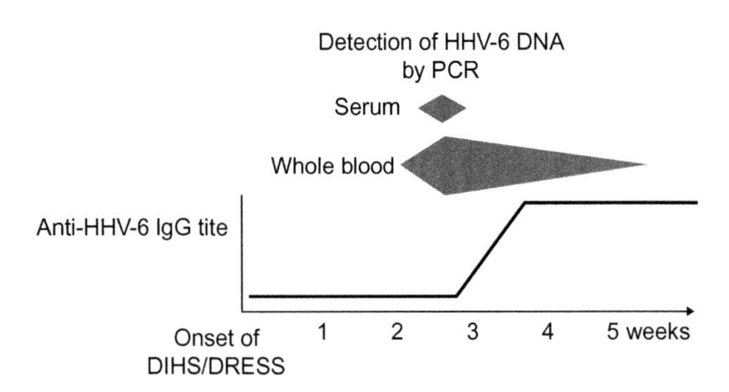

FIGURE 11.7 Detection of HHV-6 DNA and a rise of anti-HHV-6 IgG titers.

accompanies the flaring of symptoms such as fever and hepatitis.[24] The level of viral replication may influence the flaring of symptoms. Body temperature is increased during HHV-6 viremia but decreases with clearance of HHV-6 from serum. Hepatitis also relapses with HHV-6 reactivation to various degrees and is improved after the clearance of HHV-6. Thus, because HHV-6 reactivation and infection are usually self-limited in DIHS/DRESS, antiviral therapy is not required. However, the reactivation of HHV-6 can rarely cause a severe illness such as encephalitis.

Other Herpesviruses

Herpesviruses other than HHV-6, including CMV, EBV, and human herpesvirus-7 (HHV-7), may also be reactivated during DIHS/DRESS.[5,6,30] The cascade of viral reactivation initiated by HHV-6 extends to EBV and CMV (see Figure 11.7).[5,21,30] It is of interest that reactivation of these viruses has been observed in the same sequential manner as in a bone marrow transplant patient.[5] HHV-7 is detected at various periods during DIHS/DRESS[21,31] and co-reactivates with HHV-6.[30,32] In DIHS/DRESS, the reactivation of EBV and HHV-7 is thought to have no clinical relevance.

In contrast, CMV contributes to internal organ involvement and the relapse of symptoms. CMV reactivates more than 1 month after the onset of DIHS/DRESS and causes various diseases such as recurring transient fever, a skin rash, skin ulcers, hepatitis, pneumonia, and gastrointestinal bleeding.[33,34] Moreover, CMV infection might participate in the development of myocarditis or fulminant type 1 diabetes.[35,36] The detection of CMV DNA in blood has no clinical importance, as it does not indicate the existence of infectious disease. On the other hand, the detection of CMV antigenemia is useful for assessing the risk of CMV infection. In addition, positive staining for CMV antigens by immunohistochemistry can be used to make a diagnosis. Diseases caused by CMV infection such as pneumonia and gastrointestinal bleeding may lead to a fatal outcome in DIHS/DRESS.[34] Immune suppression induced by systemic corticosteroids may influence the severity of a CMV infection, although CMV can reactivate in DIHS/DRESS without systemic corticosteroids.[33] It is necessary to assess the pathological condition of a patient carefully to manage CMV infections.

VISCERAL INVOLVEMENTS AND SEQUELAE OF DIHS/DRESS

Visceral involvement may occur throughout the clinical course of DIHS/DRESS. The cause of multiorgan involvement remains unknown,[8,10,37] and these complications develop unpredictably (Table 11.2). The severity of DIHS/DRESS is commonly determined by the degree of visceral involvement. Furthermore, sequelae may occur after a disease-free interval of several months to years after resolution of the acute illness.[37]

TABLE 11.2 Visceral Involvements During the Course of DIHS/DRESS

Enterocolitis/intestinal bleeding
Fulminant type 1 diabetes mellitus
Hemophagocytic syndrome
Hepatitis/cholangitis
Limbic encephalitis/encephalopathy
Myocarditis
Nephritis
Pneumonitis/pleuritis

Visceral Involvement During the Course of DIHS/DRESS

Hepatobiliary System

Apart from the skin, the liver is the most common organ involved in DIHS/DRESS.[8] Hepatomegaly accompanied by splenomegaly is frequently observed. The presence of an underlying persistent viral infection, such as hepatitis B or hepatitis C infection, often causes a severe deterioration in liver function.[37] Although the large majority of patients recover spontaneously, massive hepatic necrosis in the setting of coagulopathy and sepsis can cause death. Prolonged prothrombin times and/or partial thromboplastin times are observed in severe cases.[38] Hepatic impairment is commonly anicteric, but if it is icteric it tends to have a poorer prognosis. The flaring of symptoms such as fever and liver dysfunction in DIHS/DRESS is closely related to HHV-6 reactivation.[24]

Renal System

Renal involvement occurs in 8 to 11% of patients with DIHS/DRESS.[8,39] Accumulated evidence from case reports suggests that severe renal insufficiency increases the risk of mortality, and mortality depends in part on the degree of renal involvement rather than hepatic involvement.[10] Renal involvement is prominent in allopurinol-induced DIHS/DRESS.[40] It is more likely to occur in patients with preexisting renal disease or those receiving diuretic therapy. Because renal function declines steadily with age, elderly patients are vulnerable to renal complications. Laboratory studies demonstrate worsening renal insufficiency, ranging from a mild elevation in serum creatinine levels to severe interstitial nephritis.[41] Granulomatous interstitial nephritis can be also noted in DIHS/DRESS.[42] Proteinuria and oliguria may also be present. No specific deposits have been detected in immunofluorescence studies of the biopsy specimens from the kidney.

Respiratory System

The prevalence of lung involvement in DIHS/DRESS is unclear because cases with less severe pulmonary involvement have been underreported, leading to a reporting bias with only severe cases being reported.[10] Pulmonary complications include abnormal pulmonary function, acute interstitial pneumonitis, lymphocytic interstitial pneumonia, Loeffler's syndrome, and acute respiratory distress syndrome (ARDS). Clinical symptoms such as a nonproductive cough and breathlessness are highly suggestive of lung involvement. Pleuritis or pleural effusions may also occur during the course of DIHS/DRESS. Apart from the life-threatening condition of ARDS,[43] most pulmonary complications resolve without sequelae. Pulmonary infections have been documented during the course of DIHS/DRESS. Infectious pneumonia caused by *Pneumocystis jiroveci* can occasionally develop during and after resolution of DIHS/DRESS.[18]

Cardiovascular System

Cardiac involvement is a rare but life-threatening complication of DIHS/DRESS. Myocarditis associated with DIHS/DRESS is observed around 3 weeks to 4 months after onset.[44,45] Symptoms/signs suggestive of myocarditis include chest pain, unexplained tachycardia, breathlessness, low blood pressure, and symptoms of congestive heart failure, but some patients may be asymptomatic. Chest radiographs may show cardiomegaly and pleural effusions, and the electrocardiogram usually shows nonspecific ST–T wave changes or ST segment elevations, sinus tachycardia, or arrhythmias. The echocardiogram may show a significant reduction in ejection fraction;[46] cardiac catheterization reveals normal coronary arteries and a decreased ejection fraction. The appearance of the above-mentioned cardiac findings in the setting of DIHS/DRESS should raise the suspicion of drug-induced myocarditis as a complication of DIHS/DRESS. The condition may resolve or progress rapidly into fulminant heart failure.[44,45] Endomyocardial biopsy—a highly invasive, seldom performed procedure—demonstrated a mixed infiltrate of lymphocytes and eosinophils with myocyte necrosis in several cases of DIHS/DRESS.[44,46] Dilated cardiomyopathy may develop as a late sequel of DIHS/DRESS.[47] Complete atrioventricular block associated with dapsone-induced DIHS/DRESS has been reported.[48]

Central Nervous System

Neurological complications such as meningitis and encephalitis in DIHS/DRESS have been documented.[49–52] Limbic encephalitis may develop approximately 2 to 4 weeks after the onset of DIHS/DRESS; the condition is suspected if there is fever accompanied by headache, altered level of consciousness, and signs of cerebral dysfunction. Focal neurological abnormalities such as anomia, dysphagia, and hemiparesis and seizure may be present. Cerebrospinal fluid (CSF) reveals an increase in the white blood cell count and protein level. Magnetic resonance imaging (MRI) has shown bilateral lesions involving the amygdala, mesial temporal lobes, insula, and cingulate gyrus.[40] The typical MRI appearance is selective involvement of the limbic system, particularly the mesial temporal lobes. An electroencephalogram shows diffuse slow waves, with an occasional solitary spike, in the frontal and temporal leads without periodic patterns. Considering the onset of limbic encephalitis and timing of HHV-6 reactivation in DIHS/DRESS, it is possible that limbic encephalitis might be due to HHV-6 reactivation. In a patient with allopurinol-induced DIHS/DRESS who developed encephalitis, an increase in anti-HHV-6 IgG titers and a detection of HHV-6 DNA by polymerase chain reaction in the CSF were noted.[50] However, viral DNA cannot be detectable in CSF samples from DIHS/DRESS patients after the onset of encephalitis, which may indicate secondary encephalitis or inappropriate timing of sample collection.[52] A case of a patient with DIHS/DRESS complicated by the syndrome of inappropriate secretion of antidiuretic hormone (SIADH) and concomitant limbic encephalitis has been documented.[52]

Gastrointestinal System

Acute, life-threatening gastrointestinal bleeding may occur during the course of DIHS/DRESS and is caused by CMV infection.[33] Despite the initial belief that only HHV-6 reactivation occur in DIHS/DRESS, other herpesviruses such as EBV, HHV-7, and CMV may also reactivate during the course of the disease.[30] Gastrointestinal ulcers due to CMV infection may be misdiagnosed as steroid-induced ulcers if CMV is not considered in the differential diagnosis. Endoscopic examination reveals arterial bleeding from "punched out" gastric ulceration.[33] Gastrointestinal ulceration is often concomitant with cutaneous CMV ulcers on the shoulders and upper trunk. Histopathological examination of the biopsy specimens from the gastric mucosa and the skin shows cytomegalic cells with characteristic "owl's eye" intranuclear inclusions in the infiltrating cells. CMV infection is usually confirmed by immunohistochemical analysis using an anti-CMV monoclonal antibody.[33] Autopsy findings of a patient with severe DIHS/DRESS revealed disseminated CMV infection involving the lung, myocardium, kidney, adrenal gland, liver, pancreas, spleen, and skin.[53]

Cytomegalovirus reactivation occurs in a predictable manner during DIHS/DRESS. In most patients, CMV DNA is detected 4 to 7 weeks after the onset of DIHS/DRESS and approximately 10 days to 3 weeks after HHV-6 reactivation.[33] Scratch dermatitis and erythematous rash on the trunk, unexplained low-grade fever, and lumbar pain during the 4- to 7-week period after the onset of DIHS/DRESS are suggestive of CMV disease. A reduction in both the platelet and white blood cell counts and a decreased serum immunoglobulin level are also useful indicators of CMV reactivation. The basis for CMV reactivation is unclear, but it appears that advanced age, particularly among those older than 60 years, and patients with antecedent high HHV-6 DNA loads are at risk for overt CMV disease.[33]

Pancreatic System

DIHS/DRESS has been associated with the development of fulminant type 1 diabetes mellitus (FT1DM).[54–56] FT1DM is a subtype of type 1 diabetes mellitus (T1DM) characterized by abrupt onset, absence of islet-related autoantibodies, and the near complete destruction of pancreatic β-cells. The initial symptoms of T1DM are vomiting and dull epigastric pain. Laboratory findings demonstrate hyperglycemia, hyperosmolarity, and metabolic acidosis, findings that are compatible with diabetic ketoacidosis. In addition, elevations in lipase and amylase consistent with acute pancreatitis have been documented at the onset of T1DM.[55]

According to a nationwide survey performed in Japan, 15 cases of FT1DM associated with DIHS/DRESS were documented between 1985 and 2010. In these cases, the mean age of FT1DM onset was 53.4 years, and the interval between the onset of DIHS/DRESS and the development of FT1DM was 39.9 days (range, 13 to 199 days). The prevalence of FT1DM in DIHS/DRESS was much greater than that found in the general Japanese population.[36] The clinical characteristics of FT1DM associated with DIHS/DRESS were similar to those unrelated to DIHS/DRESS. The etiology of FT1DM associated with DIHS/DRESS remains unknown; however, the prevalence of HLA-B62 was significantly higher in Japanese patients with FT1DM associated with DIHS/DRESS. It is likely that a genetic susceptibility may contribute to the development of FT1DM.[36] Furthermore, it seems that herpesvirus reactivation, such as HHV-6 and CMV, may contribute to the development of FT1DM in patients with DIHS/DRESS. In contrast to FT1DM, autoimmune T1DM is uncommon in patients with DIHS/DRESS. Insulin therapy should be commenced after the diagnosis of T1DM. The consequence of a missed diagnosis can be fatal.

Other Organs

Hemophagocytic syndrome (HPS) is usually associated with, and triggered by, various conditions such as viral infections (particularly EBV-related disorders), malignant tumors, or autoimmune diseases. Although rare, HPS has been noted during the course of DIHS/DRESS.[57] Reactivation of HHV-6 or EBV might be responsible for the development of HPS in DIHS/DRESS. Herpes zoster (HZ) can develop during the first 6 months after the onset of DIHS/DRESS, with HZ lesions appearing during tapering of oral corticosteroids. HZ lesions are prone to be detected on the trunk in patients with DIHS/DRESS; the cutaneous manifestations of HZ is usually mild with an uncomplicated course.[18] Telogen effluvium may develop in the absence of autoantibodies after complete recovery of DIHS/DRESS.[58]

Sequelae of DIHS/DRESS

Autoimmune sequelae can arise after resolution of DIHS/DRESS following a symptom-free interval of several months to years, with the development of autoimmune disease and/or production of autoantibodies. The link between DIHS/DRESS and autoimmune disease might be overlooked unless dermatologists perform long-term follow-up of DIHS/DRESS patients after their complete recovery.[37] These autoimmune diseases include autoimmune T1DM,[59,60] autoimmune thyroiditis,[61,62] sclerodermoid graft-versus-host disease (GVHD)-like lesions,[61] lupus erythematosus,[63] and autoimmune polyglandular syndrome.[64] Several autoimmune diseases can develop sequentially in a single patient.[61] These cases suggest that DIHS/DRESS may trigger the development of autoimmune disease. Interestingly, some of these autoimmune diseases are similar to those seen after bone marrow transplantation.

The development of autoimmune T1DM is uncommon in DIHS/DRESS. Various autoantibodies including anti-glutamic acid decarboxylase (GAD) and anti-islet cell antibodies are present in autoimmune T1DM. The coexistence of autoimmune T1DM and autoimmune thyroiditis has been observed in a patient with DIHS/DRESS. In this case, various autoantibodies, including anti-GAD, anti-thyroid peroxidase, anti-thyroglobulin, anti-nuclear, and anti-SSA, were detected.[59]

Graves' disease, characterized by anti-thyroid stimulating hormone receptor autoantibodies, may develop after the resolution of DIHS/DRESS.[59] The interval between discontinuation of the causative drug and onset of Graves' disease is approximately 2 to 4 months. Painless thyroiditis and Hashimoto's thyroiditis—characterized by anti-thyroid peroxidase and anti-thyroglobulin antibodies—may also develop after resolution of DIHS/DRESS. Anti-thyroid antibodies have been detected in asymptomatic patients without functional alteration of the thyroid gland following the resolution of DIHS/DRESS.[18] Rarely, sclerodermoid GVHD-like lesions may appear 3 to 4 years after DIHS/DRESS following the development of autoimmune thyroiditis.[61] Lupus erythematosus can also develop in DIHS/DRESS.[62]

DIFFERENTIAL DIAGNOSIS

Differentiating DIHS/DRESS from viral eruptions can be challenging, as DIHS/DRESS can resemble viral infections (Table 11.3). Thus, careful history taking, including examination of drug history, and a thorough physical examination are important for making the correct diagnosis. The diagnosis of DIHS/DRESS is usually not difficult if there is a long period of drug intake (e.g., anticonvulsants). Many patients have been initially misdiagnosed with a viral

TABLE 11.3 Differential Diagnosis

Atopic dermatitis
Infectious mononucleosis
Kawasaki disease
Measles
Pseudolymphoma
Serum sickness-like reaction

illness, such as infectious mononucleosis or measles; however, viral infections may be identified by the lack of eosinophilia and/or hypoimmunoglobulinemia in the laboratory findings. Infectious mononucleosis develops most commonly in teenagers or young adults, while DIHS/DRESS occurs in middle-aged and elderly individuals with a long period of administration of specific drugs. Kawasaki disease can be excluded by using the established diagnostic criteria and laboratory testing. The serum sickness-like reaction is characterized by the presence of urticarial lesions and the absence of internal organ involvement. The clinical manifestations of DIHS/DRESS can be indistinguishable from atopic erythroderma with bacterial infection, but hepatic and/or renal involvement is not characteristic of this condition. In a substantial number of patients with DIHS/DRESS, small pustules on the face and trunk are observed at an early stage that may resemble acute generalized exanthematous pustulosis (AGEP); however, duration of drug intake is less than a week and internal organ involvement is uncommon in AGEP. Pseudolymphoma needs to be differentiated because lymphadenopathy is frequently observed at an early stage of DIHS/DRESS. Pseudolymphomas have been reported to develop in association with drugs such as phenytoin and carbamazepine.[65] A diagnosis of drug-induced pseudolymphoma is usually based on histologic findings and clinical presentation, ranging from solitary nodules to multiple infiltrative papules or plaques without evidence of extracutaneous lymphoma and resolution of the skin eruption after drug cessation.

To identify the drug responsible for the development of DIHS/DRESS, *in vivo* and *in vitro* testing such as patch tests and lymphocyte transformation tests (LTTs) are available. Positive LTT reactions are obtained if more than 4 weeks have elapsed after the onset of DIHS/DRESS, and strong positive reactions may be seen even a year after discontinuation of the causative drug.[66]

MANAGEMENT AND TREATMENT

Early recognition of clinical symptoms associated with DIHS/DRESS is essential in improving patient outcomes. Although the management of DIHS/DRESS is not well established, prompt cessation of the culprit drug is necessary. Because a dramatic deterioration of clinical symptoms, such as fever, skin rashes, and facial edema, often develops 3 to 4 days after withdrawal of the causative drug, a suspicion of underlying infection may be raised. However, empirical treatment with antibiotics or nonsteroidal anti-inflammatory drugs should be avoided as these agents often worsen the early stage of DIHS/DRESS.[67] In contrast to toxic epidermal necrolysis, factors that predict the outcome of DIHS/DRESS have not been identified; however, advanced age, allopurinol- or minocycline-induced DIHS/DRESS, and the presence of comorbid conditions appear to be poor prognostic factors.[8,25]

Patients with moderate disease often recover within 3 to 4 weeks if given supportive care without systemic corticosteroids.[18] Oral corticosteroids are the mainstay of treatment, despite the absence of randomized control trials confirming its efficacy. Given that herpesvirus reactivation is implicated in DIHS/DRESS, the corticosteroid dose should be given carefully. The usual prednisolone dosage is 0.5 to 1.0 mg/kg/day. After commencement of oral corticosteroids, fever and skin rashes usually subside within a week. It is necessary to taper the dose down gradually to avoid the re-emergence of symptoms. The mean duration of corticosteroid treatment is 6 to 8 weeks.[67]

The risk of CMV reactivation is significantly higher in DIHS/DRESS patients under corticosteroid treatment than those managed without corticosteroid therapy. Thus, physicians should be vigilant for signs of CMV disease in these patients. Given the high mortality rate associated with gastrointestinal CMV ulceration in patients with DIHS/DRESS, early intervention with emergency endoscopic clipping and blood transfusion is usually required. If CMV antigenemia is present during the course of DIHS/DRESS, administration of ganciclovir is recommended to continue until CMV antigenemia is absent in order to avoid CMV disease progression.

Various infections have been noted in patients undergoing corticosteroid treatment in the acute early phase of DIHS/DRESS, including herpesvirus disease and *Pneumocystis jiroveci* pneumonia. Infections are frequent and most appear within 3 months of oral corticosteroid therapy. For this reason, a minimum of 3 months of follow-up is recommended after resolution of DIHS/DRESS.[18]

The development of infections in DIHS/DRESS during corticosteroid treatment may be analogous to the pathomechanism of immune reconstitution inflammatory syndrome (IRIS).[67,68] Infections in the early stage of DIHS/DRESS such as herpes zoster, CMV disease, and *Pneumocystis jiroveci* pneumonia are similar to the illnesses in acquired immune deficiency syndrome patients with IRIS after highly active antiretroviral therapy.[67]

It has been shown that intravenous immunoglobulin (IVIG) compensates for the decreased immunoglobulin concentration, provides anti-inflammatory effects, and regulates the immune response in autoimmune diseases. Beneficial effects of IVIG treatment have been observed in a patient with DIHS/DRESS;[69] however, a separate report documented severe adverse events resulting from IVIG that required systemic corticosteroid therapy.[70]

CONCLUSION

DIHS and DRESS show a wide range of clinical manifestations and laboratory findings. The association between severe visceral involvement and herpesvirus reactivation has been clarified, and the pathomechanisms have been partially elucidated. A better understanding of the interplay between drug allergy and viral reactivations will allow more efficient management of this disease. Furthermore, the immunological aspects of DIHS/DRESS may serve as excellent tools for investigating the autoimmune diseases after viral infections.

ACKNOWLEDGMENT

This work was supported by the Research Project on Measures for Intractable Diseases; a matching fund subsidy from the Ministry of Health, Labour and Welfare (to M.T. and T.S.); and by the Ministry of Education, Culture, Sports, Science and Technology (to M.T., T.S., and Y.K.).

REFERENCES

1. Bocquet H, Bagot M, Roujeau JC. Drug-induced pseudolymphoma and drug hypersensitivity syndrome (drug rash with eosinophilia and systemic symptoms: DRESS). *Semin Cutan Med Surg* 1996;**15**:250–7.
2. Shiohara T, Iijima M, Ikezawa Z, et al. The diagnosis of a DRESS syndrome has been sufficiently established on the basis of typical clinical features and viral reactivations. *Br J Dermatol* 2007;**156**:1083–4.
3. Kardaun SH, Sidoroff A, Valeyrie-Allanore L, et al. Variability in the clinical pattern of cutaneous side-effects of drugs with systemic symptoms: does a DRESS syndrome really exist?. *Br J Dermatol* 2007;**156**:609–11.
4. Ushigome Y, Kano Y, Hirahara K, et al. Human herpesvirus 6 reactivation in drug-induced hypersensitivity syndrome and DRESS validation score. *Am J Med* 2012;**125**:e9–10.
5. Kano Y, Hiraharas K, Sakuma K, et al. Several herpesviruses can reactivate in a severe drug-induced multiorgan reaction in the same sequential order as in graft-versus-host disease. *Br J Dermatol* 2006;**155**:301–6.
6. Picard D, Janela B, Descamps V, et al. Drug reaction with eosinophilia and systemic symptoms (DRESS): a multiorgan antiviral T cell response. *Sci Transl Med* 2010;**2**:46–62.
7. Shear NH, Spielberg SP. Anticonvulsant hypersensitivity syndrome. *In vitro* assessment of risk. *J Clin Invest* 1988;**82**:1826–32.
8. Cacoub P, Musette P, Descamps V, et al. The DRESS syndrome: a literature review. *Am J Med* 2011;**124**:588–97.
9. Aihara M. Pharmacogenetics of cutaneous adverse drug reactions. *J Dermatol* 2011;**38**:246–54.
10. Shiohara T, Kano Y, Takahashi R, Mizukawa Y. Drug-induced hypersensitivity syndrome: recent advances in the diagnosis, pathogenesis and management. *Chem Immunol Allergy* 2012;**97**:122–38.
11. Tohyama M, Yahata Y, Yasukawa M, et al. Severe hypersensitivity syndrome due to sulfasalazine associated with reactivation of human herpesvirus 6. *Arch Dermatol* 1998;**134**:1113–7.
12. Descamps V, Valance A, Edlinger C, et al. Association of human herpesvirus 6 infection with drug reaction with eosinophilia and systemic symptoms. *Arch Dermatol* 2001;**137**:301–4.
13. Shiohara T, Kano Y. A complex interaction between drug allergy and viral infection. *Clin Rev Allergy Immunol* 2007;**33**:124–33.
14. Mardivirin L, Descamps V, Lacroix A, et al. Early effects of drugs responsible for DRESS on HHV-6 replication *in vitro*. *J Clin Virol* 2009;**46**:300–2.
15. Pompeu YA, Stewart JD, Mallal S, et al. The structural basis of HLA-associated drug hypersensitivity syndromes. *Immunol Rev* 2012;**250**:158–66.
16. Takahashi R, Kano Y, Yamazaki Y, et al. Defective regulatory T cells in patients with severe drug eruptions: timing of the dysfunction is associated with the pathological phenotype and outcome. *J Immunol* 2009;**182**:8071–9.

17. Descamps V, Avenel-Audran M, Valeyrie-Allanore L, et al. Saliva polymerase chain reaction assay for detection and follow-up of herpesvirus reactivation in patients with drug reaction with eosinophilia and systemic symptoms (DRESS). *JAMA Dermatol* 2013;**149**(5):565–9.
18. Ushigome Y, Kano Y, Ishida T, Hirahara K, Shiohara T. Short- and long-term outcomes of 34 patients with drug-induced hypersensitivity syndrome in a single institution. *J Am Acad Dermatol* 2013;**68**(5):721–8.
19. Walsh S, Diaz-Cano S, Higgins E, et al. Drug reaction with eosinophilia and systemic symptoms: is cutaneous phenotype a prognostic marker for outcome? A review of clinicopathological features of 27 cases. *Br J Dermatol* 2013;**168**:391–401.
20. Chen YC, Chiu HC, Chu CY. Drug reaction with eosinophilia and systemic symptoms: a retrospective study of 60 cases. *Arch Dermatol* 2010;**146**:1373–9.
21. Tohyama M, Hashimoto K. New aspects of drug-induced hypersensitivity syndrome. *J Dermatol* 2011;**38**:222–8.
22. Hashimoto K, Tohyama M, Shiohara T. Drug reactions with eosinophilia and systemic symptoms(DIHS/DRESS). In: Revuz J, Roujeau JC, Kerdel FA, Valeyrie-Allanore L, editors. *Life-Threatening Dermatoses and Emergencies in Dermatology*. Berlin: Springer-Verlag; 2009. p. 97–102.
23. Criado PR, Criado RF, Avancini Jde M, et al. Drug reaction with eosinophilia and systemic symptoms (DRESS)/drug-induced hypersensitivity syndrome (DIHS): a review of current concepts. *An Bras Dermatol* 2012;**87**:435–49.
24. Tohyama M, Hashimoto K, Yasukawa M, et al. Association of human herpesvirus 6 reactivation with the flaring and severity of drug-induced hypersensitivity syndrome. *Br J Dermatol* 2007;**157**:934–40.
25. Eshki M, Allanore L, Musette P, et al. Twelve-year analysis of severe cases of drug reaction with eosinophilia and systemic symptoms: a cause of unpredictable multiorgan failure. *Arch Dermatol* 2009;**145**:67–72.
26. Agrawal S, Agarwalla A. Dapsone hypersensitivity syndrome: a clinico-epidemiological review. *J Dermatol* 2005;**32**:883–9.
27. Saltzstein SL, Ackerman LV. Lymphadenopathy induced by anticonvulsant drugs and mimicking clinically pathologically malignant lymphomas. *Cancer* 1959;**12**:164–82.
28. Schnetzke U, Bossert T, Scholl S, et al. Drug-induced lymphadenopathy with eosinophilia and renal failure mimicking lymphoma disease: dramatic onset of DRESS syndrome associated with antibiotic treatment. *Ann Hematol* 2011;**90**:1353–5.
29. Rim MY, Hong J, Yo I, et al. Cervical lymphadenopathy mimicking angioimmunoblastic T-cell lymphoma after dapsone-induced hypersensitivity syndrome. *Korean J Pathol* 2012;**46**:606–10.
30. Seishima M, Yamanaka S, Fujisawa T, et al. Reactivation of human herpesvirus (HHV) family members other than HHV-6 in drug-induced hypersensitivity syndrome. *Br J Dermatol* 2006;**155**:344–9.
31. Morito H, Kitamura K, Fukumoto T, et al. Drug eruption with eosinophilia and systemic syndrome associated with reactivation of human herpesvirus 7, not human herpesvirus 6. *J Dermatol* 2012;**39**:669–70.
32. Yagami A, Yoshikawa T, Asano Y, et al. Drug-induced hypersensitivity syndrome due to mexiletine hydrochloride associated with reactivation of human herpesvirus 7. *Dermatology* 2006;**213**:341–4.
33. Asano Y, Kagawa H, Kano Y, et al. Cytomegalovirus disease during severe drug eruptions: report of 2 cases and retrospective study of 18 patients with drug-induced hypersensitivity syndrome. *Arch Dermatol* 2009;**145**:1030–6.

34. Sano S, Ueno H, Yamagami K, et al. Isolated ileal perforation due to cytomegalovirus reactivation during management of terbinafine hypersensitivity. *World J Gastroenterol* 2010;**16**:3339–42.
35. Sekiguchi A, Kashiwagi T, Ishida-Yamamoto A, et al. Drug-induced hypersensitivity syndrome due to mexiletine associated with human herpes virus 6 and cytomegalovirus reactivation. *J Dermatol* 2005;**32**:278–81.
36. Onuma H, Tohyama M, Imagawa A, et al. High frequency of HLA B62 in fulminant type 1 diabetes with the drug-induced hypersensitivity syndrome. *J Clin Endocrinol Metab* 2012;**97**:E2277–81.
37. Kano Y, Ishida T, Hirahara K, et al. Visceral involvements and long-term sequelae in drug-induced hypersensitivity syndrome. *Med Clin North Am* 2010;**94**:743–59.
38. Newell BD, Moinfar M, Mancini AJ, et al. Retrospective analysis of 32 pediatric patients with anticonvulsant hypersensitivity syndrome (ACHSS). *Pediatr Dermatol* 2009;**26**:536–46.
39. Roujeau JC, Stern RS. Severe adverse cutaneous reactions to drugs. *N Engl J Med* 1994;**331**:1272–85.
40. Kano Y, Shiohara T. The Variable clinical picture of drug-induced hypersensitivity syndrome (DIHS)/drug rash with eosinophilia and systemic symptoms (DRESS) in relation to the eliciting drug. *Immunol Allergy Clin North Am* 2009;**29**:481–501.
41. Augusto J-F, Sayegh J, Simon A, et al. A case of sulphasalazine-induced DRESS syndrome with delayed acute interstitial nephritis. *Nephrol Dial Transplant* 2009;**24**:2940–2.
42. Eguchi E, Shimazu K, Nishiguchi K, et al. Granulomatous interstitial nephritis associated with atypical drug-induced hypersensitivity syndrome induced by carbamazepine. *Clin Exp Nephrol* 2012;**16**:168–72.
43. Roca B, Calvo B, Ferrer D. Minocycline hypersensitivity reaction with acute respiratory distress syndrome. *Intensive Care Med* 2003;**29**:338.
44. Bourgeois GP, Cafardi JA, Groysman V, et al. Fulminant myocarditis as a late sequel of DRESS: two cases. *J Am Acad Dermatol* 2011;**65**:889–90.
45. Shaughnessy KK, Bouchard SM, Mohr MR, et al. Minocycline-induced drug reaction with eosinophilia and systemic symptoms (DRESS) syndrome with persistent myocarditis. *J Am Acad Dermatol* 2010;**62**:315–8.
46. Teo RY, Tay YK, Tan CH, et al. Presumed dapson-induced drug hypersensitivity syndrome causing reversible hypersensitivity myocarditis and thyrotoxicosis. *Ann Acad Med Singapore* 2006;**35**:833–6.
47. Lo MH, Huang CF, Chang LS, et al. Drug reaction with eosinophilia and systemic symptoms syndrome associated myocarditis: a survival experience after extracorporeal membrane oxygenation support. *J Clin Pharm Ther* 2013;**38**:172–4.
48. Zhu KJ, He FT, Jin N, et al. Complete atrioventricular block associated with dapson therapy: a rare complication of dapson-induced hypersensitivity syndrome. *J Clin Pharm Ther* 2009;**34**:489–92.
49. Fujino Y, Nakajima M, Inoue H, et al. Human herpesvirus 6 encephalitis associated with hypersensitivity syndrome. *Ann Neurol* 2002;**51**:771–4.
50. Masaki T, Fukunaga A, Tohyama M, et al. Human herpes virus 6 encephalitis in allopurinol-induced hypersensitivity syndrome. *Acta Derm Venereol* 2003;**83**:128–31.
51. Descamps V, Collot S, Houhou N, et al. Human herpesvirus-6 encephalitis associated with hypersensitivity syndrome. *Ann Neurol* 2003;**53**:280.
52. Sakuma K, Kano Y, Fukuhara M, et al. Syndrome of inappropriate secretion of antidiuretic hormone associated with limbic encephalitis in a patient with drug-induced hypersensitivity syndrome. *Clin Exp Dermatol* 2008;**33**:287–90.

53. Arakawa M, Kakuto Y, Ichikawa K, et al. Allopurinol hypersensitivity syndrome associated with systemic cytomegalovirus infection and systemic bacteremia. *Intern Med* 2001;**40**:331–5.
54. Sekine N, Motokura T, Oki T, et al. Rapid loss of insulin secretion in a patient with fulminant type 1 diabetes mellitus and carbamazepine hypersensitivity syndrome. *JAMA* 2001;**285**:1153–4.
55. Sommers LM, Schoene RB. Allopurinol hypersensitivity syndrome associate with pancreatic exocrine abnormalities and new-onset diabetes mellitus. *Arch Intern Med* 2002;**162**:1190–2.
56. Chiou CC, Chung WH, Hung SI, Yang LC, Hong HS. Fulminant type 1 diabetes mellitus caused by drug hypersensitivity syndrome with human herpesvirus 6. *J Am Acad Dermatol* 2006;**54**(Suppl. 2):S14–7.
57. Descamps V, Bouscarat F, Laglenne S, et al. Human herpesvirus 6 infection associated with anticonvulsant hypersensitivity syndrome and reactive haemophagocytic syndrome. *Br J Dermatol* 1997;**137**:605–8.
58. Wongkitisophon P, Chanprapaph K, Pattanakaemakorn P, et al. Six-year retrospective review of drug reaction with eosinophilia and systemic symptoms. *Acta Derm Venereol* 2012;**92**:200–5.
59. Brown RJ, Rother KI, Artman H, et al. Minocycline-induced drug hypersensitivity syndrome followed by multiple autoimmune sequelae. *Arch Dermatol* 2009;**145**:63–6.
60. Ozaki N, Miura Y, Sakakibara A, et al. A case of hypersensitivity syndrome induced by methimazole for Graves' disease. *Thyroid* 2005;**15**:1333–6.
61. Kano Y, Sakuma K, Shiohara T. Sclerodermoid graft-versus-host disease-like lesions occurring after drug-induced hypersensitivity syndrome. *Br J Dermatol* 2007;**156**:1061–3.
62. Minegaki Y, Higashida Y, Ogawa M, et al. Drug-induced hypersensitivity syndrome complicated with concurrent fulminant type 1 diabetes mellitus and Hashimoto's thyroiditis. *Int J Dermatol* 2013;**52**:355–7.
63. Aota N, Hirahara K, Kano Y, Fukuoka T, Yamada A, Shiohara T. Systemic lupus erythematosus presenting with Kikuchi–Fujimoto's disease as a long-term sequelae of drug-induced hypersensitivity syndrome, a possible role of Epstein–Barr virus reactivation. *Dermatology* 2009;**218**:275–7.
64. Van den Driessche A, Eenkhoom V, Van Gaal L, De Block C. Type 1 diabetes and autoimmune polyglandular syndrome: a clinical review. *Neth J Med* 2009;**67**:376–87.
65. Callot V, Roujeau JC, Bagot M, et al. Drug-induced pseudolymphoma and hypersensitivity syndrome: two different clinical entities. *Arch Dermatol* 1996;**132**:1315–21.
66. Kano Y, Hirahara K, Mitsuyama Y, Takahashi R, Shiohara T. Utility of the lymphocyte transformation test in the diagnosis of drug sensitivity: dependence on its timing and type of drug eruption. *Allergy* 2007;**62**:1439–44.
67. Shiohara T, Kurata M, Mizukawa Y, Kano Y. Recognition of immune reconstitution syndrome necessary for better management of patients with severe drug eruption and those under immunosuppressive therapy. *Allegol Int* 2010;**59**:333–43.
68. Leohloenya R, Meintjes G. Dermatologic manifestations of the immune reconstitution inflammatory syndrome. *Dermatol Clin* 2006;**24**:549–70.
69. Kito Y, Ito T, Tokura Y, Hashizume H. High-dose intravenous immunoglobulin monotherapy for drug-induced hypersensitivity syndrome. *Acta Derm Venereol* 2012;**92**:100–1.
70. Joly P, Janela B, Tetart F, et al. Poor benefit/risk balance of intravenous immunoglobulins in DRESS. *Arch Dermatol* 2012;**148**:543–4.

Chapter 12

HHV-6A and HHV-6B in Solid Organ Transplantation

Irmeli Lautenschlager[a] **and Raymund R. Razonable**[b]

[a]Helsinki University Hospital and University of Helsinki, Helsinki, Finland, [b]College of Medicine, Mayo Clinic, Rochester, Minnesota

INTRODUCTION

Infections due to the two human herpesviruses (HHV-6A and HHV-6B) are common in humans, and HHV-6B causes the majority of symptomatic primary infections of infants and children.[1] HHV-6 is classically associated with roseola infantum, an illness characterized by high-grade fever, irritability, and maculopapular rash in children.[2] The clinical symptoms of primary HHV-6 infections are usually mild and self-limited, but complications such as seizures, otitis, respiratory or gastrointestinal symptoms, encephalitis, and hepatitis have been described.[2,3] HHV-6 seroconversion generally occurs before the age of 2 years, and the seroprevalence in the adult population exceeds 95%.[4,5]

After primary infection, HHV-6 persists in the state of latency in the host. During latency, the HHV-6 genome is harbored as circular DNA in various cells, though mainly in those of monocyte and macrophage origin.[6] In 1% of individuals, the HHV-6 DNA becomes incorporated in the host chromosome (known as chromosomally integrated HHV-6, or ciHHV-6),[7] and this can be inherited or transmitted vertically in a Mendelian manner. Latent HHV-6 may reactivate later in life, especially during periods of immunosuppression such as after organ and tissue transplantation.

Human herpesvirus 6 is a lymphotropic virus that replicates efficiently in CD4+ T lymphocytes. The virus uses the CD46 molecule as its cellular receptor.[8] The virus has also been shown to infect other cell types, such as monocytes and macrophages, astrocytes, fibroblasts, and cells of endothelial or epithelial origin.[9] These cells have widespread distribution; thus, HHV-6 has been shown to infect various tissues including the brain, salivary glands, tonsils, lungs, kidneys, and liver. The cellular and tissue tropism of the two viruses HHV-6A and HHV-6B vary, with HHV-6A being considered as the more neurotropic virus.[10] In solid organ recipients, most infections have been

Human Herpesviruses HHV-6A, HHV-6B & HHV-7.

attributed to HHV-6B, which has been detected in various tissues, such as liver, kidney, intestines, lungs, and heart.[11]

HHV-6 INFECTIONS IN SOLID ORGAN TRANSPLANTATION

Epidemiology

Due to the high seroprevalence and presence of preexisting immunity of the majority of humans, HHV-6 infections in organ transplant recipients are mostly asymptomatic reactivations. In adult patients, where seroprevalence may exceed 95%, primary HHV-6 infections are expectedly rare. Primary HHV-6 infections are more common in pediatric transplant recipients.[12] In a study of adult solid organ transplant recipients with high HHV-6 seroprevalence (96.4%), only one patient developed symptomatic primary HHV-6 infection after transplantation.[13] In another series of adult living related liver transplant recipients, for whom the HHV-6 seroprevalence was reported at 97%, none of the few HHV-6 seronegative patients demonstrated HHV-6 seroconversion after transplantation.[14] The incidence of HHV-6 reactivation peaks at 2 to 4 weeks after transplantation,[15–18] although late infections have been demonstrated to occur months or years after transplantation.[19]

The overall incidence of HHV-6 infection after transplantation varies widely, depending on the study design, patient population, and method of testing, from as low as 0% to as high as 82% of solid organ transplant recipients.[11,20] Most HHV-6 infections in solid organ transplant recipients were reported among liver and kidney transplant recipients, and the incidence is estimated to be between 22 and 55%.[16,17,21–28] HHV-6 infections have also been reported after kidney–pancreas[29] and intestinal[20] transplantation. The reported incidence after heart and lung transplantation is as high as 66 to 91%, but these are mostly asymptomatic infections.[30–34]

Clinical Perspectives

Clinical Syndromes

Human herpesvirus 6 infections in solid organ transplant patients are usually asymptomatic. The vast majority of patients with HHV-6 infection have viral DNA detected in the blood but no clinical manifestations. In about 1% of transplant patients with HHV-6 infection, clinical symptoms and complications have been reported. Fever is the most common clinical manifestations of HHV-6 infection, but other clinical symptoms, such as rash, graft dysfunction, pneumonitis, gastrointestinal infection, hepatitis, and neurological disorders and encephalitis, have also been reported sporadically.[15,18,19,24,35–38]

Most clinically symptomatic HHV-6 infections in solid organ transplantation have been reported in liver transplant recipients. A recent study of living donor liver transplant recipients demonstrated the significant association between a higher rate of mortality among those who developed HHV-6 reactivation compared to those without viral reactivation.[14] Although HHV-6 was not the direct cause of death in this study, it implies the potential of HHV-6 to indirectly influence clinically relevant outcomes after solid organ transplantation.

The first reports of symptomatic HHV-6 disease in transplant recipients described a syndrome of febrile dermatosis with thrombocytopenia and encephalopathy.[15,24,36] Thereafter, several studies have suggested the clinical significance of HHV-6 as a potential pathogen in liver recipients.[16,18,19,39] In one study of 200 liver transplant recipients, HHV-6 was found to be a causative agent of febrile disease in 1% of cases, when other pathogens had been excluded.[18] Gastrointestinal HHV-6 involvement may also occur.[38] In a study of 90 liver transplant recipients undergoing gastroscopic examination for dyspeptic symptoms, HHV-6B antigen-positive cells were demonstrated in the biopsy specimens in gastroduodenal mucosa of 21 (23%) of the patients.[38] Furthermore, HHV-6 may infect the transplanted liver and cause allograft dysfunction,[19] although no direct loss of a transplant allograft due to HHV-6 has been reported.

Posttransplant allograft hepatitis[35] and intrahepatic HHV-6 infections associated with liver dysfunction have been described.[19] In a retrospective analysis of 121 liver recipients, HHV-6 was thought to be an etiological agent in eight cases (6.7%) of allograft hepatitis, and HHV-6 was demonstrated in six available liver biopsies.[19] The hepatic pathogenicity of HHV-6 was further demonstrated in a prospective series of 51 adult liver transplant recipients, where eight out of 11 patients with HHV-6 antigenemia demonstrated significant graft dysfunction, and three of these had HHV-6 antigens in the liver biopsy specimens (5.9%).[17] All cases of HHV-6 antigenemia were due to HHV-6B; however, infection with HHV-6A may occur, as others have reported a case of giant cell hepatitis caused by HHV-6A.[40] Pretransplant HHV-6B infection with acute liver failure was reported to be a risk factor for the development of posttransplant HHV-6 infection of the liver allograft, although the intragraft HHV-6 infection had no appreciable long-term impact on the transplant outcome in this study.[41,42] However, another study of 170 liver transplant recipients indicates that high intrahepatic HHV-6 loads (but neither CMV nor Epstein–Barr virus) are associated with decreased allograft survival following diagnosis of allograft hepatitis.[43]

Human herpesvirus 6 infections after kidney transplantation are also common but usually asymptomatic, and the virus has been detected by either serology or viral culture from peripheral blood.[21–24,44] Frequent detection of HHV-6 DNA in peripheral blood mononuclear cells from asymptomatic kidney transplant patients has been reported.[45] In general, the clinical effect of HHV-6 is suggested to be minimal after kidney transplantation;[46] however, symptomatic disease due to HHV-6 infections has also been reported rarely in kidney recipients, and these are usually nonspecific manifestations, including fever or gastrointestinal symptoms. Unlike liver transplant recipients, HHV-6 hepatitis is much less common after kidney transplantation. HHV-6 has been associated with a higher risk of CMV disease after kidney transplantation, and concomitant or recent CMV infection may contribute to the clinical symptoms.[47,48] Isolated HHV-6 infections (in the absence of CMV or HHV-7) are limited to small series and case reports describing fever, elevated creatinine levels, liver dysfunction, and colitis.[49,50] In a recent study, HHV-6B was frequently found in the gastrointestinal tract of 67 kidney transplant patients undergoing gastrointestinal endoscopy for gastrointestinal symptoms.[51] HHV-6B antigens were demonstrated both in gastroduodenal (34%) and colorectal (36%) mucosa, but the more intense infection demonstrated by immunohistochemistry was seen in the upper gastrointestinal tract. The role of HHV-6 in these cases was difficult to interpret, as some patients had CMV-positive findings from the same specimens. The clinical symptoms were mild and no antiviral treatment was administered.[51] Only a few fatal cases of HHV-6 disease have been described, and all have been due to HHV-6A causing hemophagocytic syndrome, pancytopenia, encephalitis, severe hepatitis, or colitis.[52,53]

Human herpesvirus 6-specific antigens have been demonstrated by immunohistochemistry in kidney biopsy material, and this has been associated with pathological conditions, such as acute and chronic rejection or cyclosporine-related nephropathy.[22,54] HHV-6 antigens have been found in the distal tubular epithelial cells and in a few infiltrating lymphocytes, but also in the glomeruli and vascular endothelia.[54] HHV-6 has also been isolated from renal tissues without concurrent viremia.[22] An association between HHV-6 and kidney allograft rejection has been reported, and viral antigens were detected in the tubular epithelium of 5 of 9 patients with acute rejection. HHV-6 antigens have also been found in some intraoperative biopsy specimens.[54,55] In one study, the HHV-6 isolate was found to be similar for the donor and the kidney transplant recipient;[23] this suggests that HHV-6 may be transmitted with the donor kidney allograft and that reactivation after transplantation could be due to the HHV-6 strain of either the recipient or donor origin. HHV-6 may persist in kidney allografts.[56] In an analysis of 22 kidney transplant biopsies of patients with previous CMV infection, HHV-6B antigens were detected in 7 patients, including one with both HHV-6B and CMV antigens detected in the graft. High viral loads have been detected in the allograft tissue of pediatric kidney transplant recipients with significant illness due to HHV-6 infection.[57]

Human herpesvirus 6 has been detected in the blood and bronchoalveolar lavage (BAL) specimens of lung transplant patients.[33,34,58] The overall frequency of HHV-6 DNA detection in BAL was about 20%, and it may be detected together with other opportunistic herpesviruses such as CMV, HHV-7, and Epstein–Barr virus.[59–61] In a recent report of 27 lung transplant recipients, HHV-6 was found in 21% of BAL specimens and in one transbronchial (TBB) specimen.[61] The clinical significance of HHV-6 detection in the lung remains unclear, probably due to the coexistence of other viral infections. In heart and heart–lung transplant recipients, HHV-6 is mostly found in association with CMV (and HHV-7), and the clinical symptoms have been concluded to be due mainly to CMV infection.[31,34] On the other hand, another report has described a higher mortality rate among lung and heart–lung transplant patients with detectable HHV-6 compared to those without HHV-6 infections.[33] Cases of giant cell hepatitis and encephalitis have been reported after heart transplantation.[37,62]

Encephalitis is the most prominent severe clinical disease due to HHV-6 infection in allogeneic hematopoietic stem cell transplant recipients, but this entity is much less observed after solid organ transplantation. There have been a few reports that describe neurological complications,[28,63] but no HHV-6-associated neurological symptoms were recorded in the large series of liver transplant patients.[18] Only a single case of HHV-6 pneumonia has been described after solid organ transplantation,[36] and few cases of encephalitis or colitis have been reported.[34,64]

Indirect Sequelae

Human herpesvirus 6 is an immunomodulatory virus that may facilitate superinfections with fungal or other opportunistic infections.[15,28,65,66] HHV-6 reactivations are often related to CMV and HHV-7 infections and recurrent hepatitis C, and interactions between these viruses have been suggested.[24,46,67–72] Concurrent intragaft infections of HHV-6 and CMV have been found in liver and kidney transplants.[19,56] HHV-6 has been associated with liver allograft rejection and chronic allograft nephropathy.[16,19,73–77] In liver transplantation, intragraft HHV-6B infection has been associated with portal lymphocyte infiltration and increase of vascular adhesion molecules, such as ICAM-1 and VCAM-1, known to be important in leukocyte extravasation and lymphocyte activation.[75] Recently, it has been reported that both HHV-6 and HHV-7 infections are associated with the development of chronic allograft nephropathy.[77] Demonstration of HHV-6 in BAL has also been suggested as a risk factor for bronchiolitis obliterans syndrome, which is a form of chronic rejection in lung transplant recipients.[58] However, no association between beta-herpesviruses, including HHV-6, and bronchiolitis obliterans syndrome have been demonstrated in the era of antiviral prophylaxis.[60] The HHV-6-associated direct and indirect effects in solid organ transplantation are summarized in Table 12.1.

CHROMOSOMALLY INTEGRATED HHV-6 (CIHHV-6) IN ORGAN TRANSPLANT PATIENTS

In the minority of individuals, the HHV-6 genome becomes integrated into the host chromosome (ciHHV-6).[7,78–80] The incidence of ciHHV-6 is between 0.2 and 1% based on a study of blood donors from the United Kingdom.[81] In a cohort study of 47 kidney transplant recipients, ciHHV-6 was detected in one patient, giving a calculated prevalence of 2.1%.[82] In a large cohort of 548 liver transplant recipients, ciHHV-6 was detected in seven patients, and

TABLE 12.1 Direct and Indirect Effects of HHV-6 Infection in Solid Organ Transplant Recipients

Organ	Direct Effects	Indirect Effects
Kidney	Allograft infection Gastrointestinal infection Fever and myelosuppression	Association with CMV infection Acute and chronic rejection
Liver	Encephalitis and neurologic disorders Allograft hepatitis Gastrointestinal infection Pneumonia Fever, rash, myelosuppression	Higher incidence of fungal infections Higher incidence of CMV infection More severe HCV recurrent infections Acute and chronic rejection
Lung	Encephalitis Pneumonia Colitis Fever	Association with CMV Bronchiolitis obliterans
Heart	Encephalitis Hepatitis Fever	Association with CMV

the prevalence was calculated at 1.3%.[83] Both HHV-6A and HHV-6B can be chromosomally integrated, although the majority of reported cases have been due to ciHHV-6B.

The clinical significance of ciHHV-6 after solid organ transplantation has not been defined. Reactivation of ciHHV-6 has been demonstrated in experimental settings, and this poses a theoretical concern during periods of immunosuppression such as after transplantation when reactivations of latent viruses are common and may cause clinical disease.[84] In a recent study of liver transplant recipients, bacterial infections were significantly more common in the ciHHV-6 group compared to the group without HHV-6 (71.4% versus 31.4%; $p = 0.04$). A higher rate of allograft rejection was observed in the ciHHV-6 group compared to the group of patients with low-level HHV-6 DNA load (71.4% versus 37.1%; $p = 0.12$) and those without HHV-6 DNA infection (71.4% versus 42.9%; $p = 0.25$), although these differences did not reach statistical significance. These data suggest that patients with ciHHV-6 may be at increased risk of indirect HHV-6 effects after transplantation, but this will require confirmation using a larger cohort of transplant recipients.[83] The single kidney transplant patient with ciHHV-6 developed neurologic illness and thrombosis after kidney transplantation and died from pulseless electrical activity. It was unclear if this was related to ciHHV-6 reactivation and disease.[82]

The greatest impact of ciHHV-6 has been in the difficulty in the interpretation of nucleic acid testing for diagnosis.[7,80] Because ciHHV-6 is present in every cell, individuals with ciHHV-6 have exceedingly high levels (most likely over a million copies per milliliter of blood) of the HHV-6 genome in their blood samples. Several cases of ciHHV-6 in transplant recipients have been misdiagnosed as true HHV-6 infections and treated unnecessarily.

Diagnosis of HHV-6 Infection

The laboratory diagnosis of HHV-6A and HHV-6B infections is described in detail elsewhere in this book. The clinical diagnosis of significant HHV-6 infection in solid organ transplant recipients is not easy for several reasons. First, most infections are asymptomatic. Second, the detection of HHV-6 DNA or antigens in the clinical specimen does not necessarily implicate the virus as the etiology of a specific illness. Third, differentiation between latent and active HHV-6 infection is not always possible.

Serology is of very limited diagnostic value due to high seroprevalence rate. Viral culture of HHV-6 is laborious and not used routinely for laboratory diagnosis. HHV-6 antigenemia assay, which detects the viral antigens in peripheral blood mononuclear cells (PBMCs) by monoclonal antibodies and immunoperoxidase staining or immunofluorescence, demonstrates an active infection with viral proteins and can discriminate between HHV-6A and HHV-6B.[9,17,23,85] Quantitation of HHV-6 antigen-positive cells is also possible, and this could be a marker of disease severity.[86] HHV-6B antigenemia has been demonstrated to have a good correlation with quantitative polymerase chain reaction (PCR) methods.[87] Demonstration of HHV-6-specific antigens in tissue or cellular specimens by immunostaining is suggestive of tissue-invasive HHV-6 disease (Figure 12.1).[19,38,54]

Molecular assays are the most common laboratory methods for the detection of HHV-6 reactivation and replication in solid organ transplant patients. Quantitative nucleic acid amplification assays are preferred over qualitative assays as they can indicate the severity of infection and can guide the efficacy of antiviral therapies. Qualitative assays often cannot differentiate latent from active viral infection, with few exceptions. Qualitative assays that detect HHV-6 RNA, which indicates active viral replication, may be useful. Likewise, qualitative assay demonstrating viral DNA in the relatively acellular cerebrospinal fluid may be suitable for the diagnosis of HHV-6 encephalitis, although quantitative methods are still preferred. Quantitative HHV-6 PCR assays are recommended for the diagnosis of systemic HHV-6 infection. Quantitative tests have been developed to monitor viral load in plasma, serum, whole blood, or mononuclear cells, mostly by real-time PCR, and some PCR tests are able to differentiate HHV-6A and HHV-6B.[88–91] It has been recently demonstrated that HHV-6 DNA in plasma reflects the presence of infected blood cells rather than circulating viral particles, and quantification of viral DNA in whole blood has been found preferable in the diagnosis of active HHV-6 infection.[92] In addition, quantitative PCR methods can also be used to demonstrate HHV-6 in BAL and transbronchial biopsies.[61,93] Thus,

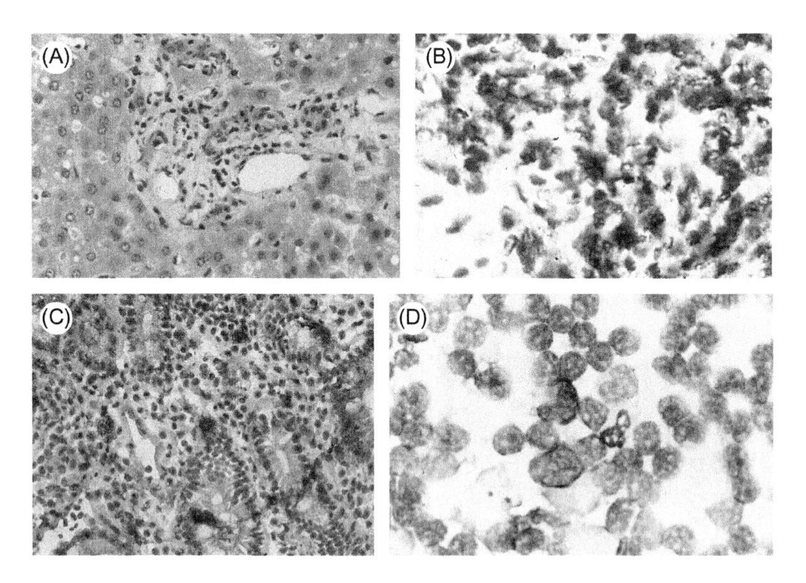

FIGURE 12.1 (A) HHV-6 infection of the liver allograft with portal inflammatory infiltrate (H&E staining). (B) HHV-6B-positive cells in the portal area demonstrated by immunohistochemistry (indirect immunoperoxidase staining). (From Härmä M, et al. *Transplantation* 2006; **81**: 367–72. With permission of Wolters Kluwer Health.) (C) HHV-6B-positive cells in gastric mucosa demonstrated by immunohistochemistry. (Photograph courtesy of J. Arola) (D) HHV-6B-positive lymphocytes in the cytocentrifuge preparation of mononuclear cells (HHV-6 antigenemia test) demonstrated by immunocytochemistry. *(Photograph by I. Lautenschlager).*

the quantification of HHV-6 DNA by means of real-time PCR is currently the most common method for the diagnosis of HHV-6, but these methods are not standardized, the results are not interchangeable between assays, and there are no clear cutoff levels to guide diagnosis and treatment.

The interpretation of active HHV-6 infection on the basis of nucleic acid testing that amplifies and detects HHV-6 DNA should take into account the possibility that one may be detecting ciHHV-6. Because ciHHV-6 is present in every nucleated cell, HHV-6 DNA can be demonstrated by qualitative and quantitative nucleic acid amplification tests. Quantitative PCR testing of cell-containing samples such as whole blood will often yield exceedingly high viral genome copies. In a cohort study of 548 liver transplant recipients, ciHHV-6 was detected in seven patients, and these were characterized by having HHV-6 loads of over a million copies per milliliter of whole blood.[83] A HHV-6 quantitative PCR test with over 1 million genomic copies per milliliter should be highly suspected of having ciHHV-6. In many instances, the detection of high level viral load in a patient with ciHHV-6 has been misinterpreted as an active HHV-6 infection in solid organ transplant patients and treated unnecessarily.[94,95] The diagnostic questions concerning ciHHV-6 are also discussed in a recent review.[7]

Antiviral Treatment

The antiviral treatment of HHV-6 infections are described in detail elsewhere in this book. Based on *in vitro* studies, the current antiviral drugs with efficacy against CMV, such as ganciclovir, foscarnet, and cidofovir, also have antiviral activity against HHV-6.[4,96,97] The most effective compounds against HHV-6 *in vitro* are cidofovir and foscarnet, but the toxicity of these compounds may limit their use in patients with mild infections.[4,96,97] The activity of ganciclovir is highly superior to that of acyclovir; however, ganciclovir is comparatively more effective against CMV than HHV-6B and HHV-6A. There is some indication *in vitro* that the antimalaria drug artesunate is active against HHV-6.[98] The novel formulation of cidofovir, hexadecyloxypropyl cidofovir (HDP-CDV, or CMX001), has also been shown to be active against HHV-6.[99]

Antiviral treatment is recommended after transplantation in patients with HHV-6 encephalitis and other severe syndromes of HHV-6. The degree of immunosuppression should also be assessed, and reduced accordingly. The efficacy of antiviral drugs for HHV-6 has not been subjected to rigorous clinical trials in transplant patients. Data to support their use in the clinical setting are based on observational series and case reports. It was shown that antiviral prophylaxis with ganciclovir or valganciclovir decreased the incidence and HHV-6 viremia in solid organ transplant patients.[100] Ganciclovir prophylaxis has also been reported to delay and shorten HHV-6 viremia in renal transplant recipients.[101] During ganciclovir treatment of CMV disease in liver transplant patients, there is also a decline in concurrent HHV-6 antigenemia although at a slower rate compared to the decline in CMV levels.[69] However, in a large randomized trial comparing ganciclovir and valganciclovir in the treatment of CMV disease in solid organ recipients, there was no clear effect of the antiviral drugs on concomitant HHV-6 and HHV-7 infections.[102] The guidelines from the American Society of Transplantation Infectious Disease Community of Practice do not recommend specific antiviral prophylaxis or preemptive therapy for HHV-6 infection, but intravenous ganciclovir and foscarnet are considered to be first-line agents for treatment of established HHV-6 disease, especially with encephalitis.[103]

CONCLUDING REMARKS

Human herpesvirus 6 is a common infection, primarily caused by HHV-6B, after solid organ transplantation. The reactivation rate is high, but symptomatic clinical disease or graft dysfunction is estimated to occur in only 1 to 6.7% of patients. Fever, myelosuppression, hepatitis, gastrointestinal disease, and encephalitis have been associated with HHV-6. In addition, a higher rate of acute and chronic graft rejection, opportunistic infections such as CMV infection, and invasive fungal disease have been recorded. The clinical significance of chromosomally integrated HHV-6, occurring approximately in 1% of solid organ transplant recipients, is not clear. The diagnosis of HHV-6 is now generally based on quantitative nucleic acid testing, although this can be difficult to interpret due to the frequent asymptomatic reactivations. Antiviral prophylaxis and preemptive therapy are not recommended for HHV-6. Treatment of HHV-6 is indicated in encephalitis and established end-organ disease. Foscarnet, ganciclovir, and cidofovir have been used for treatment, together with reassessment of the level of pharmacologic immunosuppression.

REFERENCES

1. Dewhurst DH, McIntyre JC, Schnabel L, Hall CB. Human herpesvirus 6 (HHV-6) variant B accounts for the majority of symptomatic primary HHV-6 infections in a population of U.S. infants. *J Clin Microbiol* 1993;**31**:416–8.
2. Asano Y, Yoshikawa T, Suga S, et al. Clinical features of infants with primary human herpesvirus 6 infection (exanthem subitum, roseola infantum). *Pediatrics* 1994;**93**:104–8.
3. Hall CB, Long CE, Schnabel KC, et al. Human herpesvirus-6 infection in children. A prospective study of complications and reactivation. *N Engl J Med* 1994;**331**:432–8.
4. De Bolle L, Naesens L, De Clercq E. Update of human herpesvirus 6 biology, clinical features, and therapy. *Clin Microbiol Rev* 2005;**18**:217–45.
5. Okuno T, Takahaski K, Balachandra K, et al. Seroepidemiology of human herpesvirus 6 infection in normal children and adults. *J Clin Microbiol* 1989;**27**:651–3.
6. Kondo K, Kondo T, Okuno T, Takahashi M, Yamanishi K. Latent human herpesvirus 6 infection of human monocytes/macrophages. *J Gen Virol* 1991;**72**:1401–8.
7. Pellett PE, Ablashi DV, Ambros PF, et al. Chromosomally integrated human herpesvirus 6: questions and answers. *Rev Med Virol* 2012;**22**:144–55.
8. Santoro F, Kennedy PE, Locatelli G, Malnati MS, Berger EA, Lusso P. CD46 is a receptor for human herpesvirus 6. *Cell* 1999;**99**:817–27.
9. Flamand L, Komaroff AL, Arbuckle JH. Review part 1. Human herpesvirus 6: basic biology, diagnostic testing, and antiviral efficacy. *J Med Virol* 2010;**82**:1560–8.
10. Yao K, Crawford JR, Komaroff AL, Ablashi DV, Jacobson S. Reviewpart 2. Human herpesvirus 6 in central nervous system diseases. *J Med Virol* 2010;**82**:1669–78.
11. Lautenschlager I, Razonable RR. Human herpesvirus-6 infections in kidney, liver, lung, and heart transplantation: review. *Transplant Int* 2012;**25**:493–502.
12. Yoshikawa T, Ihira M, Suzuki K, et al. Primary human herpesvirus 6 infection in liver transplant recipients. *J Pediatr* 2001;**138**:921–5.
13. Cervera C, Marcos MA, Linares L, et al. A prospective survey of human herpesvirus 6 primary infection in solid organ transplant recipients. *Transplantation* 2006;**82**:979–82.
14. Ohashi M, Sugata K, Ihara M, et al. Human herpesvirus 6 infection in adult living related liver transplant recipients. *Liver Transplant* 2006;**14**:100–9.
15. Sigh N, Carrigan DR, Gayowski T, Marino IR. Human herpesvirus-6 in liver transplant recipients: documentation of pathogenicity. *Transplantation* 1997;**64**:674–8.
16. Griffiths PD, Ait-Khaled M, Bearcroft CP, et al. Human herpesvirus 6 and 7 as potential pathogens after liver transplant: prospective comparison with the effect of cytomegalovirus. *J Med Virol* 1999;**59**:496–501.
17. Lautenschlager I, Linnavuori K, Höckerstedt K. Human herepesvirus-6 antigenemia after liver transplantation. *Transplantation* 2000;**69**:2561–6.
18. Humar A, Kumar D, Caliendo AM, et al. Clinical impact of human herpesvirus 6 infection after liver transplantation. *Transplantation* 2002;**73**:599–604.

19. Lautenschlager I, Höckerstedt K, Linnavuori K, Taskinen E. Human herpesvirus-6 infection after liver transplantation. *Clin Infect Dis* 1998;**26**:702–7.
20. Razonable RR, Paya CV. The impact of human herpesvirus-6 and -7 infection on the outcome of liver transplantation. *Liver Transplant* 2002;**8**:651–8.
21. Morris DJ, Littler E, Arrand JR, Jordan D, Mallick NP, Johnson RW. Human herpesvirus 6 infection in renal-transplant recipients. *N Engl J Med* 1989;**320**:1560–1.
22. Okuno T, Higashi K, Shiraki K, et al. Human herpesvirus 6 infection in renal transplantation. *Transplantation* 1990;**49**:519–22.
23. Yoshikawa T, Suga S, Asano Y, et al. A prospective study of human herpesvirus-6 infection in renal transplantation. *Transplantation* 1992;**54**:879–83.
24. Herbein G, Srasswimmer J, Altieri M, Woehl-Jaegle ML, Wolf P, Obert G. Longitudinal study of human herpesvirus 6 infection in organ transplant recipients. *Clin Infect Dis* 1996;**22**:171–3.
25. Schmidt CA, Wilbron F, Weiss K, et al. A prospective study of human herpes type 6 detected by polymerase chain reaction after liver transplantation. *Transplantation* 1996;**61**:662–4.
26. Dockrell DH, Prada J, Jones MF, et al. Seroconversion to human herpesvirus 6 following liver transplantation is a marker of cytomegalovirus disease. *J Infect Dis* 1997;**176**:1135–40.
27. Kidd IM, Clark DA, Sabin CA, et al. Prospective study of human betaherpesviruses after renal transplantation: association of human herpesvirus 7 and cytomegalovirus co-infection with cytomegalovirus disease and increased rejection. *Transplantation* 2000;**69**:2400–4.
28. Rogers J, Rohal S, Carrigan DR, et al. Human herpesvirus-6 in liver transplant recipients: role in pathogenesis of fungal infections neurologic complications, and outcome. *Transplantation* 2000;**69**:2566–73.
29. Benito N, Ricart MJ, Pumarola T, Marcos MA, Oppenheimer F, Camacho AM. Infection with human herpesvirus 6 after kidney-pancreas transplant. *Am J Transplant* 2004;**4**:1197–9.
30. Petrisli E, Chiereghin A, Gabrielli L, et al. Early and late virological monitoring of cytomegalovirus, Epstein–Barr virus, and human herpesvirus 6 infections in small bowel/multivisceral transplant recipients. *Transplant Proc* 2010;**42**:74–8.
31. Moschettini D, De Milito A, Catucci M, et al. Detection of human herpesviruses 6 and 7 in heart transplant recipients by a multiplex polymerase chain reaction method. *Eur J Microbiol Infect Dis* 1998;**17**:117–9.
32. De Ona M, Melon S, Rodriguez JL, Sanmartin JC, Bernardo MJ. Association between human herpesvirus type 6 and type 7, and cytomegalovirus disease in heart transplant recipients. *Transplant Proc* 2002;**34**:75–6.
33. Jacobs F, Knoop C, Brancart F, Gilot P, Brussels Heart and Lung Transplantation Group Human herpesvirus-6 infection after lung and heart-lung transplantation: a prospective longitudinal study. *Transplantation* 2003;**75**:1996–2001.
34. Lehto JT, Halme M, Tukiainen P, Harjula A, Sipponen J, Lautenschlager I. Human herpesvirus-6 and -7 after lung and heart–lung transplantation. *J Heart Lung Transplant* 2007;**26**:41–7.
35. Ward KN, Gray JJ, Efstathiou S. Brief report: primary human herpesvirus 6 infection in a patient following liver transplantation from a seropositive donor. *J Med Virol* 1989;**28**:69–72.
36. Singh N, Carrigan DR, Gayowski T, Singh J, Marino IR. Variant B human herpesvirus-6 associated febrile dermatosis with thrombocytopenia and encephalopathy in a liver transplant recipient. *Transplantation* 1995;**60**:1355–7.

37. Nash PJ, Avery RK, Tang WHW, Starling RC, Taege A, Yamani MH. Encephalitis owing to human herpesvirus-6 after cardiac transplant. *Am J Transplant* 2004;**4**:1200–3.
38. Halme L, Arola J, Höckerstedt K, Lautenschlager I. Human herpesvirus 6 infection of the gastroduodenal mucosa. *Clin Infect Dis* 2008;**46**:434–9.
39. Yoshikawa T, Iharu M, Suzuki K, et al. Human herpesvirus 6 infection after living related liver transplantation. *J Med Virol* 2000;**62**:52–9.
40. Potenza L, Luppi M, Barozzi P, et al. HHV-6A in syncytial giant-cell hepatitis. *N Engl J Med* 2008;**359**:593–602.
41. Härmä M, Höckerstedt K, Lautenschlager I. Human herpesvirus-6 and acute liver failure. *Transplantation* 2003;**76**:536–9.
42. Härmä M, Höckerstedt K, Krogerus L, Lautenschlager I. Pretransplant human herpesvirus 6 infection of the patients with acute liver failure is a risk factor for posttransplant human herpesvirus 6 infection of the liver. *Transplantation* 2006;**81**:367–72.
43. Pischke S, Gösling J, Engelmann I, et al. High intrahepatic HHV-6 virus loads but neither CMV nor EBV are associated with decreased graft survival after diagnosis of graft hepatitis. *J Hepatol* 2012;**56**:1063–9.
44. Merlino C, Ciacchino F, Segloni GP, Ponzi AN. Human herpesvirus-6 infection and renal transplantation. *Transplantation* 1990;**49**:519–22.
45. Kikuta H, Itami N, Matsumoto S, Chikaraishi T, Togashi M. Frequent detection of human herpesvirus 6 DNA in peripheral blood mononuclear cells from kidney transplant patients. *J Infect Dis* 1991;**163**:925.
46. Caïola D, Karras A, Flandre P, et al. Confirmation of the low clinical effect of human herpesvirus-6 and -7 infections after renal transplantation. *J Med Virol* 2012;**84**:450–6.
47. DesJardin JA, Gibbons L, Cho E, et al. Human herpesvirus reactivation is associated with cytomegalovirus infection and syndromes in kidney transplant recipients at risk for primary infection. *J Infect Dis* 1998;**178**:1783–6.
48. Ratnamohan VM, Chapman J, Howse H, et al. Cytomegalovirus and human herpesvirus 6 both cause viral disease after renal transplantation. *Transplantation* 1998;**66**:877–82.
49. Delbridge MS, Karim MS, Shrestha BM, McKane W. Colitis in a renal transplant patient with human herpesvirus-6 infection. *Transplant Infect Dis* 2006;**8**:226–8.
50. Koukourgianni F, Pichalt V, Liutkus A, et al. HHV-6 infection in a pediatric kidney transplant patient. *Pediatr Nephrol* 2009;**24**:2445–8.
51. Lempinen M, Halme L, Arola J, Honkanen E, Salmela K, Lautenschlager I. HHV-6B is frequently found in the gastrointestinal tract in kidney transplantation patients. *Transplant Int* 2012;**25**:776–82.
52. Rossi C, Delforge ML, Jacobs F, et al. Fatal primary infection due to human herpesvirus 6 variant A in a renal transplant recipient. *Transplantation* 2001;**71**:288–92.
53. Pilmore H, Collins J, Dittmer I, et al. Fatal human herpesvirus-6 infection after renal transplantation. *Transplantation* 2009;**88**:762–5.
54. Hoshino K, Nishi T, Adachi H, et al. Human herpesvirus-6 infection in renal allografts: retrospective immunohistochemical study in Japanese recipients. *Transplant Int* 1995;**8**:169–73.
55. Asano Y, Yoshikawa T, Suga S, et al. Human herpesvirus 6 harbouring in kidney [letter]. *Lancet* 1989;**2**(8676):1391.
56. Helanterä I, Loginov R, Koskinen P, Lautenschlager I. Demonstration of HHV-6 antigens in biopsies of kidney transplant recipients with cytomegalovirus infection. *Transplant Int* 2008;**21**:980–4.
57. Gupta M, Diaz-Mitoma F, Feber J, Shaw L, Forget C, Filler G. Tissue HHV6 and 7 determination in pediatric solid organ recipients: a pilot study. *Pediatr Transplant* 2003;**7**:458–63.

58. Neurohr C, Huppmann P, Leuchte H, for the Munich Lung Transplant Group Human herpesvirus 6 in bronchoalveolar lavage fluid after lung transplantation: a risk factor for bronchiolitis obliterans syndrome?. *Am J Transplant* 2005;**5**:2982–91.
59. Bauer CC, Jaksch P, Aberle SW, et al. Relationship between cytomegalovirus DNA load in epithelial lining fluid and plasma of lung transplant recipients and analysis of coinfection with Epstein–Barr virus and human herpesvirus 6 in the lung compartment. *J Clin Microbiol* 2007;**45**:324–8.
60. Manuel O, Kumar D, Moussa G, et al. Lack of association between beta-herpesvirus infection and bronchiolitis obliterans syndrome in lung transplant recipients in the era of antiviral prophylaxis. *Transplantation* 2009;**87**:719–25.
61. Costa C, Delsedimide L, Solidoro P, et al. Herpesviruses detection by quantitative real-time polymerase chain reaction in bronchoalveolar lavage and transbronchial biopsy in lung transplant: viral infections and histopathological correlation. *Transplant Proc* 2010;**42**:1270–4.
62. Randhawa PS, Jenkins FJ, Nalesnik MA, et al. Herpesvirus 6 variant A infection after heart transplantation with giant cell transformation in bile ductular and gastroduodenal epithelium. *Am J Surg Pathol* 1997;**21**:847–53.
63. Montejo M, Ramon Ferenandez J, et al. Encephalitis caused by human herepesvirus-6 in a liver transplant recipient. *Eur Neurol* 2002;**48**:234–5.
64. Lamoth F, Jaynet PY, Aubert JD, et al. Case report: human herpesvirus 6 reactivation associated with colitis in a lung transplant recipient. *J Med Virol* 2008;**80**:1804–7.
65. Flamand L, Gosselin J, D'Addario M, et al. Human herpesvirus 6 induces interleukin-1 beta and tumor necrosis factor alpha, but not interleukin-6, in peripheral blood mononuclear cell cultures. *J Virol* 1991;**65**:5105–10.
66. Flamand L, Gosselin J, D'Addario M, et al. Immunosuppressive affect of human herpesvirus 6 on T-cell functions: suppression of interleukin-2 synthesis and cell proliferation. *Blood* 1995;**85**:1263–71.
67. Mendez JC, Dockrell DH, Espy MJ, et al. Human beta-herpesvirus interactions in solid organ transplant recipients. *J Infect Dis* 2001;**183**:179–84.
68. Humar A, Kumar D, Raboud J, et al. Interactions between cytomegalovirus, human herpesvirus-6 and the recurrence of hepatitis C after liver transplantation. *Am J Transplant* 2002;**2**:461–6.
69 Lautenschlager I, Lappalainen M, Linnavuori K, Suni J, Höckerstedt K. CMV infection is usually associated with concurrent HHV-6 and HHV-7 antigenemia in liver transplant patients. *J Clin Virol* 2002;**25**(Suppl. 2):S57–61.
70. Singh N, Husain S, Carrigan DR, et al. Impact of human herpesvirus-6 on the frequency and severity of recurrent hepatitis C virus hepatitis in liver transplant recipients. *Clin Transplant* 2002;**16**:92–6.
71. Razonable RR, Rivero A, Brown RA, et al. Detection of simultaneous beta-herpesvirus infections in clinical syndromes due to defined cytomegalovirus infection. *Clin Transplant* 2003;**17**:114–20.
72. Härmä M, Höckerstedt K, Lyytikäinen O, Lautenschlager I. HHV-6 and HHV-7 antigenemia related to CMV infection after liver transplantation. *J Med Virol* 2006;**78**:800–5.
73. Tong CY, Bakran A, Peiris JS, Muir P, Herrington CS. The association of viral infection and chronic allograft nephropathy with graft dysfunction after renal transplantation. *Transplantation* 2002;**74**:576–8.
74. Deborska-Materkowska D, Sadowska A, Wesolowska A, Perkowska-Ptasisnka A, Durlik M. Human herpesvirus 6 variant A infection in renal transplant recipients. *Nephrol Dial Transplant* 2005;**20**:2294.

75. Lautenschlager I, Härmä M, Höckerstedt K, Linnavuori K, Loginov R, Taskinen E. Human herpesvirus-6 infection is associated with adhesion molecule induction and lymphocyte infiltration in liver allografts. *J Hepatol* 2002;**37**:648–54.
76. Feldstein AE, Razonable RR, Boyce TG, et al. Prevalence and clinical significance of human herpesviruses 6 and 7 active infection in pediatric liver transplant patients. *Pediatr Transplant* 2003;**7**:125–9.
77. Chapenko S, Folkmane I, Ziedina I, et al. Association of HHV-6 and HHV-7 reactivation with the development of chronic allograft nephropathy. *J Clin Virol* 2009;**46**:29–32.
78. Luppi M, Marasca R, Barozzi P, et al. Three cases of human herpesvirus-6 latent infection: integration of viral genome in peripheral blood mononuclear cell DNA. *J Med Virol* 1993;**40**:44–52.
79. Luppi M, Barozzi P, Bosco R, et al. Human herpesvirus 6 latency characterized by high viral load: chromosomal integration in many, but not all, cells. *J Infect Dis* 2006;**194**:1020–1.
80. Ward KN, Leong HN, Nacheva EP, et al. Human herpesvirus 6 chromosomal integration in immunocompetent patients results in high levels of viral DNA in blood, sera, and hair follicles. *J Clin Microbiol* 2006;**44**(4):1571–4.
81. Leong HN, Tuke PW, Tedder RS, et al. The prevalence of chromosomally integrated human herpesvirus 6 genomes in the blood of UK blood donors. *J Med Virol* 2007;**79**:45–51.
82. Lee SO, Brown RA, Eid AJ, Razonable RR. Chromosomally integrated human herpesvirus-6 in kidney transplant recipients. *Nephrol Dial Transplant* 2011;**26**:2391–3.
83. Lee SO, Brown RA, Razonable RR. Clinical significance of pretransplant chromosomally integrated humanherpesvirus-6 in liver transplant recipients. *Transplantation* 2011;**92**:224–9.
84. Ablashi DV, Levin CL, Yoshikawa T, et al. Review part 3. Human hepesvirus-6 in multiple non-neurological diseases. *J Med Virol* 2010;**82**:1903–10.
85. Sampaio AM, Thomasini RL, Guardia AC, et al. Human herpesvirus-6 and human herpasvirus-7 in adult transplant recipients: diagnosis based on antigenemia. *Transplant Proc* 2011;**43**:1357–9.
86. Loginov R, Höckerstedt K, Ablashi D, Lautenschlager I. Quantitative HHV6B antigenemia test for the monitoring of transplant patients. *Eur J Clin Microbiol Infect Dis* 2010;**29**:881–6.
87. Karlsson T, Mannonen L, Loginov R, Lappalainen M, Höckerstedt K, Lautenschlager I. Development of a new quantitative real-time HHV-6–PCR and monitoring of HHV-6 DNAaemia after liver transplantation. *J Virol Methods* 2012;**181**:25–36.
88. Locatelli G, Santoro F, Veglia F, Gobbi A, Lusso P, Malnati MS. Real-time quantitative PCR for human herpesvirus 6 DNA. *J Clin Microbiol* 2000;**38**:4042–8.
89. Safronetz D, Humar A, Tipples GA. Differentiation and quantitation of human herpesvirus 6A, 6B and 7 by real-time PCR. *J Virol Methods* 2003;**112**:99–105.
90. Boutolleau D, Duros C, Bonnafous P, et al. Identification of human herpesvirus 6 variants A and B by primer-specific real-time PCR may help to revisit their respective role in pathology. *J Clin Virol* 2006;**35**:257–63.
91. Flamand L, Gravel A, Boutolleau D, et al. Multicenter comparison of PCR assays for detection of human herpesvirus 6 DNA in serum. *J Clin Microbiol* 2008;**46**:2700–6.
92. Achour A, Boutolleau D, Slim A, Agut H, Gautheret-Dejean A. Human herpesvirus-6 (HHV-6) DNA in plasma reflects the presence of infected blood cells rather than circulating viral particles. *J Clin Virol* 2007;**38**:280–5.

93. Costa C, Curtoni A, Bergallo M, et al. Quantitative detection of HHV-6 and HHV-7 in transbronchial biopsies from lung transplant recipients. *New Microbiologica* 2011;**34**:275–80.
94. Clark DA, Ward KN. Importance of chromosomally integrated HHV-6A and -6B in the diagnosis of active HHV-6 infection. *Herpes* 2008;**15**:28–32.
95. Potenza L, Barozzi P, Masetti M, et al. Prevalence of human herpesvirus-6 chromosomal integration (ciHHV-6) in Italian solid organ and allogenic stem cell transplant patients. *Am J Transplant* 2009;**9**:1690–7.
96. Yoshida M, Yamada M, Tsukazaki T, et al. Comparison of antiviral compounds against human herpesvirus 6 and 7. *Antiviral Res* 1998;**40**:73–84.
97. De Clercq E, Naesens L, De Bolle L, Schols D, Zhang Y, Neyts J. Antiviral agents active against human herpesviruses HHV-6, HHV-7 and HHV-8. *Rev Med Virol* 2001;**11**:381–95.
98. Milbradt J, Auerochs S, Korn K, Marschall M. Sensitivity of human herpesvirus 6 and other herpesviruses to the broad-spectrum antiinfective drug artesunate. *J Clin Virol* 2009;**46**:24–8.
99. Bonnafous P, Bogaert S, Godet AN, Agut H. HDP-CDV as an alternative for treatment of human herpesvirus-6 infections. *J Clin Virol* 2013;**56**(2):175–6.
100. Razonable RR, Brown RA, Humar A, Covington E, Alecock E, Paya C, The PV 16000 Study Group Herpesvirus infections in solid organ transplant patients at high risk of primary cytomegalovirus disease. *J Infect Dis* 2005;**192**:1331–9.
101. Galarraga MC, Gomez E, De Ona M, et al. Ganciclovir prophylaxis and β-herpesvirus in renal transplant recipients. *Transplant Int* 2005;**18**:1016–7.
102. Humar A, Åsberg A, Kumar D, On behalf of the VICTOR Study Group An assessment of herpesvirus co-infections in patients with CMV disease: correlation with clinical and virologic outcomes. *Am J Transplant* 2009;**9**:374–81.
103. Razonable RR, Zerr DM, The AST Infectious Diseases Community of Practice HHV-6, HHV-7 and HHV-8 in solid organ transplant recipients. *Am J Transplant* 2009;**9**(Suppl. 4):S97–100.

Chapter 13

HHV-6A and HHV-6B in Recipients of Hematopoietic Cell Transplantation

Danielle M. Zerr[a] and Masao Ogata[b]
[a]*University of Washington and Seattle Children's Research Institute, Seattle, Washington,*
[b]*Oita University Hospital, Hasama-machi, Yufu-city, Oita, Japan*

INTRODUCTION

Primary infection due to human herpesvirus 6 (HHV-6) occurs during early childhood. Thus, most recipients of hematopoietic cell transplantation (HCT) have already been infected with HHV-6 prior to transplantation. With HCT, however, HHV-6 becomes active again or "reactivates" in 30 to 50% of patients between 2 and 4 weeks after transplantation.[1–5] HHV-6B accounts for most reactivation events, with HHV-6A accounting for fewer than 3% of cases.[3,6,7] HHV-6 is an important cause of encephalitis after HCT. In addition, HHV-6 has been associated with mortality as well as a number of important clinical syndromes in HCT recipients. Whether these associations are causal, however, remains undetermined.

EPIDEMIOLOGY

HHV-6 Reactivation

Both demographic and clinical factors may influence the risk of HHV-6 reactivation. For example, recipients of allogeneic HCT are at higher risk of HHV-6 reactivation than recipients of autologous HCT, and among allogeneic HCT recipients those who receive transplants from unrelated or human leukocyte antigen (HLA)-mismatched donors are at particularly increased risk.[4,5,8,9] The source of stem cells also influences the risk of reactivation; recipients of umbilical cord blood stem cells experience reactivation rates of 70 to 90%, compared to rates of 30 to 50% in recipients of peripheral blood or bone marrow stem cells.[8–11] Younger age and underlying disease have variably been identified as pretransplantation risk factors for HHV-6 reactivation occurring after transplantation,[2,3] whereas receipt of glucocorticoids has been identified as a posttransplantation risk factor.[3,5]

Disease Associations

Encephalitis

The accumulated evidence supports a causal association between HHV-6 and encephalitis in HCT recipients. The Bradford Hill criteria of strength of association, consistency, specificity, temporality, biological gradient (dose–response relationship), plausibility, coherence, analogy, and experimental evidence provide a structure by which to assess epidemiological evidence for causality.[12,13] For example, fulfilling criteria for the strength and consistency of the association, multiple case reports[14–17] and observational studies[1,3,5,8,9,18,19] have linked HHV-6 with encephalitis in HCT recipients. In fact, HHV-6 has been observed to be the most common cause of encephalitis in this setting, demonstrating some level of specificity.[20] Addressing temporality, given the early age of primary infection, it is likely that HCT recipients with HHV-6 encephalitis were first infected with the virus well before the episode of encephalitis. In addition, a few studies have also demonstrated the presence of HHV-6 reactivation before the onset of encephalitis.[18] Theoretical plausibility for HHV-6 as a cause of encephalitis is supported by the fact that HHV-6 can infect the cells of the central nervous system and other similar information. The criterion of coherence is fulfilled in that the proposed causal association does not conflict with what is known about HHV-6 or encephalitis. Finally, analogy exists in that other similar viruses (herpesviruses) are known to cause encephalitis. The one criterion that has not been clearly demonstrated is experimental evidence that the condition can be prevented or ameliorated by an appropriate antiviral regimen.

Of the reported cases of HHV-6 encephalitis where the virus was typed, approximately 90% of them have been due to HHV-6B.[14–17] Case reports also provide information regarding characteristics of the patient population most at risk of developing HHV-6 encephalitis. Approximately 85% of reported HHV-6 encephalitis cases have occurred in recipients of unrelated or HLA-mismatched-related allogeneic HCT.[17] Recipients of umbilical cord transplantations also make up a significant proportion of case reports.[17] In a population-based single-center study, the incidence of HHV-6 encephalitis was 0.7% in recipients of transplantations from adult donors versus 9.9% in recipients of umbilical cord blood transplantations.[19]

As previously mentioned, several observational studies have demonstrated associations between systemic HHV-6 reactivation and encephalitis or central nervous system dysfunction in HCT recipients.[1,3,5,8,9,18] In a study of 50 allogeneic HCT recipients, HHV-6 reactivation was detected in 24 patients (48%), and four patients (8%) developed HHV-6 encephalitis.[5] The HHV-6 DNA levels were significantly higher in the plasma of those who developed encephalitis compared with those who did not. Similarly, in a study of 111 HCT recipients conducted by the same group, 60 (54%) patients were demonstrated to have HHV-6 reactivation.[18] Within these 60 patients, high plasma HHV-6 DNA levels ($\geq 10^4$ copies per mL) were associated with encephalitis; none of the 36 patients with low-level HHV-6 reactivation developed encephalitis compared to 8 of 24 patients with high-level HHV-6 reactivation ($p=0.0003$). Another study of 315 allogeneic HCT recipients demonstrated an independent association between HHV-6 reactivation and subsequent delirium, as indicated by neuropsychiatric screening. HHV-6 reactivation was also independently associated with subsequent neurocognitive decline at approximately 3 months after HCT, as measured by baseline and follow-up neurocognitive testing. An important limitation of this study is that a cerebrospinal fluid (CSF) sample was obtained from only 4 of 19 patients with delirium, 2 of whom had HHV-6 DNA detected by polymerase chain reaction. In addition, brain magnetic resonance imaging (MRI) was obtained in only 9 of 19 patients with delirium, and none had the typical abnormalities associated with HHV-6 encephalitis. Because MRI and CSF HHV-6 testing were not obtained in all patients with delirium, it is not possible to determine how many of the patients with delirium had HHV-6 encephalitis. This study suggests that HHV-6 may lead to central nervous system dysfunction in the absence of encephalitis, but further studies are necessary to prove this. For further discussion of the clinical aspects of HHV-6 encephalitis, see the Encephalitis section below.

Bone Marrow Suppression

Clinical studies demonstrate an association between HHV-6 reactivation and bone marrow suppression in HCT recipients.[1,3,4,21–25] Case reports further suggest that the suppressive effects of HHV-6 may even cause secondary graft failure.[26,27] In addition, a single center study demonstrated that among five patients who underwent umbilical cord blood transplantation and died of HHV-6 encephalitis, four of the patients never engrafted.[19] Clinical and *in vitro* studies provide biological evidence supporting the clinical association between HHV-6 and bone marrow suppression. HHV-6 has been shown to latently infect early bone marrow progenitors *in vivo*,[28] and *in vitro* studies have demonstrated that HHV-6 infection of bone marrow progenitor cells suppresses the proliferation of granulocyte–macrophage, erythroid, and megakaryocyte cell lines.[29,30] In addition, HHV-6 has been shown to inhibit granulocyte–macrophage and erythrocyte progenitors in human bone marrow.[21] Taken together, these data suggest that HHV-6 may be considered as a differential diagnosis of secondary graft failure after HCT.

Other Disease Associations

Other important clinical manifestations, such as acute graft-versus-host disease (aGVHD), cytomegalovirus (CMV) reactivation, pneumonitis, and gastrointestinal disease, have also been associated with HHV-6 reactivation. However, as discussed below, these associations have not been consistently demonstrated and potential pathways of pathogenesis are not well defined.

Many previous researchers have reached conflicting conclusions regarding the role of HHV-6 reactivation in aGVHD;[3,4,10,22,31–33] however, two recent large studies have produced concordant results. The first study analyzed 235 consecutive patients and revealed a significant association between aGVHD and HHV-6 reactivation.[34] This association was stronger for high-grade aGVHD and remained significant in the patients who experienced HHV-6 reactivation before initiation of salvage immunosuppression for aGVHD. In this study, the authors suggest several potential explanations of how HHV-6 may trigger aGVHD, including through the host's inflammatory responses to HHV-6 reactivation, an effect of HHV-6 against the host immune system through various mechanisms, and an HHV-6-induced activation of alloreactive T cells. The other large prospective study included 315 HCT recipients and revealed an association between high-level HHV-6 reactivation and subsequent grade II–IV aGVHD after controlling for potential confounding factors (adjusted hazard ratio [HR], 2.4; $p<0.001$). In this study, the authors speculated that HHV-6 reactivation causes a proinflammatory or type I immune response, which could play a role in the development of aGVHD. Further research is required to confirm a causal association between HHV-6 and acute GVHD and to elucidate the pathogenesis of HHV-6-associated aGVHD.

Several reports have shown association between HHV-6 reactivation and CMV reactivation.[35–37] A recent prospective study[35] analyzing 315 HCT recipients demonstrated that HHV-6 reactivation was independently associated with subsequent CMV reactivation (HR, 1.9; $p=0.002$). This study also demonstrated that high-level HHV-6 reactivation was strongly associated with increased risk of subsequent high-level CMV reactivation ($p=0.002$). Whether HHV-6 induces CMV reactivation through its immunosuppressive effect, or whether severe immunosuppression after HCT accounts for the reactivation of both HHV-6 and CMV remains uncertain.

Human herpesvirus 6 may cause pneumonitis in HCT recipients.[4,38–40] In a study of lung biopsy specimens from 15 HCT patients with pneumonitis, both high HHV-6 DNA levels in lung tissue and low HHV-6 antibody levels at the time of the lung biopsy were associated with idiopathic pneumonitis.[38] In some patients with pneumonia, immunohistochemical staining of lung tissue demonstrated HHV-6 antigens without any other identifiable pathogen.[39] However, the association with HHV-6 has been less certain in other reports because of the isolation of additional potential pathogens such as cytomegalovirus, *Pneumocystis jirovecii*, and adenovirus.[41,42] Further study is necessary to demonstrate causality for each of these possible associations.

Human herpesvirus 6 may also play a pathogenic role for gastrointestinal diseases in HCT recipients. In HCT recipients with gastrointestinal symptoms, HHV-6 DNA has been detected by nucleic acid amplification in gastroduodenal mucosa.[4] In some patients with severe diarrhea, HHV-6B DNA was demonstrated in the nuclei of goblet cells by *in situ* hybridization.[43] Histological findings in the study indicated GVHD in two patients, but not in the other two patients. A recent retrospective single-center analysis including 50 patients with severe vomiting and diarrhea after HCT showed that detection of HHV-6 DNA in biopsy specimens was not associated with overall survival, and antiviral therapy against HHV-6 provided no beneficial effects.[44] Nucleic acid amplification for HHV-6 DNA in biopsy specimens may be too sensitive to diagnose clinically relevant HHV-6 infection. Further studies investigating HHV-6 antigen expression on pathological mucosa are needed to elucidate whether HHV-6 causes gastrointestinal disease in HCT recipients, as seen in CMV infection.

Human herpesvirus 6 appears to cause some cases of chronic hepatitis or fulminant hepatitis in infants.[45,46] A few cases of liver dysfunction associated with HHV-6 have been reported in HCT recipients.[1,47] In one patient with liver dysfunction, HHV-6 was detected in a liver biopsy along with histopathological changes suggestive of viral hepatitis, high levels of systemic HHV-6, and no other explanations for the hepatitis.[1] However, demonstrating HHV-6 as a causative pathogen for liver dysfunction in HCT recipients is difficult, given the many potential causes of liver dysfunction, including aGVHD, drug side effects, and infectious agents. Carefully designed studies are needed to determine the roles of HHV-6 in hepatic damage after HCT.

Human herpesvirus 6 reactivation has been associated with an increased risk of mortality in HCT recipients.[3,32,35] In a study in which plasma samples were collected prospectively from 315 allogeneic HCT recipients and tested serially for HHV-6 DNA, HHV-6 reactivation (>1000 copies per mL) was associated with nonrelapse mortality (adjusted hazard ratio [aHR], 2.7; 95% CI, 1.2–6.3) and subsequent aGVHD (aHR, 2.4; 95% CI, 1.6–3.6). Whether HHV-6 is just a marker for a patient at higher risk of mortality or whether it directly or indirectly (through other associated diseases or syndromes) causes mortality has not been determined.

PATHOGENESIS

As most individuals are infected with HHV-6 during early childhood, detectable virus in the plasma or serum of an HCT recipient is typically a result of reactivation of a latent infection. Multiple cell types can support HHV-6 infection, including monocytes/macrophages[48,49] and cells of the central nervous system, such as astrocytes.[14,50] It is unclear whether HHV-6 isolated or detected in the setting of encephalitis originated from latent virus associated with monocytes in the blood or cells of the central nervous system, or both.

Both clinical and experimental evidence suggests that HHV-6 may interfere with normal functioning of the immune system through a variety of mechanisms.[51] The virus can infect several types of immune cells, and infection has been associated with a selective suppression of interleukin-12, an important cytokine in the T helper cell type 1 (Th1)-polarized antiviral immune response.[52,53] Such suppression of antiviral immune responses could promote the HHV-6 infection itself as well as other viral infections, which may explain the documented associations between HHV-6 reactivation and subsequent cytomegalovirus reactivation in allogeneic HCT recipients.[36,37]

Seemingly in contrast to the concept of an HHV-6-induced suppression of the Th1-polarized antiviral immune response, results from other studies suggest that HHV-6 appears to activate a Th1-polarized response.[18,54,55] In several studies, Th1-type cytokines have been associated with acute graft-versus-host disease (aGVHD).[56–59] Although few studies exploring potential associations between the HHV-6 immune response and aGVHD have been performed, data from six patients with HHV-6 reactivation suggest a temporal association between HHV-6 reactivation, elevated cytokines, and rash or aGVHD.[54]

ENCEPHALITIS

Given the strength of the association between HHV-6 and encephalitis as well as the associated morbidity and mortality, a more in-depth discussion of HHV-6 encephalitis is warranted. A discussion of HHV-6 encephalitis is also provided in Chapter 5. As presented above, numerous case reports and case series describing HHV-6 encephalitis in allogeneic HCT recipients have been published.[14–17,60,61] Typically in such cases, HHV-6 was identified in the cerebrospinal fluid (CSF) without another cause identified to explain the central nervous system abnormalities. In these reports, the symptoms and signs of encephalitis typically presented between 2 and 6 weeks after transplantation and were characterized by confusion and anterograde amnesia. Overt seizures have been reported in 40 to 70% of patients,[17,60] and when electroencephalograms (EEGs) are used an even higher proportion of patients are observed to have seizures.[15,16,62] Among patients with clinically apparent seizures, some have generalized seizures, whereas others have partial seizures.[15]

One group has termed the encephalitis associated with HHV-6 in HCT recipients "post-transplant acute limbic encephalitis," or "PALE," describing it as a distinct syndrome of anterograde amnesia, syndrome of inappropriate antidiuretic hormone, mild cerebrospinal fluid pleocytosis, temporal electroencephalogram abnormalities often reflecting clinical or subclinical seizures, and MRI hyperintensities in the limbic system.[15] In this series of nine patients, engraftment preceded the onset of the neurologic syndrome in all patients, and the median time to onset of neurologic symptoms was 29 days following HCT (range, 14 to 61 days).[15] The syndrome began as confusion in most patients and progressed to dense anterograde amnesia with patchy retrograde amnesia within several days in all patients. Fever was present in only two patients and could have been due to other processes that were present, such as graft-versus-host disease or concurrent infection.

Cerebrospinal fluid findings are typically normal or only mildly abnormal in patients with HHV-6 encephalitis.[17] The most common abnormalities are elevations in CSF protein concentration and white blood cell count. In a series of nine allogeneic HCT recipients with HHV-6 encephalitis, the median CSF white blood cell count was 5 leukocytes per mm^3 (range, 1 to 41) with lymphocyte predominance; the median protein concentration was 48 mg/dL (range, 19 to 189).[15] The glucose concentration was normal in all patients.

Abnormal findings are noted on brain MRI in most patients with HHV-6 encephalitis.[17,60] Abnormalities typically involve the medial temporal lobes, particularly the amygdala and hippocampus; hyperintensities in these regions are visualized on T2, fluid-attenuated inversion recovery (FLAIR), and diffusion-weighted imaging (DWI) sequences.[15,19,60,63] Patients with HHV-6 encephalitis may also have abnormalities in limbic structures outside the medial temporal lobes.[64] Computed tomography (CT) of the brain, especially when obtained early in the course of illness, is often normal.[63]

As noted above, seizures are common in patients with HHV-6 encephalitis. On EEG, focal abnormalities are often observed over the temporal or frontotemporal regions.[15] Abnormalities may include epileptiform activity, including electrographic seizures, periodic lateralized epileptiform discharges, and/or sporadic interictal discharges. Diffuse slowing can also occur.

Autopsy studies of patients who died with HHV-6 encephalitis demonstrate lesions involving the white and/or gray matter and injuries characterized by necrosis, neuronal loss, demyelination, and astrogliosis.[14–16,65,66] Earlier in the course of infection, edema and inflammation may be seen.[67] Consistent with the clinical findings of memory impairment and focal findings on brain imaging, the hippocampus is the area most commonly involved on pathology,[14–16,65,66] but other areas of the brain may also be involved.[66,67] In the few published studies that have attempted to correlate detection of HHV-6 with pathology, high levels of HHV-6 mRNA and HHV-6 antigen have been documented in diseased areas of the brain, with astrocytes being the predominant cells involved.[14,66]

Differential Diagnosis

Although HHV-6 is the most common cause of viral encephalitis in HCT recipients,[20] a broad differential diagnosis, including other viral, bacterial, fungal, and parasitic pathogens, must be considered. Noninfectious etiologies should also be considered in allogeneic HCT recipients with altered mental status, as a wide range of causes can contribute to encephalopathy in such patients.

DIAGNOSIS

Please see Chapter 2 for a complete discussion of HHV-6 diagnosis. In HCT recipients, direct detection of HHV-6 using nucleic acid amplification is the preferred method for diagnosis because serological methods are unreliable in these severely immunocompromised hosts. Detection of HHV-6 DNA in plasma or serum using nucleic acid amplification correlates well with viremia and seroconversion[68,69] and is commonly used to detect reactivation. HHV-6 DNA detection in whole blood has also been used to identify HHV-6 reactivation after HCT, with similar frequencies of detection and similar clinical associations as those seen in studies using plasma or serum.[70] In contrast, detection of viral DNA in peripheral blood mononuclear cells by nucleic acid amplification can be difficult to interpret since the mononuclear cell is a site of virus latency.

In the absence of another identified etiology, detection of HHV-6 in the CSF is considered diagnostic of HHV-6 encephalitis. The HHV-6 viral load from the CSF varies widely in patients with HHV-6 encephalitis. For example, in a series of eight patients with HHV-6 encephalitis, the peak median CSF viral load was 16,600 copies per mL and ranged from 600 to 288,975 copies per mL.[61] In another series that included two patients with HHV-6 encephalitis who underwent quantitative nucleic acid amplification of the CSF at the time of onset, the HHV-6 DNA levels were 203,000 copies per mL in one patient and >999,000 copies per mL in the other.[15]

It should be noted that young children have detectable HHV-6 in their CSF around the age when primary HHV-6 infection occurs.[71] Thus, when evaluating young HCT recipients (under the age of 3 years), the possibility of a recent primary infection should be considered. It is unknown what proportion of HCT recipients without central nervous system abnormalities has detectable HHV-6 in the CSF at the time of HHV-6 reactivation, as reactivation occurs early after transplantation, a time when many HCT recipients have not yet engrafted and their physicians are reluctant to perform lumbar punctures. HHV-6 was not detected in a study of 29 CSF samples from HCT recipients without encephalitis; however, CSF samples were obtained a median of 51 days after HCT, a time after when most reactivation occurs.[15] In another study of 22 HCT recipients with central nervous system symptoms and 107 immunocompromised controls without symptoms, HHV-6 DNA was detected in CSF of five (23%) of the HCT recipients with symptoms compared to only one (0.9%) of the controls without symptoms, although it should be noted that few of the controls in this study were HCT recipients.[67]

Few studies have investigated the viral dynamics in patients with HHV-6 encephalitis. In a study of 11 HCT recipients with HHV-6 detected by nucleic acid amplification in the CSF and serum (8 of whom had HHV-6 encephalitis) who were treated with foscarnet and/or ganciclovir, the concentration of HHV-6 in the CSF declined more slowly than in the serum.[61] This suggests that the inability to detect HHV-6 DNA in a blood specimen should not rule out HHV-6 encephalitis. Furthermore, it has been demonstrated in three patients who died with or after HHV-6 encephalitis that active HHV-6 infection can still be detected in the brain tissue even after HHV-6 DNA has become undetectable in the serum and CSF,[14] demonstrating that resolution of CSF HHV-6 DNA alone should not be used to shorten therapeutic courses for HHV-6 encephalitis.

Chromosomal Integration

Human herpesvirus 6 chromosomal integration, or inherited HHV-6, can present challenges when attempting to diagnose HHV-6-associated disease. Chromosomal integration occurs when HHV-6 is integrated into a germ cell of an individual, and the resulting offspring and subsequent generations have at least one copy of HHV-6 DNA in every cell of their body. Consequently, high levels of HHV-6 DNA are detectable in blood, CSF, and tissue samples of affected individuals. Population-based studies have estimated chromosomally integrated HHV-6 to be present in about 0.2 to 0.8% of people.[72,73] In the HCT setting, high levels of HHV-6 DNA are detectable in the blood of HCT recipients with integrated HHV-6 as well as in HCT recipients whose donors have integrated HHV-6.[74–76]

When the HCT donor has inherited HHV-6, HHV-6 levels will increase in the recipient with engraftment, and very high levels are often detected, particularly when whole blood is tested.[74,76] As an example, in one HCT recipient who received a transplant from a donor with chromosomal integration, persistent HHV-6 concentrations of 2.5×10^6 to 10×10^6 copies per mL were detected in whole blood following engraftment.[74] In another report, two HCT recipients who received transplants from donors with chromosomal integration had 6.1×10^6 and 9.7×10^5 copies per mL in the whole blood following engraftment but only 3600 and 15,400 copies per mL in the plasma, respectively.[76] Other features of chromosomal integration in donor cells are that HHV-6 DNA levels remain persistently and stably elevated and do not decline with time or antiviral therapy.[77] Consequently, serial testing and quantification of HHV-6 DNA copies per cell assayed can provide evidence of inherited HHV-6 if at least one copy of HHV-6 per cell assayed is consistently detected over time.

In contrast to the pattern observed when the HCT recipient receives a transplant from a donor with chromosomal integration with HHV-6, when the HCT recipient has inherited HHV-6, high levels of HHV-6 DNA can be detected in the blood prior to and immediately after transplantation, and then levels decrease with successful engraftment.

Distinguishing inherited HHV-6 from HHV-6-associated disease can be challenging, especially when dealing with a single positive result. This is true even when interpreting results from CSF specimens. As noted above, high levels of HHV-6 DNA can be detected in the CSF in individuals with chromosomal integration. For example, among 21 patients with presumed chromosomal integration, HHV-6 DNA was detected in the CSF with a mean level of 4.0 $\log_{10}$ copies per mL (95% CI, 3.5–4.5).[78] This was higher than the mean level detected in immunocompetent children with primary HHV-6 infection (2.4 $\log_{10}$ copies per mL, 95% CI, 1.0–3.7), but similar to levels detected in HCT recipients with HHV-6 encephalitis.

Despite these challenges, there are strategies to help distinguish the integrated state from reactivation. Fluorescent *in situ* hybridization (FISH) with a specific HHV-6 probe performed on metaphase chromosome preparations from peripheral blood will demonstrate integrated HHV-6.[74,79] In addition, in contrast to individuals with latent HHV-6 infection, individuals with integrated HHV-6 have detectable HHV-6 in their hair follicles.[80] Thus, if FISH is not available, nucleic acid amplification testing of hair follicles for HHV-6 DNA can be performed to identify an individual with integrated HHV-6. This would mean testing the donor's hair follicles in the case of an HCT recipient whose donor is suspected to have integrated HHV-6. When there is concern for integrated HHV-6 in the HCT donor cells, testing of donor serum or whole blood with HHV-6 nucleic acid amplification may provide supporting evidence for integrated HHV-6.[76,80] Both HHV-6A and HHV-6B can be integrated and inherited. Because the vast majority of reactivation events following HCT are due to HHV-6B, detection of HHV-6A can be a clue to chromosomally integrated HHV-6.[3,6,7]

Whether HCT recipients with integrated HHV-6 can also reactivate HHV-6 and suffer complications such as encephalitis or central nervous system dysfunction is unknown. A case of encephalitis occurring in an immunocompetent patient with integrated HHV-6 has been reported.[81] Although a thorough evaluation was performed and an alternative etiology was not identified, causality remains uncertain. In addition, another case report describes two siblings with integrated HHV-6 and cognitive dysfunction. Both patients appeared to respond to antiviral therapy.[82]

The possibility of chromosomally integrated or inherited HHV-6 should be considered in the setting of high and persistent levels of HHV-6 DNA in the peripheral blood, particularly if the patient has no clinical signs of encephalitis and/or the level of viremia does not decline following initiation of antiviral therapy with activity against HHV-6.

Approach to Diagnosis

Any allogeneic HCT recipient with clinical findings suggestive of HHV-6 encephalitis (confusion, anterograde amnesia, short-term memory loss, seizures) should undergo lumbar puncture and MRI. A peripheral blood sample and a CSF sample should be tested for HHV-6 using nucleic acid amplification in addition to routine CSF studies (e.g., cell count and differential, protein, glucose, Gram stain, aerobic culture). Other potential etiologies should be pursued. The diagnosis of HHV-6 encephalitis is established in allogeneic HCT recipients with the characteristic clinical findings in whom HHV-6 DNA is detected in the CSF without other causes identified.

Human herpesvirus 6 nucleic acid amplification from the peripheral blood can be helpful for establishing that HHV-6 reactivation has occurred and should be obtained when considering a diagnosis of HHV-6 encephalitis. However, detection of HHV-6 DNA in the blood alone is not specific for HHV-6 encephalitis, as only a small proportion of patients with HHV-6 reactivation develop encephalitis. In addition, HHV-6 viral load may become undetectable in the blood before it resolves from CSF.[61]

Testing a blood specimen can also be considered when determining the cause of other entities that have been associated with HHV-6, such as bone marrow suppression or pneumonitis. In these cases it is especially important to test for other entities that are known causes of these conditions.

TREATMENT

No antiviral agent has been approved by the U.S. Food and Drug Administration for the treatment of HHV-6 infection. However, *in vitro* studies demonstrate that foscarnet, ganciclovir, and cidofovir have antiviral activity against HHV-6.[83] There are also small case series suggesting a potential anti-HHV-6 effect of foscarnet and ganciclovir. Studies of patients with HHV-6 encephalitis demonstrate an association between foscarnet or ganciclovir treatment and a reduction of viral DNA levels in serum and CSF.[61] However, these studies did not have a comparison group of patients with HHV-6 encephalitis who were not treated with these antivirals. Although the optimal therapy is unknown, based on the available data foscarnet or ganciclovir is recommended for the treatment of HHV-6 encephalitis in HCT recipients.[84–86] Cidofovir has been proposed as a second-line agent, but it is avoided as a first-line agent because it is highly nephrotoxic.[84] Generally, full doses of foscarnet (180 mg/kg/day) or ganciclovir (10 mg/kg/day) should be used when treating HHV-6 encephalitis. There is no evidence to guide the optimal duration of therapy; however, many clinicians would plan for at least 3 weeks of antiviral therapy at full dose.

PREVENTION

Although small studies exploring the safety of preemptive or prophylactic antiviral therapy directed against HHV-6 in HCT patients have been conducted, large studies demonstrating the efficacy and safety of HHV-6 prophylaxis or preemptive therapy have not been performed. Therefore, there are insufficient data to recommend prophylactic antiviral strategies or preemptive monitoring and treatment aimed at reducing the risk of HHV-6-associated disease, such as HHV-6 encephalitis.[85]

Studies of recipients of solid organ transplants who received CMV-directed prophylactic ganciclovir suggest a possible effect on HHV-6.[87,88] Although ganciclovir prophylaxis was not associated with a decreased frequency of HHV-6 reactivation in a study of 134 renal transplant recipients, it was associated with delayed onset and shortened duration of HHV-6 detection.[87] Similarly, several small retrospective studies in HCT recipients suggest a potential benefit of prophylactic ganciclovir in reducing HHV-6 activity and its associated manifestations.[89–91]

Two very small (6 to 8 patients receiving drug) prospective studies have explored preemptive ganciclovir or foscarnet therapy in HCT recipients.[92,93] These studies highlight the challenge of initiating preemptive therapy after a viral threshold has been reached but prior to the onset of manifestations of HHV-6 disease in high-risk patients. In a study of prophylactic foscarnet in 10 HCT patients, the drug was generally well tolerated.[94]

A larger study of HCT recipients of unrelated bone marrow or cord blood stem cells compared consecutive cohorts of patients not receiving prophylaxis (cohort 1, $n=51$) with those receiving foscarnet prophylaxis (cohort 2, $n=67$).[95] The prophylactic regimen was 50 mg/kg/day of foscarnet for 10 days after engraftment. High-level reactivation was defined as the primary endpoint. No significant reduction of high-level reactivation (HHV-6 DNA $\geq 10^4$ copies per mL) by day 70 was seen in Cohort 2 (19.4%) compared with Cohort 1 (33.8%, $p=0.095$). A trend was identified toward fewer high-level HHV-6 reactivations in Cohort 2 among recipients of unrelated bone marrow transplantation ($p=0.067$), but no difference in incidence was observed among recipients of cord blood transplantations ($p=0.75$). Breakthrough HHV-6 encephalitis occurred following foscarnet prophylaxis in three patients, and the incidence of HHV-6 encephalitis did not differ between Cohort 1 (9.9%) and Cohort 2 (4.5%, $p=0.24$). In conclusion, 50 mg/kg/day of foscarnet for 10 days after engraftment did not effectively suppress HHV-6 reactivation and did not prevent all cases of HHV-6 encephalitis.

OUTCOMES

Outcomes of HHV-6 encephalitis have been difficult to evaluate because allogeneic HCT recipients often have complex comorbidities. It appears that outcomes vary widely, with some patients recovering full neurologic function and other patients being left with residual neurologic deficits. Similarly, some patients with HHV-6 encephalitis appear to have died from encephalitis, but many have died of other identifiable causes.[14–16,65,66]

In a review of case reports and case series of patients with HHV-6 encephalitis, information was provided on the neurologic course and outcomes in 44 cases.[17] Although difficult to determine with certainty, 11 patients (25%) had a progressive course and died within 1 to 4 weeks of diagnosis. Eight patients (18%) improved but were left with residual neurologic compromise. Nineteen patients (43%) made a full recovery, although for some, the recovery process lasted several weeks and required rehabilitation services. Six patients (14%) initially showed improvement but then succumbed to respiratory failure, multiorgan failure, or other documented infections. Whether these conditions were directly or indirectly a result of the HHV-6 infection remains unclear. Similar observations were made in a surveillance study of HHV-6 encephalitis in HCT recipients in Japan.[60]

Little information has been available about the long-term neurocognitive outcomes of HHV-6 encephalitis in survivors. A retrospective study showed that four of five surviving patients were unable to return to society because of neuropsychological disorders at the end of follow-up, including anterograde amnesia and seizures.[96] In that report, prominent hippocampal atrophy in the late phase was demonstrated on MRI.

CONCLUSION

Human herpesvirus 6B commonly reactivates after HCT and has been associated with a number of important outcomes. The evidence supports a causal association between HHV-6B and encephalitis. In fact, HHV-6 appears to be the most common cause of encephalitis following HCT. HCT recipients with signs and symptoms of encephalitis should have CSF tested for HHV-6, and empiric therapy with foscarnet or ganciclovir should be initiated. Detection of HHV-6 in the CSF in the absence of other causes of encephalitis should lead to a definitive course of therapy. Further research is needed to establish whether other outcomes are causally associated with HHV-6 and if so the pathogenesis of those relationships.

REFERENCES

1. Ljungman P, Wang FZ, Clark DA, et al. High levels of human herpesvirus 6 DNA in peripheral blood leucocytes are correlated to platelet engraftment and disease in allogeneic stem cell transplant patients. *Br J Haematol* 2000;**111**:774–81.
2. Yoshikawa T, Asano Y, Ihira M, et al. Human herpesvirus 6 viremia in bone marrow transplant recipients: clinical features and risk factors. *J Infect Dis* 2002;**185**:847–53.
3. Zerr DM, Corey L, Kim HW, Huang ML, Nguy L, Boeckh M. Clinical outcomes of human herpesvirus 6 reactivation after hematopoietic stem cell transplantation. *Clin Infect Dis* 2005;**40**:932–40.
4. Hentrich M, Oruzio D, Jager G, et al. Impact of human herpesvirus-6 after haematopoietic stem cell transplantation. *Br J Haematol* 2005;**128**:66–72.
5. Ogata M, Kikuchi H, Satou T, et al. Human herpesvirus 6 DNA in plasma after allogeneic stem cell transplantation: incidence and clinical significance. *J Infect Dis* 2006;**193**:68–79.
6. Reddy S, Manna P. Quantitative detection and differentiation of human herpesvirus 6 subtypes in bone marrow transplant patients by using a single real-time polymerase chain reaction assay. *Biol Blood Marrow Transplant* 2005;**11**:530–41.
7. Wang LR, Dong LJ, Zhang MJ, Lu DP. The impact of human herpesvirus 6B reactivation on early complications following allogeneic hematopoietic stem cell transplantation. *Biol Blood Marrow Transplant* 2006;**12**:1031–7.
8. Yamane A, Mori T, Suzuki S, et al. Risk factors for developing human herpesvirus 6 (HHV-6) reactivation after allogeneic hematopoietic stem cell transplantation and its association with central nervous system disorders. *Biol Blood Marrow Transplant* 2007;**13**:100–6.

9. Zerr DM, Fann JR, Breiger D, et al. HHV-6 reactivation and its effect on delirium and cognitive functioning in hematopoietic cell transplantation recipients. *Blood* 2011;**117**:5243–9.
10. Chevallier P, Hebia-Fellah I, Planche L, et al. Human herpes virus 6 infection is a hallmark of cord blood transplant in adults and may participate to delayed engraftment: a comparison with matched unrelated donors as stem cell source. *Bone Marrow Transplant* 2010;**45**:1204–11.
11. Sashihara J, Tanaka-Taya K, Tanaka S, et al. High incidence of human herpesvirus 6 infection with a high viral load in cord blood stem cell transplant recipients. *Blood* 2002;**100**:2005–11.
12. Hill AB. The environment and disease: association or causation?. *Proc R Soc Med* 1965;**58**:295–300.
13. Weed DL. On the use of causal criteria. *Int J Epidemiol* 1997;**26**:1137–41.
14. Fotheringham J, Akhyani N, Vortmeyer A, et al. Detection of active human herpesvirus-6 infection in the brain: correlation with polymerase chain reaction detection in cerebrospinal fluid. *J Infect Dis* 2007;**195**:450–4.
15. Seeley WW, Marty FM, Holmes TM, et al. Post-transplant acute limbic encephalitis: clinical features and relationship to HHV6. *Neurology* 2007;**69**:156–65.
16. Wainwright MS, Martin PL, Morse RP, et al. Human herpesvirus 6 limbic encephalitis after stem cell transplantation. *Ann Neurol* 2001;**50**:612–9.
17. Zerr DM. Human herpesvirus 6 and central nervous system disease in hematopoietic cell transplantation. *J Clin Virol* 2006;**37**(Suppl. 1):S52–6.
18. Ogata M, Satou T, Kawano R, et al. Correlations of HHV-6 viral load and plasma IL-6 concentration with HHV-6 encephalitis in allogeneic stem cell transplant recipients. *Bone Marrow Transplant* 2010;**45**:129–36.
19. Hill JA, Koo S, Guzman Suarez BB, et al. Cord-blood hematopoietic stem cell transplant confers an increased risk for human herpesvirus-6-associated acute limbic encephalitis: a cohort analysis. *Biol Blood Marrow Transplant* 2012;**18**(11):1638–48.
20. Schmidt-Hieber M, Schwender J, Heinz WJ, et al. Viral encephalitis after allogeneic stem cell transplantation: a rare complication with distinct characteristics of different causative agents. *Haematologica* 2011;**96**:142–9.
21. Drobyski WR, Dunne WM, Burd EM, et al. Human herpesvirus-6 (HHV-6) infection in allogeneic bone marrow transplant recipients: evidence of a marrow-suppressive role for HHV-6 *in vivo*. *J Infect Dis* 1993;**167**:735–9.
22. Imbert-Marcille BM, Tang XW, Lepelletier D, et al. Human herpesvirus 6 infection after autologous or allogeneic stem cell transplantation: a single-center prospective longitudinal study of 92 patients. *Clin Infect Dis* 2000;**31**:881–6.
23. Maeda Y, Teshima T, Yamada M, et al. Monitoring of human herpesviruses after allogeneic peripheral blood stem cell transplantation and bone marrow transplantation. *Br J Haematol* 1999;**105**:295–302.
24. Savolainen H, Lautenschlager I, Piiparinen H, Saarinen-Pihkala U, Hovi L, Vettenranta K. Human herpesvirus-6 and -7 in pediatric stem cell transplantation. *Pediatr Blood Cancer* 2005;**45**:820–5.
25. Wang FZ, Dahl H, Linde A, Brytting M, Ehrnst A, Ljungman P. Lymphotropic herpesviruses in allogeneic bone marrow transplantation. *Blood* 1996;**88**:3615–20.
26. Johnston RE, Geretti AM, Prentice HG, et al. HHV-6-related secondary graft failure following allogeneic bone marrow transplantation. *Br J Haematol* 1999;**105**:1041–3.
27. Lagadinou ED, Marangos M, Liga M, et al. Human herpesvirus 6-related pure red cell aplasia, secondary graft failure, and clinical severe immune suppression after allogeneic

hematopoietic cell transplantation successfully treated with foscarnet. *Transpl Infect Dis* 2010;**12**:437–40.

28. Luppi M, Barozzi P, Morris C, et al. Human herpesvirus 6 latently infects early bone marrow progenitors *in vivo*. *J Virol* 1999;**73**:754–9.
29. Isomura H, Yamada M, Yoshida M, et al. Suppressive effects of human herpesvirus 6 on *in vitro* colony formation of hematopoietic progenitor cells. *J Med Virol* 1997;**52**:406–12.
30. Isomura H, Yoshida M, Namba H, et al. Suppressive effects of human herpesvirus-6 on thrombopoietin-inducible megakaryocytic colony formation *in vitro*. *J Gen Virol* 2000;**81**:663–73.
31. Appleton AL, Sviland L, Peiris JS, et al. Human herpes virus-6 infection in marrow graft recipients: role in pathogenesis of graft-versus-host disease. Newcastle Upon Tyne Bone Marrow Transport Group. *Bone Marrow Transplant* 1995;**16**:777–82.
32. de Pagter PJ, Schuurman R, Visscher H, et al. Human herpes virus 6 plasma DNA positivity after hematopoietic stem cell transplantation in children: an important risk factor for clinical outcome. *Biol Blood Marrow Transplant* 2008;**14**:831–9.
33. Wang LR, Dong LJ, Zhang MJ, Lu DP. Correlations of human herpesvirus 6B and CMV infection with acute GVHD in recipients of allogeneic haematopoietic stem cell transplantation. *Bone Marrow Transplant* 2008;**42**:673–7.
34. Dulery R, Salleron J, Dewilde A, et al. Early human herpesvirus type 6 reactivation after allogeneic stem cell transplantation: a large-scale clinical study. *Biol Blood Marrow Transplant* 2012;**18**:1080–9.
35. Zerr DM, Boeckh M, Delaney C, et al. HHV-6 reactivation and associated sequelae after hematopoietic cell transplant. *Biol Blood Marrow Transplant* 2012;**18**(11):1700–8.
36. Tormo N, Solano C, de la Camara R, et al. An assessment of the effect of human herpesvirus-6 replication on active cytomegalovirus infection after allogeneic stem cell transplantation. *Biol Blood Marrow Transplant* 2010;**16**:653–61.
37. Wang FZ, Larsson K, Linde A, Ljungman P. Human herpesvirus 6 infection and cytomegalovirus-specific lymphoproliferative responses in allogeneic stem cell transplant recipients. *Bone Marrow Transplant* 2002;**30**:521–6.
38. Cone RW, Hackman RC, Huang ML, et al. Human herpesvirus 6 in lung tissue from patients with pneumonitis after bone marrow transplantation. *N Engl J Med* 1993;**329**:156–61.
39. Carrigan DR, Drobyski WR, Russler SK, Tapper MA, Knox KK, Ash RC. Interstitial pneumonitis associated with human herpesvirus-6 infection after marrow transplantation. *Lancet* 1991;**338**:147–9.
40. Cone RW, Huang ML, Hackman RC. Human herpesvirus 6 and pneumonia. *Leuk Lymphoma* 1994;**15**:235–41.
41. Carrigan DR. Human herpesvirus-6 and bone marrow transplantation. In: Ablashi DV, Krueger GRF, Salahuddin SZ, editors. *Human Herpesvirus-6: Epidemiology, Molecular Biology, and Clinical Pathology*. Amsterdam: Elsevier; 1992. p. 281–97.
42. Vuorinen T, Kotilainen P, Lautenschlager I, Kujari H, Krogerus L, Oksi J. Interstitial pneumonitis and coinfection of human herpesvirus 6 and *Pneumocystis carinii* in a patient with hypogammaglobulinemia. *J Clin Microbiol* 2004;**42**:5415–8.
43. Amo K, Tanaka-Taya K, Inagi R, et al. Human herpesvirus 6B infection of the large intestine of patients with diarrhea. *Clin Infect Dis* 2003;**36**:120–3.
44. Mousset S, Martin H, Berger A, et al. Human herpesvirus 6 in biopsies from patients with gastrointestinal symptoms after allogeneic stem cell transplantation. *Ann Hematol* 2012;**91**: 737–42.
45. Tajiri H, Tanaka-Taya K, Ozaki Y, Okada S, Mushiake S, Yamanishi K. Chronic hepatitis in an infant, in association with human herpesvirus-6 infection. *J Pediatr* 1997;**131**:473–5.

46. Asano Y, Yoshikawa T, Suga S, Yazaki T, Kondo K, Yamanishi K. Fatal fulminant hepatitis in an infant with human herpesvirus-6 infection. *Lancet* 1990;**335**:862–3.
47. Kuribayashi K, Matsunaga T, Iyama S, et al. Human herpesvirus-6 hepatitis associated with cyclosporine-A encephalitis after bone marrow transplantation for chronic myeloid leukemia. *Intern Med* 2006;**45**:475–8.
48. Kondo K, Kondo T, Okuno T, Takahashi M, Yamanishi K. Latent human herpesvirus 6 infection of human monocytes/macrophages. *J Gen Virol* 1991;**72**(Pt 6):1401–8.
49. Kondo K, Kondo T, Shimada K, Amo K, Miyagawa H, Yamanishi K. Strong interaction between human herpesvirus 6 and peripheral blood monocytes/macrophages during acute infection. *J Med Virol* 2002;**67**:364–9.
50. He J, McCarthy M, Zhou Y, Chandran B, Wood C. Infection of primary human fetal astrocytes by human herpesvirus 6. *J Virol* 1996;**70**:1296–300.
51. Lusso P. HHV-6 and the immune system: mechanisms of immunomodulation and viral escape. *J Clin Virol* 2006;**37**(Suppl. 1):S4–10.
52. Smith A, Santoro F, Di Lullo G, Dagna L, Verani A, Lusso P. Selective suppression of IL-12 production by human herpesvirus 6. *Blood* 2003;**102**:2877–84.
53. Smith AP, Paolucci C, Di Lullo G, Burastero SE, Santoro F, Lusso P. Viral replication-independent blockade of dendritic cell maturation and interleukin-12 production by human herpesvirus 6. *J Virol* 2005;**79**:2807–13.
54. Fujita A, Ihira M, Suzuki R, et al. Elevated serum cytokine levels are associated with human herpesvirus 6 reactivation in hematopoietic stem cell transplantation recipients. *J Infect* 2008;**57**:241–8.
55. Mayne M, Cheadle C, Soldan SS, et al. Gene expression profile of herpesvirus-infected T cells obtained using immunomicroarrays: induction of proinflammatory mechanisms. *J Virol* 2001;**75**:11641–50.
56. Fujii N, Hiraki A, Aoe K, et al. Serum cytokine concentrations and acute graft-versus-host disease after allogeneic peripheral blood stem cell transplantation: concurrent measurement of ten cytokines and their respective ratios using cytometric bead array. *Int J Mol Med* 2006;**17**:881–5.
57. Ju XP, Xu B, Xiao ZP, et al. Cytokine expression during acute graft-versus-host disease after allogeneic peripheral stem cell transplantation. *Bone Marrow Transplant* 2005;**35**:1179–86.
58. Mohty M, Blaise D, Faucher C, et al. Inflammatory cytokines and acute graft-versus-host disease after reduced-intensity conditioning allogeneic stem cell transplantation. *Blood* 2005;**106**:4407–11.
59. Shaiegan M, Iravani M, Babaee GR, Ghavamzadeh A. Effect of IL-18 and sIL2R on aGVHD occurrence after hematopoietic stem cell transplantation in some Iranian patients. *Transpl Immunol* 2006;**15**:223–7.
60. Muta T, Fukuda T, Harada M. Human herpesvirus-6 encephalitis in hematopoietic SCT recipients in Japan: a retrospective multicenter study. *Bone Marrow Transplant* 2009;**43**:583–5.
61. Zerr DM, Gupta D, Huang ML, Carter R, Corey L. Effect of antivirals on human herpesvirus 6 replication in hematopoietic stem cell transplant recipients. *Clin Infect Dis* 2002;**34**:309–17.
62. Tiacci E, Luppi M, Barozzi P, et al. Fatal herpesvirus-6 encephalitis in a recipient of a T-cell-depleted peripheral blood stem cell transplant from a 3–loci mismatched related donor. *Haematologica* 2000;**85**:94–7.
63. Noguchi T, Mihara F, Yoshiura T, et al. MR imaging of human herpesvirus-6 encephalopathy after hematopoietic stem cell transplantation in adults. *AJNR Am J Neuroradiol* 2006;**27**:2191–5.

64. Provenzale JM, van Landingham KE, Lewis DV, Mukundan Jr. S, White LE. Extra-hippocampal involvement in human herpesvirus 6 encephalitis depicted at MR imaging. *Radiology* 2008;**249**:955–63.
65. De Almeida Rodrigues G, Nagendra S, Lee CK, De Magalhaes-Silverman M. Human herpes virus 6 fatal encephalitis in a bone marrow recipient. *Scand J Infect Dis* 1999;**31**:313–5.
66. Drobyski WR, Knox KK, Majewski D, Carrigan DR. Brief report: fatal encephalitis due to variant B human herpesvirus-6 infection in a bone marrow-transplant recipient. *N Engl J Med* 1994;**330**:1356–60.
67. Wang FZ, Linde A, Hagglund H, Testa M, Locasciulli A, Ljungman P. Human herpesvirus 6 DNA in cerebrospinal fluid specimens from allogeneic bone marrow transplant patients: does it have clinical significance?. *Clin Infect Dis* 1999;**28**:562–8.
68. Huang LM, Kuo PF, Lee CY, Chen JY, Liu MY, Yang CS. Detection of human herpesvirus-6 DNA by polymerase chain reaction in serum or plasma. *J Med Virol* 1992;**38**:7–10.
69. Yoshikawa T, Ihira M, Suzuki K, et al. Human herpesvirus 6 infection after living related liver transplantation. *J Med Virol* 2000;**62**:52–9.
70. Betts BC, Young JA, Ustun C, Cao Q, Weisdorf DJ. Human herpesvirus 6 infection after hematopoietic cell transplantation: is routine surveillance necessary?. *Biol Blood Marrow Transplant* 2011;**17**:1562–8.
71. Hall CB, Caserta MT, Schnabel KC, et al. Persistence of human herpesvirus 6 according to site and variant: possible greater neurotropism of variant A. *Clin Infect Dis* 1998;**26**:132–7.
72. Tanaka-Taya K, Sashihara J, Kurahashi H, et al. Human herpesvirus 6 (HHV-6) is transmitted from parent to child in an integrated form and characterization of cases with chromosomally integrated HHV-6 DNA. *J Med Virol* 2004;**73**:465–73.
73. Leong HN, Tuke PW, Tedder RS, et al. The prevalence of chromosomally integrated human herpesvirus 6 genomes in the blood of UK blood donors. *J Med Virol* 2007;**79**:45–51.
74. Clark DA, Nacheva EP, Leong HN, et al. Transmission of integrated human herpesvirus 6 through stem cell transplantation: implications for laboratory diagnosis. *J Infect Dis* 2006;**193**:912–6.
75. Jeulin H, Guery M, Clement L, et al. Chromosomally integrated HHV-6: slow decrease of HHV-6 viral load after hematopoietic stem-cell transplantation. *Transplantation* 2009;**88**:1142–3.
76. Kamble RT, Clark DA, Leong HN, Heslop HE, Brenner MK, Carrum G. Transmission of integrated human herpesvirus-6 in allogeneic hematopoietic stem cell transplantation. *Bone Marrow Transplant* 2007;**40**:563–6.
77. Jeulin H, Salmon A, Gautheret-Dejean A, et al. Contribution of human herpesvirus 6 (HHV-6) viral load in whole blood and serum to investigate integrated HHV-6 transmission after haematopoietic stem cell transplantation. *J Clin Virol* 2009;**45**:43–6.
78. Ward KN, Leong HN, Thiruchelvam AD, Atkinson CE, Clark DA. Human herpesvirus 6 DNA levels in cerebrospinal fluid due to primary infection differ from those due to chromosomal viral integration and have implications for diagnosis of encephalitis. *J Clin Microbiol* 2007;**45**:1298–304.
79. Daibata M, Taguchi T, Nemoto Y, Taguchi H, Miyoshi I. Inheritance of chromosomally integrated human herpesvirus 6 DNA. *Blood* 1999;**94**:1545–9.
80. Ward KN, Leong HN, Nacheva EP, et al. Human herpesvirus 6 chromosomal integration in immunocompetent patients results in high levels of viral DNA in blood, sera, and hair follicles. *J Clin Microbiol* 2006;**44**:1571–4.
81. Troy SB, Blackburn BG, Yeom K, Caulfield AK, Bhangoo MS, Montoya JG. Severe encephalomyelitis in an immunocompetent adult with chromosomally integrated human herpesvirus 6 and clinical response to treatment with foscarnet plus ganciclovir. *Clin Infect Dis* 2008;**47**:e93–6.

82. Montoya JG, Neely MN, Gupta S, et al. Antiviral therapy of two patients with chromosomally-integrated human herpesvirus-6A presenting with cognitive dysfunction. *J Clin Virol* 2012;**55**(1):40–5.
83. De Bolle L, Naesens L, De Clercq E. Update on human herpesvirus 6 biology, clinical features, and therapy. *Clin Microbiol Rev* 2005;**18**:217–45.
84. Dewhurst S. Human herpesvirus type 6 and human herpesvirus type 7 infections of the central nervous system. *Herpes* 2004;**11**(Suppl. 2):105A–111AA.
85. Tomblyn M, Chiller T, Einsele H, et al. Guidelines for preventing infectious complications among hematopoietic cell transplantation recipients: a global perspective. *Biol Blood Marrow Transplant* 2009;**15**:1143–238.
86. Tunkel AR, Glaser CA, Bloch KC, et al. The management of encephalitis: clinical practice guidelines by the Infectious Diseases Society of America. *Clin Infect Dis* 2008;**47**:303–27.
87. Galarraga MC, Gomez E, de Ona M, et al. Influence of ganciclovir prophylaxis on cytomegalovirus, human herpesvirus 6, and human herpesvirus 7 viremia in renal transplant recipients. *Transplant Proc* 2005;**37**:2124–6.
88. Razonable RR, Brown RA, Humar A, Covington E, Alecock E, Paya CV. Herpesvirus infections in solid organ transplant patients at high risk of primary cytomegalovirus disease. *J Infect Dis* 2005;**192**:1331–9.
89. Cheng FW, Lee V, Leung WK, et al. HHV-6 encephalitis in pediatric unrelated umbilical cord transplantation: a role for ganciclovir prophylaxis?. *Pediatr Transplant* 2010;**14**:483–7.
90. Rapaport D, Engelhard D, Tagger G, Or R, Frenkel N. Antiviral prophylaxis may prevent human herpesvirus-6 reactivation in bone marrow transplant recipients. *Transpl Infect Dis* 2002;**4**:10–16.
91. Tokimasa S, Hara J, Osugi Y, et al. Ganciclovir is effective for prophylaxis and treatment of human herpesvirus-6 in allogeneic stem cell transplantation. *Bone Marrow Transplant* 2002;**29**:595–8.
92. Ishiyama K, Katagiri T, Hoshino T, Yoshida T, Yamaguchi M, Nakao S. Preemptive therapy of human herpesvirus-6 encephalitis with foscarnet sodium for high-risk patients after hematopoietic SCT. *Bone Marrow Transplant* 2011;**46**:863–9.
93. Ogata M, Satou T, Kawano R, et al. Plasma HHV-6 viral load-guided preemptive therapy against HHV-6 encephalopathy after allogeneic stem cell transplantation: a prospective evaluation. *Bone Marrow Transplant* 2008;**41**:279–85.
94. Ishiyama K, Katagiri T, Ohata K, et al. Safety of pre-engraftment prophylactic foscarnet administration after allogeneic stem cell transplantation. *Transpl Infect Dis* 2012;**14**:33–9.
95. Ogata M, Satou T, Inoue Y, et al. Foscarnet against human herpesvirus (HHV)-6 reactivation after allo-SCT: breakthrough HHV-6 encephalitis following antiviral prophylaxis. *Bone Marrow Transplant* 2013;**48**(2):257–64.
96. Sakai R, Kanamori H, Motohashi K, et al. Long-term outcome of human herpesvirus-6 encephalitis after allogeneic stem cell transplantation. *Biol Blood Marrow Transplant* 2011;**17**:1389–94.

Chapter 14

The Immune Response to HHV-6

J. Mauricio Calvo-Calle and Lawrence J. Stern
University of Massachusetts Medical School, Worcester, Massachusetts

INTRODUCTION

The human herpesvirus 6 was discovered in 1986 by Dharam Ablashi, Robert Gallo, and Zaki Salahuddin.[1] In subsequent years, two closely related variants were recognized and designated as HHV-6A and HHV-6B. In 2011, these two variants were reclassified as distinct virus species in the genus *Roseolavirus* of the subfamily Betaherpesvirinae. In this review, HHV-6 will be used to refer collectively to these viruses. The specific virus species will be indicated whenever possible. The group of HHV-6 viruses is characterized by its immunosuppressive and immunomodulatory functions. Although recently there have been advances in the identification of some HHV-6 genes implicated in immunomodulation, very little is known about the underlying mechanisms. During the past few years, significant advances have been made in understanding both the innate and adaptive immune responses to HHV-6, which are the focus of this review. However, as we will discuss later, many responses to HHV-6 antigens can barely be detected *ex vivo* in peripheral blood samples, and more studies are necessary to acquire a basic understanding of the immunodulatory process.

HHV-6 AND INNATE IMMUNE RESPONSES

Herpesviruses elicit innate immune responses that control viral replication before adaptive responses are mounted.[2] Innate immune responses are elicited by pattern recognition receptors that sense the presence of microbial molecular signatures. Upon ligand binding, the receptor relays signals that lead to secretion of cytokines such as type I interferon (IFN). For HHV-6, some specific recognition pathways have been identified.

Human Herpesviruses HHV-6A, HHV-6B & HHV-7.

Toll-Like Receptors and Type I Interferon Responses

The expression and signaling of Toll-like receptors (TLRs) in HHV-6-infected cells have been investigated in two studies, one in CD4+ T cells and one in dendritic cells (DCs). In HHV-6A-infected CD4+ T cells, TLR-9 was found to be upregulated with concomitant signal transduction and apoptosis.[3] Although the expression of TLR1-10 was not altered in DCs infected with HHV-6B, cytokine responses associated with stimulation of these receptors were impaired.[4] Production of type I IFN is central to innate immune responses, and pathways responsible are frequently targeted by pathogens. Early studies demonstrated that both HHV-6 species were sensitive to the type I IFN antiviral effects when the cytokine was added to cultures before HHV-6 inoculation.[5,6] However, when cells were infected before exposure to type I IFN, only HHV-6B was able to overcome the antiviral effects of type I IFN.[7] In a series of elegant experiments, Jaworska et al.[7] demonstrated that immediate-early 1 (IE1) protein from HHV-6B is a potent inhibitor of IFN-stimulated gene (ISG) activation. ISG inhibition was mapped to 41 amino acids in the C-terminal segment of IE1 that differs between HHV-6A and HHV-6B, which inhibits ISG activation by sequestrating signal transducer and activator of transcription 2 (STAT-2).[7]

NK Cell Responses

Natural killer (NK) cells are a key component of the innate immune response and play a major role in the control of tumors and microbial infections. Patients with deficient NK-cell responses are highly susceptible to herpesviruses.[8] NK cells respond quickly to challenge, and their activity is greatly enhanced by cytokines. Type I IFN and interleukin 12 (IL-12) support NK effector functions, and IL-15 supports NK cell expansion.

Children undergoing primary HHV-6B infection have NK-cell responses that peak in the acute phase and continuously decline toward the convalescence phase, in which the virus load falls and T-cell responses become prominent.[6,9] These findings suggest that NK cell responses are key in controlling the HHV-6B expansion during the early stage of primary infection.[9] An *in vitro* study has shown that NK-cell cytotoxicity toward HHV-6A-infected peripheral blood mononuclear cells (PBMCs) was mediated by IL-15, produced presumably by monocytes and NK cells.[10] However, IL-15 also stimulated synthesis of IFN-γ by CD4+ T cells and by NK cells[11] and also rendered monocytes more susceptible to HHV-6 infection.[12] NK cell activity was enhanced in cultures supplemented with IL-2 and IL-12.[13] Little is known about the mechanisms used by NK cells to eliminate HHV-6B-infected cells. It was suggested that the killing of HHV-6A infected cells by NK cells did not require a complete absence of MHC class I, but instead depended on the expression of NK ligands on the infected cells.[14] Human and murine cytomegaloviruses, other chronic viruses in the same betaherpesvirus family as HHV-6, are known to encode several proteins that either act as ligands for NK receptors or modulate the surface expression of endogenous NK ligands,[15] but whether HHV-6 encodes such proteins is not known.

Modulation of the Chemokine Network by HHV-6

The chemokine network is crucial in activating and regulating innate and adaptive immune responses. This network consists of chemokines and G-protein-coupled receptors that mediate cell recruitment, activation, and other processes. HHV-6 has evolved two modulators of the chemokine network, a secreted virus-encoded chemokine U83, and two virus-encoded chemokine receptors, U12 and U51. HHV-6 U83 was initially found in the media of cells infected with HHV-6B and designated as U83B. U83B is a chemoattractant for CCR2+ CD14^{hi} CD16-(classical) monocytes.[16,17] Analysis of a U83A protein product, the HHV-6A homolog of U83B, has shown that this protein has two versions: a shorter version called U83A-N that is produced early after cell infection and a full version produced later. The short version is an antagonist that inhibits chemotaxis mediated by several receptors including CCR5.[18,19] The long version binds to CCR5 with affinities higher than the human ligands for this receptor. These findings suggest that early in infection HHV-6 might be attracting classical monocytes (CCR2+) that have low antimicrobial properties and inhibiting migration of nonclassical and intermediate monocytes (CCR5+) that have high antimicrobial properties. These observations also suggest that U83 might contribute to viral dissemination late during infection.[16–18]

The genome of HHV-6 viruses encodes two chemokine receptors, U12 and U51. U12 is a late product that encodes for a calcium-mobilizing receptor that induces calcium mobilizations and binds RANTES, macrophage inflammatory protein 1α (MIP-1α) and MIP-1β, and monocyte chemoattractant protein 1 (MCP-1). U51A (HHV-6A, U51) binds five inflammatory chemokines with similar affinities to their human receptors.[20] U51A acts as a constitutive relay signal[21] and probably enhances cell fusion.[22] Binding of U51A to the human chemokines XCL1 and CCL-19 suggests that this receptor might participate in the mobilization of infected cells to sites of high cytokine production—for example, lymph nodes where the viruses will be in close proximity to their preferred host cells, the T cells. HHV-6 also induces downregulation of the host chemokine receptor CXCR4 affecting chemotaxis.[23] CXCR4 is used by HIV to infect CD4+ T cells; therefore, it has been suggested that HHV-6 could protect T cells from HIV infection.[24]

ADAPTIVE IMMUNE RESPONSE TO HHV-6

Adaptive immune responses to herpesviruses are critical in the control of virus replication during primary infection and reactivation. Characterization of these responses has allowed the development of immunotherapies for human cytomegalovirus (HCMV) and Epstein–Barr virus (EBV).[25,26] By comparison with these viruses, the immune response to HHV-6 is poorly characterized. This has hindered the development of immunotherapies for HHV-6 reactivation and also our understanding of the pathogenesis of HHV-6 in human diseases in general. Characterization of the T-cell response to HHV-6 is complicated by its tropism for CD4+ T cells[27,28] and by its cytopathic effects on infected cells.[29–31] In addition, the almost universal exposure of humans to HHV-6 precludes access to nonexposed donors required to determine basal responses in noninfected donors.

Antibody Response to HHV-6

The human antibody response to HHV-6 has been the subject of a previous review by Wang and Pellett.[32] Here we will describe briefly major features of the antibody response and recent advances in the development of a HHV-6 variant-specific determination. Primary HHV-6B infection induces production of IgM antibodies that reach a peak by the third week and is followed by the production of low-avidity IgG antibodies that increase in affinity over the years (reviewed in Braun et al.[33]). Antibodies to HHV-6 have been studied mainly by indirect immunofluorescence assay (IFA) and by western blot (WB). Although IFA is considered a gold standard, at present this method cannot differentiate HHV-6A and HHV-6B species-specific antibody responses. WB studies have demonstrated antibody responses to at least 30 different bands.[34] A major target of the antibody response revealed by western blot is U11,[35] a major virion tegument phosphoprotein that constitutes almost 25% of the total mass viral particle (Becerra-Artiles et al., in preparation). Because all individuals positive by IFA to HHV-6 also have antibody responses by WB to U11, antibody response to U11 has been considered a primary candidate for a HHV-6 serological marker (reviewed in Wang et al.). Studies of antibody and cellular responses indicate that HHV-6 shares antigenic determinants with its close relative HHV-7; therefore, a diagnostic method must also discriminate between HHV-6 and HHV-7. Studies by Yamamoto et al.[36] and Black et al.[35] have suggested the presence of HHV-6 and HHV-7 virus-specific antibody determinants in the respective U11 homologs.

Two recent reports have described progress toward a HHV-6 species-specific WB method based on recognition of U11. The first of these reports, by Thäder-Voigt et al.,[37] used a microblot that includes HHV-6A and HHV-6B versions of recombinant fragments of U11, U47, and glycoproteins B and L. The second report, by Higashimoto et al.,[38] used only recombinant versions of U11 produced in prokaryotic expression systems in a regular western blot format. Both methods revealed species-specific antibody responses to HHV-6B using serum samples from children. These studies also suggest a higher seroprevalence of HHV-6B; however, a lack of serum samples with primary HHV-6A impaired reaching conclusions about the specificity of the HHV-6A reported responses.

Modulation of T-Cell Responses

The first published study of the T-cell response to HHV-6 showed robust proliferative responses to UV-irradiated virus in PBMCs in most healthy seropositive adults;[39] however, a second study seemed to contradict the first report and found that HHV-6 actually induced inhibition of antigen-specific proliferative responses.[29,40] Subsequent studies have corroborated both inferences. Today it is known that inhibition of T-cell proliferation includes effects in T cells as well as in antigen-presenting cells (APCs) and is not limited to HHV-6-infected cells but also includes effects in bystander cells.

Direct Modulation in Infected Cells

Cell proliferation and effector functions of T cells are significantly affected by HHV-6 infection. HHV-6A induces marked T-cell receptor (TCR) downregulation,[41,42] apoptosis,[43] inhibition of IL-2 synthesis,[29] and cell-cycle arrest.[31,44] HHV-6B induces a slight decrease in CD3 levels,[41] apoptosis,[43] and cell-cycle arrest.[44] Apoptosis seems to occur primary in noninfected cells.[43] Professional antigen-presenting cells are severely affected by HHV-6 infection. Studies in infected monocytes have demonstrated that both HHV-6A and HHV-6B viruses induced a pronounced inhibition of interleukin-12p70 production without affecting IFN/lipopolysaccharide (LPS) signaling.[45] Monocyte-derived DCs infected with HHV-6B have reduced presentation of exogenous antigens;[46] however, monocyte-derived DCs infected with HHV-6B were able to respond to LPS and produce large quantities of IL-10 and IL-12p70.[47] As expected, DCs infected with either HHV-6A or HHV-6B had reduced ability to trigger T cells.[45] HHV-6B-infected monocytes showed decreased levels of CD14 (LPS receptor), CD64 (Fc receptor), and major histocompatibility complex (MHC) class II molecules and were less effective in eliciting T-cell responses.[48]

Indirect Modulation in Noninfected Cells

Inhibition of T-cell proliferation by HHV-6 has been attributed mainly to IL-10 that is secreted into the cultures upon priming of CD4+ T cells from seropositive but not seronegative donors.[49] The induction of IL-10 production did not require virus replication and was observed with both HHV-6 viruses.[40,49] IL-10 is known to impair antigen presentation, T-cell development, and differentiation, in addition to modulating production of proinflammatory cytokines and chemokines. However, IL-10 also promotes immune responses by promoting proliferation of B cells and granulocytes and by activating NK and CD8+ T cells. HCMV and EBV encode IL-10-like virokines that bind human IL-10 receptor and play important roles in viral persistence.[50] However, in contrast to HCMV and EBV, HHV-6 does not appear to express a virally encoded IL-10 analog, so HHV-6-stimulated IL-10 responses would appear to be due to induction of host IL-10 responses.

Interleukin 10 can be produced by a variety of cell types, including monocytes, macrophages, DCs, B cells, and T cells.[51] Although CD4+ T cells have been considered an important source of IL-10, the context of the response dictates the major IL-10 cell source in a response.[52–54] Little is known about the source of IL-10 in cultures of cells infected with HHV-6. Production of IL-10 and MCP-1 by HHV-6-infected monocytes has been reported.[55] A likely source of IL-10 in PBMCs challenged with HHV-6 are regulatory T cells (Tregs) or Type 1 regulatory T cells (Tr1).[52] Tregs express the transcription factor Foxp3 and have been shown to exhibit suppressor function both *in vitro* and *in vivo*. Tr1 cells are a "functional" regulatory CD4+ T-cell population that lacks specific markers but is characterized by a high level of IL-10 secreted upon co-engagement of CD46 and TCR or by vitamin D3 and dexamethasone. Because HHV-6 uses CD46 as a cell receptor, several researchers have suggested a role of Tr1 in the HHV-6-mediated IL-10 induction. Induction of IL-10 producing Tr1-like CD4+ T cells has already been reported for *Streptococcus* using whole bacteria or just M protein, the bacterial ligand for CD46.[56] However, to date, neither Foxp3 positive cells nor Tr1 cells have been reported following HHV-6 infection.

T-Cell Responses to HHV-6

Studies of T-cell response to HHV-6 are limited to humans and close relatives because of the highly restricted host range of this virus. Rodent cells are not productively infected,[57] and even if virus does enter the cell viral replication is very limited.[58] In humans, some studies have reported primary responses during infection in early childhood or secondary responses upon virus reactivation, but most studies have concentrated on responses in latently infected adults where virus levels (and responding T-cell frequencies) are relatively low.

T-Cell Response to HHV-6 in Children

Although only a handful of studies have investigated T-cell responses to HHV-6 in children, these have provided valuable information about the immune response to HHV-6. T-cell responses to HHV-6B develop in most children within 2 weeks of the first symptoms of virus infection (fever); however, responses to HHV-6B are delayed when compared with those elicited by other herpesviruses.[9] The reasons for the delay are not clear, but it is not due to generalized T-cell dysfunctions, because proliferative responses to a T-cell mitogen are not significantly affected.[9] Proliferative memory responses to HHV-6 seen in adults are thought to have been developed between the ages of 2 and 12 years.[59] Analysis of HHV-6B-expanded PBMCs by flow cytometry showed similar frequencies of CD4+ and CD8+ T cells in cultures of healthy adults and children post-infection.[59,60]

T-Cell Response to HHV-6 in Transplant Recipients

Robust CD4+ and CD8+ T-cell proliferative responses to HHV-6 in allogeneic hematopoietic stem cell transplantation (allo-HSCT) patients are observed after 3 months of the engraftment.[61,62] The protective role of HHV-6-specific T-cell responses was inferred by the higher incidence of persistent HHV-6 viremia in patients without proliferative responses.[62] However, proliferative responses to HHV-6 in allo-HSCT patients with HHV-6 reactivation have been shown to be associated with severe graft-versus-host disease.[63] The abundance of the HHV-6-specific T cells in the overall T-cell populations has been analyzed in allo-HSCT patients in one study and was shown to be low (0.05%) and comparable to the levels described for healthy adults.[61] Interestingly, patients with allo-reactive disease had a 3.5-fold increase in the frequency of CD8+ T-cell responses to HHV-6B-infected PBMCs.[64] HHV-6 reactivation has been reported in liver transplant recipients and may be associated with posttransplant complications[65,66] and suppressed HHV-6T-cell response.[67] Responses to HCMV and mitogen recovered within weeks of the transplant, but the response to HHV-6 only partially recovered 1 year after the transplantation, suggesting a prolonged suppression of the HHV-6-specific but not overall memory T-cell responses after liver transplantation.

T-Cell Responses in Patients with Multiple Sclerosis

Multiple sclerosis (MS) is an autoimmune disease of the nervous system with genetic, environmental, and microbial involvements. Numerous lines of evidence, although controversial, indicate that HHV-6 might be one of the pathogenic microbial components.[68] Indirect evidence from immunopathology studies suggests that HHV-6 infection may be a trigger for MS onset. Mechanistically, there are two major scenarios: harmful inflammatory responses to HHV-6 and molecular mimicry.

In support of the harmful inflammatory response theory, it has been reported that MS patients have more frequent and stronger responses to HHV-6A than HHV-6B.[69] This finding is significant for three reasons: First, HHV-6A is more neurotropic than HHV-6B,[28] which helps in locating the virus in the nervous system. Second, the responses in healthy individuals are more frequent to HHV-6B than HHV-6A[69,70] and follow the epidemiological pattern of higher prevalence of HHV-6B. Finally, responses to HHV-6A in MS patients are more prevalent and stronger than those observed in healthy individuals to HHV-6B.[69] A shortcoming of using complete virus for such studies is the possibility of cross-reactive responses with other related viruses—for example HHV-7 and HCMV, which have several regions of high sequence homology with HHV-6. HHV-7 in particular has also been found in tissue of MS patients. To overcome this problem, Tejada-Simon et al.[71] studied T-cell response to a series of recombinant fragments of the HHV-6B U11 protein, excluding regions that have significant homology to HCMV or HHV-7.

MS patients had a cytokine profile characterized by high IFN-γ, low IL-4, and low IL-10 compatible with proinflammatory responses; healthy donors have responses characterized by high IL-10 and low IFN-γ. Although these findings are circumstantial, Wang et al.[49] have reported that HHV-6 in healthy donors induces arrest of proliferation mediated by IL-10, whereas low IL-10 responses seem to be a characteristic of T-cell responses in MS patients.[72,73]

Support for molecular mimicry was provided by a study that evaluated T-cell responses to a HHV-6 U24 and a host myelin basic protein 13-mer peptide that share seven residues in the MHC-binding cores of the peptide sequences.[74] Over 50% of the clones recognized both peptide sequences, and MS patients had a higher frequency of cross-reactive precursor T cells. However, a second study that addressed the cross-reactivity by raising T-cell lines to UV-irradiated HHV-6A virus and commercial bovine MBP protein reached an opposite conclusion,[75] that the frequency of cross-reactive lines in MS and healthy controls were similar.

T-Cell Responses to HHV-6 in Healthy Adults

Multiple studies have demonstrated robust proliferation of CD4+ as well as CD8+ T cells to HHV-6A and HHV-6B in PBMCs from most healthy adults,[49,60,71] with higher frequencies of cells responding to HHV-6B.[69,70] Proliferation was shown to be restricted mainly to effector memory cells defined by the expression of surface markers.[30] In sharp contrast to HCMV, in which T-cell responses could reach over 4% of the total PBMC T-cell population,[76] T-cell responses to HHV-6 antigens are low. It was reported that IFN-γ CD4+ T-cell frequencies were below 0.12% using purified and disrupted HHV-6 viruses.[77] Screening with pools of overlapping peptides covering HHV-6B protein homologs for five major immunodominant HCMV antigens showed a total IFN-γ+ T-cell frequency below 0.1%.[61]

Antigen Specificity of the T-Cell Response to HHV-6

Three studies have reported fine characterization of the antigen specificity of human T-cell responses to HHV-6 in blood samples from healthy donors.[61,77,78] In each case, sets of synthetic peptides were used to map targets of the T-cell response, and IFN-γ was used as a readout. Nastke et al.[77] used a T-cell epitope prediction algorithm to select 9-mer sequences with high scores for binding to MHC class II DR1 (DRB1*0101) allele. Because MHC class II motifs are promiscuous, peptides that bind to DR1 are also expected to bind to many other MHC class II alleles. The screening included a set of peptides from the entire translated viral HHV-6B genome as well as another set from six virion proteins found by proteomic analysis to be abundantly expressed in HHV-6B infected cells. Eleven peptides were found to induce CD4 responses in PBMCs of several donors. Only two of these peptides belong to the set of peptides predicted in the whole genome, whereas nine peptides were derived from virion proteins: five from the major capsid protein U57, two from tegument protein U11 and U14, and two from the glycoproteins U38 and U48. Earlier studies had found that capsid and tegument antigens induced strong T-cell proliferation.[70,71] Gerdemann et al.[61] restricted the assessment to five proteins chosen because they are homologs of the major HCMV immunodominant antigens: the myristylated virion protein U711, the transactivator U90, and the tegument proteins U11, U14, and U54. Finally, the study by Martin et al.[78] evaluated the CD8 response to 12 HLA-A*0201 predicted peptides selected for high binding scores, derived from U90, U11, and U54.

Functional Characterization of CD4+ T-Cell Responses to HHV-6

The study of T-cell responses to HHV-6 in PBMCs is complicated by the low frequency of antigen-specific T cells. For this reason, *in vitro* expanded T cells have been used for the characterization of T-cell specificities. Nastke et al.[77] showed that a single *in vitro* expansion with disrupted HHV-6A or HHV-6B virus preparations resulted in expansion of CD4+ antigen-specific responses that could reach over 25% of the total population. Studies in expanded T cells corroborated responses to all 11 candidate epitopes originally identified in PBMCs. Anti-peptide T-cell lines demonstrated that these epitopes were generated by processing of HHV-6 antigens. Further characterization of the T-cell lines showed that most CD4+ T cells, in addition to secreting IFN-γ, exhibited cytolytic potential as suggested by the expression of CD107a/b.[79] This is consistent with an earlier study showing that anti-HHV-6 IFN-γ producing CD4+ T-cell clones also had cytolytic capabilities.[80] Gerdemann et al.[61] also observed robust expansion of T cells, with responses in all donors to U90, U14, U54, and U11 and in approximately 60% of donors to U71. Good correlation between IFN-γ and cytotoxicity was found in this study.

Cross-Reactivity Between HHV-6A and HHV-6B Epitopes

Although HHV-6A and HHV-6B share an overall 90% homology,[81] these viruses differ in multiple aspects of their biology[82] and in the outcome of primary infections and possibly in reactivation. As expected from the homology, less than 10% of CD4+ T-cell clones are species specific.[83] Studies of proliferative responses of primary T cells to capsid antigens, however, suggest some species-specific T-cell responses,[70] although all five defined capsid-derived CD4+ T-cell epitopes reported to date do not seem to be species specific.[77]

Detection of HHV-6 Peptide-Specific CD4+ T-Cell Responses Ex Vivo Directly in PBMC

Direct *ex vivo* analysis of responses using HHV-6-specific peptides is simple and easily implemented but is of limited use due to the low frequency of peptide-specific CD4+ T-cell responses in PBMCs (0.02%) in healthy donors.[77] However, in transplant patients where frequencies of T cells to HHV-6 are higher,[61] this technique might be more feasible. Staining of T cells with HHV-6–MHC tetramers is widely used to identify low-frequency antigen-specific T cells.[84] As expected from *ex vivo* peptide responses, tetramer staining of PBMCs showed relatively low signals;[77] however, enrichment of tetramer positive cells has significantly improved the signal-to-noise ratio.[85,86] Our limited experimental evidence so far indicates that tetramer staining of PBMCs followed by enrichment could be used to track with confidence HHV-6 epitope-specific CD4 T-cell responses in peripheral blood,[77] although this method is restricted by individual differences in MHC haplotypes and might become practical only for common MHC alleles.

Characterization of CD8+ T-Cell Responses

Although *in vitro* stimulation of PBMCs with HHV-6 virus induces the expansion of both CD8+ and CD4+ T cells,[29,30,49,59] early studies reported only HHV-6 specific CD4+ T-cell clones. Just recently, the first HHV-6B-specific CD8+ T-cell clones and lines were described.[61,78] These two studies found that HHV-6 antigen-specific CD8+ T cells are present at frequencies below 10^{-5} in PBMCs[78] and even in *in vitro* expanded cultures are a minor fraction of the T cells.[61] Despite the low frequencies, eight defined HHV-6 CD8 T-cell epitope peptides have been reported. Five T-cell epitopes (three from U11 and two from U54) are HLA-A*0201 restricted, and three epitopes (two from transactivator U90 and one from U14) are restricted by HLA B40(60).[61] Characterization of the HLA-A*0201 CD8 epitopes using T-cell clones demonstrated that these epitopes induce polyfunctional cells that produce IFN-γ, TNF-α, and granzyme B and have cytotoxic activity for antigen-sensitized cells. All HLA-A*0201 clones specific for U54 and one for U11 recognized HHV-6B infected cells;[78] however, none of the HLA-A*0201 clones recognized HHV-6A-infected cells, perhaps due to virus-specific immunomodulatory mechanisms. Similarly to the HLA-A*0201, CD8+ T cells restricted by HLA-B*40(60) were also polyfunctional, producing IFN-γ, TNF-α, and granzyme B, and were cytotoxic.[61]

CONCLUSIONS

Our understanding of the immune response to HHV-6 is still minimal and lags well behind that of other herpesviruses such as HSV, HCMV, and EBV. Poor characterization of human immune responses elicited by HHV-6 has been a major obstacle to understanding the role of HHV-6 in various human diseases. Recent advances in HHV-6 biology and in innate and adaptive immunity can be expected to move the field forward, and it is likely that characterization of T-cell responses to HHV-6 will provide tools for developing HHV-6 immunotherapy and for understanding the role of HHV-6 in immunopathology.

ACKNOWLEDGMENTS

We gratefully acknowledge and thank Omar Dominguez-Amorocho and Lianjun Shen for helpful discussions and critical reading of the manuscript. This work was supported by National Institutes of Health grant U19-57319.

REFERENCES

1. Salahuddin SZ, Ablashi DV, Markham PD, Josephs SF, Sturzenegger S, Kaplan M, et al. Isolation of a new virus, HBLV, in patients with lymphoproliferative disorders. *Science* 1986;**234**(4776):596–601.
2. Mossman KL, Ashkar AA. Herpesviruses and the innate immune response. *Viral Immunol* 2005;**18**(2):267–81.
3. Chi J, Wang F, Li L, Feng D, Qin J, Xie F, et al. The role of MAPK in CD4(+) T cells toll-like receptor 9-mediated signaling following HHV-6 infection. *Virology* 2012;**422**(1):92–8.
4. Murakami Y, Tanimoto K, Fujiwara H, An J, Suemori K, Ochi T, et al. Human herpesvirus 6 infection impairs Toll-like receptor signaling. *Virol J* 2010;**7**:91.
5. Kikuta H, Nakane A, Lu H, Taguchi Y, Minagawa T, Matsumoto S. Interferon induction by human herpesvirus 6 in human mononuclear cells. *J Infect Dis* 1990;**162**(1):35–8.
6. Takahashi K, Segal E, Kondo T, Mukai T, Moriyama M, Takahashi M, et al. Interferon and natural killer cell activity in patients with exanthem subitum. *Pediatr Infect Dis J* 1992;**11**(5):369–73.
7. Jaworska J, Gravel A, Flamand L. Divergent susceptibilities of human herpesvirus 6 variants to type I interferons. *Proc Natl Acad Sci USA* 2010;**107**(18):8369–74.
8. Orange JS. Human natural killer cell deficiencies and susceptibility to infection. *Microbes Infect* 2002;**4**(15):1545–58.
9. Kumagai T, Yoshikawa T, Yoshida M, Okui T, Ihira M, Nagata N, et al. Time course characteristics of human herpesvirus 6 specific cellular immune response and natural killer cell activity in patients with exanthema subitum. *J Med Virol* 2006;**78**(6):792–9.
10. Flamand L, Stefanescu I, Menezes J. Human herpesvirus-6 enhances natural killer cell cytotoxicity via IL-15. *J Clin Invest* 1996;**97**(6):1373–81.
11. Gosselin J, Tomolu A, Gallo RC, Flamand L. Interleukin-15 as an activator of natural killer cell-mediated antiviral response. *Blood* 1999;**94**(12):4210–9.
12. Arena A, Merendino RA, Bonina L, Iannello D, Stassi G, Mastroeni P. Role of IL-15 on monocytic resistance to human herpesvirus 6 infection. *New Microbiol* 2000;**23**(2):105–12.
13. Kida K, Isozumi R, Ito M. Killing of human herpes virus 6-infected cells by lymphocytes cultured with interleukin-2 or -12. *Pediatr Int* 2000;**42**(6):631–6.
14. Malnati MS, Lusso P, Ciccone E, Moretta A, Moretta L, Long EO. Recognition of virus-infected cells by natural killer cell clones is controlled by polymorphic target cell elements. *J Exp Med* 1993;**178**(3):961–9.
15. Slavuljica I, Krmpotic A, Jonjic S. Manipulation of NKG2D ligands by cytomegaloviruses: impact on innate and adaptive immune response. *Front Immunol* 2011;**2**:85.
16. Luttichau HR, Clark-Lewis I, Jensen PO, Moser C, Gerstoft J, Schwartz TW. A highly selective CCR2 chemokine agonist encoded by human herpesvirus 6. *J Biol Chem* 2003;**278**(13):10928–33.
17. Sjahril R, Isegawa Y, Tanaka T, Nakano K, Yoshikawa T, Asano Y, et al. Relationship between U83 gene variation in human herpesvirus 6 and secretion of the U83 gene product. *Arch Virol* 2009;**154**(2):273–83.

18. Dewin DR, Catusse J, Gompels UA. Identification and characterization of U83A viral chemokine, a broad and potent beta-chemokine agonist for human CCRs with unique selectivity and inhibition by spliced isoform. *J Immunol* 2006;**176**(1):544–56.
19. Catusse J, Parry CM, Dewin DR, Gompels UA. Inhibition of HIV-1 infection by viral chemokine U83A via high-affinity CCR5 interactions that block human chemokine-induced leukocyte chemotaxis and receptor internalization. *Blood* 2007;**109**(9):3633–9.
20. Catusse J, Spinks J, Mattick C, Dyer A, Laing K, Fitzsimons C, et al. Immunomodulation by herpesvirus U51A chemokine receptor via CCL5 and FOG-2 down-regulation plus XCR1 and CCR7 mimicry in human leukocytes. *Eur J Immunol* 2008;**38**(3):763–77.
21. Fitzsimons CP, Gompels UA, Verzijl D, Vischer HF, Mattick C, Leurs R, et al. Chemokine-directed trafficking of receptor stimulus to different g proteins: selective inducible and constitutive signaling by human herpesvirus 6-encoded chemokine receptor U51. *Mol Pharmacol* 2006;**69**(3):888–98.
22. Zhen Z, Bradel-Tretheway B, Sumagin S, Bidlack JM, Dewhurst S. The human herpesvirus 6G protein-coupled receptor homolog U51 positively regulates virus replication and enhances cell-cell fusion *in vitro*. *J Virol* 2005;**79**(18):11914–24.
23. Yasukawa M, Hasegawa A, Sakai I, Ohminami H, Arai J, Kaneko S, et al. Down-regulation of CXCR4 by human herpesvirus 6 (HHV-6) and HHV-7. *J Immunol* 1999;**162**(9):5417–22.
24. Lusso P. HHV-6 and the immune system: mechanisms of immunomodulation and viral escape. *J Clin Virol* 2006;**37**(Suppl. 1):S4–10.
25. Leen AM, Sili U, Bollard CM, Rooney CM. Adoptive immunotherapy for herpesviruses. In: Arvin A, Campadelli-Fiume G, Mocarski E, Moore PS, Roizman B, Whitley R, editors. *Human Herpesviruses: Biology, Therapy, and Immunoprophylaxis*. Cambridge: Cambridge University Press; 2007.
26. Leen AM, Christin A, Myers GD, Liu H, Cruz CR, Hanley PJ, et al. Cytotoxic T lymphocyte therapy with donor T cells prevents and treats adenovirus and Epstein–Barr virus infections after haploidentical and matched unrelated stem cell transplantation. *Blood* 2009;**114**(19):4283–92.
27. Takahashi K, Sonoda S, Higashi K, Kondo T, Takahashi H, Takahashi M, et al. Predominant CD4 T-lymphocyte tropism of human herpesvirus 6-related virus. *J Virol* 1989;**63**(7):3161–3.
28. De Bolle L, Van Loon J, De Clercq E, Naesens L. Quantitative analysis of human herpesvirus 6 cell tropism. *J Med Virol* 2005;**75**(1):76–85.
29. Flamand L, Gosselin J, Stefanescu I, Ablashi D, Menezes J. Immunosuppressive effect of human herpesvirus 6 on T-cell functions: suppression of interleukin-2 synthesis and cell proliferation. *Blood* 1995;**85**(5):1263–71.
30. Gupta S, Agrawal S, Gollapudi S. Differential effect of human herpesvirus 6A on cell division and apoptosis among naive and central and effector memory CD4+ and CD8+ T-cell subsets. *J Virol* 2009;**83**(11):5442–50.
31. Li L, Gu B, Zhou F, Chi J, Wang F, Peng G, et al. Human herpesvirus 6 suppresses T cell proliferation through induction of cell cycle arrest in infected cells in the G2/M phase. *J Virol* 2011;**85**(13):6774–83.
32. Wang FZ, Pellett PE. HHV-6A, 6B, and 7: immunobiology and host response. In: Arvin A, Campadelli-Fiume G, Mocarski E, Moore PS, Roizman B, Whitley R, editors. *Human Herpesviruses: Biology, Therapy, and Immunoprophylaxis*. Cambridge: Cambridge University Press; 2007.
33. Braun DK, Dominguez G, Pellett PE. Human herpesvirus 6. *Clin Microbiol Rev* 1997;**10**(3):521–67.
34. Balachandran N, Amelse RE, Zhou WW, Chang CK. Identification of proteins specific for human herpesvirus 6-infected human T cells. *J Virol* 1989;**63**(6):2835–40.

35. Black JB, Schwarz TF, Patton JL, Kite-Powell K, Pellett PE, Wiersbitzky S, et al. Evaluation of immunoassays for detection of antibodies to human herpesvirus 7. *Clin Diagn Lab Immunol* 1996;**3**(1):79–83.
36. Yamamoto M, Black JB, Stewart JA, Lopez C, Pellett PE. Identification of a nucleocapsid protein as a specific serological marker of human herpesvirus 6 infection. *J Clin Microbiol* 1990;**28**(9):1957–62.
37. Thäder-Voigt A, Jacobs E, Lehmann W, Bandt D. Development of a microwell adapted immunoblot system with recombinant antigens for distinguishing human herpesvirus (HHV)6A and HHV6B and detection of human cytomegalovirus. *Clin Chem Lab Med* 2011;**49**(11):1891–8.
38. Higashimoto Y, Ohta A, Nishiyama Y, Ihira M, Sugata K, Asano Y, et al. Development of a human herpesvirus 6 species-specific immunoblotting assay. *J Clin Microbiol* 2012;**50**(4):1245–51.
39. Yakushijin Y, Yasukawa M, Kobayashi Y. T-cell immune response to human herpesvirus-6 in healthy adults. *Microbiol Immunol* 1991;**35**(8):655–60.
40. Horvat RT, Parmely MJ, Chandran B. Human herpesvirus 6 inhibits the proliferative responses of human peripheral blood mononuclear cells. *J Infect Dis* 1993;**167**(6):1274–80.
41. Furukawa M, Yasukawa M, Yakushijin Y, Fujita S. Distinct effects of human herpesvirus 6 and human herpesvirus 7 on surface molecule expression and function of CD4+ T cells. *J Immunol* 1994;**152**(12):5768–75.
42. Lusso P, Malnati M, De Maria A, Balotta C, DeRocco SE, Markham PD, et al. Productive infection of CD4+ and CD8+ mature human T cell populations and clones by human herpesvirus 6. Transcriptional down-regulation of CD3. *J Immunol* 1991;**147**(2):685–91.
43. Inoue Y, Yasukawa M, Fujita S. Induction of T-cell apoptosis by human herpesvirus 6. *J Virol* 1997;**71**(5):3751–9.
44. Mlechkovich G, Frenkel N. Human herpesvirus 6A (HHV-6A) and HHV-6B alter E2F1/Rb pathways and E2F1 localization and cause cell cycle arrest in infected T cells. *J Virol* 2007;**81**(24):13499–508.
45. Smith AP, Paolucci C, Di Lullo G, Burastero SE, Santoro F, Lusso P. Viral replication-independent blockade of dendritic cell maturation and interleukin-12 production by human herpesvirus 6. *J Virol* 2005;**79**(5):2807–13.
46. Kakimoto M, Hasegawa A, Fujita S, Yasukawa M. Phenotypic and functional alterations of dendritic cells induced by human herpesvirus 6 infection. *J Virol* 2002;**76**(20):10338–45.
47. Bertelsen LB, Petersen CC, Kofod-Olsen E, Oster B, Hollsberg P, Agger R, et al. Human herpesvirus 6B induces phenotypic maturation without IL-10 and IL-12p70 production in dendritic cells. *Scand J Immunol* 2010;**71**(6):431–9.
48. Janelle ME, Flamand L. Phenotypic alterations and survival of monocytes following infection by human herpesvirus-6. *Arch Virol* 2006;**151**(8):1603–14.
49. Wang F, Yao K, Yin QZ, Zhou F, Ding CL, Peng GY, et al. Human herpesvirus-6-specific interleukin 10-producing CD4+ T cells suppress the CD4+ T-cell response in infected individuals. *Microbiol Immunol* 2006;**50**(10):787–803.
50. Jochum S, Moosmann A, Lang S, Hammerschmidt W, Zeidler R. The EBV immunoevasins vIL-10 and BNLF2a protect newly infected B cells from immune recognition and elimination. *PLoS Pathog* 2012;**8**(5):e1002704.
51. Pestka S, Krause CD, Sarkar D, Walter MR, Shi Y, Fisher PB. Interleukin-10 and related cytokines and receptors. *Annu Rev Immunol* 2004;**22**:929–79.
52. Couper KN, Blount DG, Riley EM. IL-10: the master regulator of immunity to infection. *J Immunol* 2008;**180**(9):5771–7.

53. Lee CC, Kung JT. Marginal zone B cell is a major source of IL-10 in *Listeria monocytogenes* susceptibility. *J Immunol* 2012;**189**(7):3319–27.
54. Brockman MA, Kwon DS, Tighe DP, Pavlik DF, Rosato PC, Sela J, et al. IL-10 is upregulated in multiple cell types during viremic HIV infection and reversibly inhibits virus-specific T cells. *Blood* 2009;**114**(2):346–56.
55. Arena A, Stassi G, Speranza A, Iannello D, Mastroeni P. Modulatory effect of HHV-6 on MCP-1 production by human monocytes. *New Microbiol* 2002;**25**(3):335–40.
56. Price JD, Schaumburg J, Sandin C, Atkinson JP, Lindahl G, Kemper C. Induction of a regulatory phenotype in human CD4+ T cells by streptococcal M protein. *J Immunol* 2005;**175**(2):677–84.
57. Santoro F, Kennedy PE, Locatelli G, Malnati MS, Berger EA, Lusso P. CD46 is a cellular receptor for human herpesvirus 6. *Cell* 1999;**99**(7):817–27.
58. Mock DJ, Strathmann F, Blumberg BM, Mayer-Proschel M. Infection of murine oligodendroglial precursor cells with human herpesvirus 6 (HHV-6): establishment of a murine *in vitro* model. *J Clin Virol* 2006;**37**(Suppl. 1):S17–23.
59. Koide W, Ito M, Torigoe S, Ihara T, Kamiya H, Sakurai M. Activation of lymphocytes by HHV-6 antigen in normal children and adults. *Viral Immunol* 1998;**11**(1):19–25.
60. Haveman LM, Scherrenburg J, Maarschalk-Ellerbroek LJ, Hoek PD, Schuurman R, W de Jager, et al. T-cell response to viral antigens in adults and children with common variable immunodeficiency and specific antibody deficiency. *Clin Exp Immunol* 2010;**161**(1):108–17.
61. Gerdemann U, Keukens L, Keirnan JM, Katari UL, Nguyen CT, de Pagter AP, et al. Immunotherapeutic strategies to prevent and treat human herpesvirus 6 reactivation after allogeneic stem cell transplantation. *Blood* 2013;**121**(1):207–18.
62. Wang FZ, Linde A, Dahl H, Ljungman P. Human herpesvirus 6 infection inhibits specific lymphocyte proliferation responses and is related to lymphocytopenia after allogeneic stem cell transplantation. *Bone Marrow Transplant* 1999;**24**(11):1201–6.
63. de Pagter PJ, Schuurman R, Meijer E, van Baarle D, Sanders EA, Boelens JJ. Human herpesvirus type 6 reactivation after haematopoietic stem cell transplantation. *J Clin Virol* 2008;**43**(4):361–6.
64. de Pagter AP, Boelens JJ, Scherrenburg J, Vroom-de Blank T, Tesselaar K, Nanlohy N, et al. First analysis of human herpesvirus 6T-cell responses: specific boosting after HHV-6 reactivation in stem cell transplantation recipients. *Clin Immunol* 2012;**144**(3):179–89.
65. Singh N, Carrigan DR. Human herpesvirus-6 in transplantation: an emerging pathogen. *Ann Intern Med* 1996;**124**(12):1065–71.
66. Singh N, Husain S, Carrigan DR, Knox KK, Weck KE, Wagener MM, et al. Impact of human herpesvirus-6 on the frequency and severity of recurrent hepatitis C virus hepatitis in liver transplant recipients. *Clin Transplant* 2002;**16**(2):92–6.
67. Singh N, Bentlejewski C, Carrigan DR, Gayowski T, Knox KK, Zeevi A. Persistent lack of human herpesvirus-6 specific T-helper cell response in liver transplant recipients. *Transpl Infect Dis* 2002;**4**(2):59–63.
68. Fotheringham J, Jacobson S. Human herpesvirus 6 and multiple sclerosis: potential mechanisms for virus-induced disease. *Herpes* 2005;**12**(1):4–9.
69. Soldan SS, Leist TP, Juhng KN, McFarland HF, Jacobson S. Increased lymphoproliferative response to human herpesvirus type 6A variant in multiple sclerosis patients. *Ann Neurol* 2000;**47**(3):306–13.
70. Wang FZ, Dahl H, Ljungman P, Linde A. Lymphoproliferative responses to human herpesvirus-6 variant A and variant B in healthy adults. *J Med Virol* 1999;**57**(2):134–9.

71. Tejada-Simon MV, Zang YC, Hong J, Rivera VM, Killian JM, Zhang JZ. Detection of viral DNA and immune responses to the human herpesvirus 6 101-kilodalton virion protein in patients with multiple sclerosis and in controls. *J Virol* 2002;**76**(12):6147–54.
72. Ozenci V, Kouwenhoven M, Huang YM, Xiao B, Kivisakk P, Fredrikson S, et al. Multiple sclerosis: levels of interleukin-10-secreting blood mononuclear cells are low in untreated patients but augmented during interferon-beta-1b treatment. *Scand J Immunol* 1999;**49**(5):554–61.
73. Navikas V, Link H. Review: cytokines and the pathogenesis of multiple sclerosis. *J Neurosci Res* 1996;**45**(4):322–33.
74. Tejada-Simon MV, Zang YC, Hong J, Rivera VM, Zhang JZ. Cross-reactivity with myelin basic protein and human herpesvirus-6 in multiple sclerosis. *Ann Neurol* 2003;**53**(2):189–97.
75. Cirone M, Cuomo L, Zompetta C, Ruggieri S, Frati L, Faggioni A, et al. Human herpesvirus 6 and multiple sclerosis: a study of T cell cross-reactivity to viral and myelin basic protein antigens. *J Med Virol* 2002;**68**(2):268–72.
76. Sylwester AW, Mitchell BL, Edgar JB, Taormina C, Pelte C, Ruchti F, et al. Broadly targeted human cytomegalovirus-specific CD4+ and CD8+ T cells dominate the memory compartments of exposed subjects. *J Exp Med* 2005;**202**(5):673–85.
77. Nastke MD, Becerra A, Yin L, Dominguez-Amorocho O, Gibson L, Stern LJ, et al. Human CD4+ T-cell response to human herpesvirus 6. *J Virol* 2012;**86**(9):4776–92.
78. Martin LK, Schub A, Dillinger S, Moosmann A. Specific CD8(+) T cells recognize human herpesvirus 6B. *Eur J Immunol* 2012;**42**(11):2901–12.
79. Betts MR, Koup RA. Detection of T-cell degranulation: CD107a and b. *Methods Cell Biol* 2004;**75**:497–512.
80. Yakushijin Y, Yasukawa M, Kobayashi Y. Establishment and functional characterization of human herpesvirus 6-specific CD4+ human T-cell clones. *J Virol* 1992;**66**(5):2773–9.
81. Dominguez G, Dambaugh TR, Stamey FR, Dewhurst S, Inoue N, Pellett PE. Human herpesvirus 6B genome sequence: coding content and comparison with human herpesvirus 6A. *J Virol* 1999;**73**(10):8040–52.
82. Ablashi D, Salahuddin SZ. *Biology and Strain Variants*, 2nd ed. Amsterdam: Elsevier; 1992.
83. Yasukawa M, Yakushijin Y, Furukawa M, Fujita S. Specificity analysis of human CD4+ T-cell clones directed against human herpesvirus 6 (HHV-6), HHV-7, and human cytomegalovirus. *J Virol* 1993;**67**(10):6259–64.
84. Vollers SS, Stern LJ. Class II major histocompatibility complex tetramer staining: progress, problems, and prospects. *Immunology* 2008;**123**(3):305–13.
85. Day CL, Seth NP, Lucas M, Appel H, Gauthier L, Lauer GM, et al. *Ex vivo* analysis of human memory CD4 T cells specific for hepatitis C virus using MHC class II tetramers. *J Clin Invest* 2003;**112**(6):831–42.
86. Scriba TJ, Purbhoo M, Day CL, Robinson N, Fidler S, Fox J, et al. Ultrasensitive detection and phenotyping of CD4+ T cells with optimized HLA class II tetramer staining. *J Immunol* 2005;**175**(10):6334–43.

Chapter 15

Chromosomally Integrated HHV-6

Mario Luppi[a], Leonardo Potenza[a], Guillaume Morissette[b], and Louis Flamand[c]

[a]Università degli Studi di Modena e Reggio Emilia, Modena, Italy, [b]Axe maladies infectieuses et immunitaires, Centre de recherche du CHU de Québec, Québec, Canada, [c]Axe maladies infectieuses et immunitaires, Centre de recherche du CHU de Québec et Université Laval, Quebec, Canada

HISTORICAL PERSPECTIVES ON CIHHV-6

Officially, the history of chromosomally integrated human herpesvirus 6 (ciHHV-6) began in 1993, with two key concepts: *serendipity* and *pulsed-field gel electrophoresis* (PFGE).[1] However, to tell the truth, the story actually began in 1988, two years after the initial HHV-6 isolation, when Josephs and colleagues[2] from the United States and Jarrett and colleagues[3] from the United Kingdom reported in two different papers that HHV-6 DNA sequences could be detected at high copy numbers in affected tissues from five patients with B-cell non-Hodgkin's lymphoma (NHL) and T-cell angioimmunoblastic lymphoma (AIL), not only by polymerase chain reaction (PCR) but also by the less sensitive Southern blot analysis. At that time, people in the field believed that HHV-6 could actually play a major role in human lymphomagenesis. The authors (Torelli and Luppi), at the Experimental Hematology Laboratory in Modena, Italy, were looking for HHV-6 in Italian lymphomas and identified the first two patients with Hodgkin's disease (HD), who harbored an unexpectedly high number of HHV-6 DNA sequences both in pathologic lymph nodes and in normal peripheral blood mononuclear cells (PBMCs).[4] We and others erroneously thought that the presence of a high level of HHV-6 DNA in lymphoma patients was a major argument for an etiopathogenetic role of HHV-6, as in the case of Epstein–Barr virus infection. However, expression of HHV-6 proteins was restricted to nonmalignant lymphoproliferative disorders, such as Rosai–Dorfman disease, while the expression of viral proteins was detected in neither the neoplastic cells of NHL nor the viable neoplastic Reed–Sternberg (R-S) cells of HD cases.[5] In HHV-6 Southern blot-positive cases, an early HHV-6 protein could be detected only in the so-called "ghost" ("mummified") apoptotic R-S cells.

Human Herpesviruses HHV-6A, HHV-6B & HHV-7.

The detection by Lacroix and colleagues[6] of HHV-6B DR 7 protein in viable R-S cells in a significant proportion of HD cases has renewed general interest in a pathogenic role for HHV-6 in HD, but the possible relevance of this finding in HD cases with ciHHV-6 has not yet been investigated. At that time, while we were performing epidemiologic molecular studies and contributing to identify cases with high-level HHV-6 DNA, including not only patients with lymphoma but also patients with multiple sclerosis (MS), we were working in the lab on patients with another hematologic disease, chronic myeloid leukemia, and we were attempting to define, by PFGE, the molecular features of some unusual cases with a chromosome 22 breakpoint, outside the breakpoint cluster region. By chance, in the middle of setting up experiments utilizing PFGE and the Bio-Rad clamped-homogeneous electrical field (CHEF) apparatus, we analyzed HHV-6 (strain GS)-infected HSB-2 cells and PBMCs from the NHL, HD, and MS patients, who carried high viral copy numbers.[1] Using CHEF, HSB-2 DNA showed only one band of about 170kb, corresponding to the free linear viral genome, and hybridized with the ZVH14 sequence.[1] Three patients' PBMC DNAs hybridized with the probe in the starting well, without showing free migrating bands. Then, infected HSB-2 DNA was digested with rare cutting enzymes, such as Not I, Mlu I, and Sma I, and the restriction analysis showed fragments invariably shorter than 170kb.[1] Unexpectedly, the restriction analysis performed on the three patients' PBMC DNAs showed a pattern of fragments larger than 170kb, which is consistent with the presence of viral sequences linked to high-molecular-weight cellular DNA.[1] A similar pattern could be observed using other probes such as the ZVB70 and the HD 12 sequences at the 5′ and the 3′ ends of the viral genomes, respectively.[7] Although in the first described cases an apparently complete viral genome was present, in four further cases (one chronic fatigue syndrome, two HD, and one NHL) the genome was either deleted or grossly rearranged.[8] Of interest, in the two HD and one NHL cases with high viral copy numbers and carrying the ZVB70 and the ZVH14 sequences, the HD 12 sequence was absent.[7]

The lack of the extreme right end of the HHV-6 genome containing the adeno-associated virus type 2 rep 68/78 homologous gene indicates that mechanisms other than the rep homologous gene may mediate the HHV-6 integration, at least in some instances. In the three patients carrying an apparently complete HHV-6 genome, fluorescent *in situ* hybridization (FISH) analysis showed that the viral genome was integrated at the telomeric end of the short arm of chromosome 17, on band 17p13.3 in 50%, 77%, and 53% of the metaphase cells, respectively, derived from phytohemagglutinin (PHA)-stimulated PBMCs.[7] A more refined analysis was made in collaboration with Morris and colleagues[9] from New Zealand in 1994 to more precisely map the location of the viral integration site, relative to two known oncogenes, CRK and the more distal (relative to centromere) ABR oncogene. It was shown that the HHV-6 integration site was at least 1000kb downstream of ABR, close to or within the telomeric sequences of 17p.[9] This was significant, given that human telomeric-like repeats flank the terminal ends of the HHV-6 genome, and it is presumed that, through homologous recombination between these viral sequences and the telomeres, the HHV-6 genome or part of it integrate into the human chromosomes.[9] Thus, at that time, we were thinking that 17p13 was a preferential site of integration, because the telomere of the p arm of chromosome 17 is shorter, on average, than that of any other human chromosomes and may be vulnerable to rearrangements and to have a propensity to integrate viral DNA.

Moreover, we also demonstrated HHV-6 integration in bone marrow (BM) progenitor cells, suggesting that HHV-6 infects BM progenitors and is transmitted longitudinally to cells that differentiate along the committed pathways, but further analyses of hair follicles or other body cells or tissues, such as fibroblasts, were not performed.[10] Thus, we suggested that the HHV-6 genome could be found integrated in only a proportion of host cells, given that analysis by two different observers of unstimulated interphase PBMCs showed a single discrete integration signal in no more than 45% and 52% of interphase cells, respectively, and that PCR analysis of sorted peripheral blood cell fractions showed different viral loads among the various cell fractions.[11] However, these arguments were not convincing enough to formally rule out the phenomenon of inheritance of ciHHV-6, subsequently described by Daibata and colleagues from Japan.[12–14] In 1998, they were the first to demonstrate vertical transmission of HHV-6 DNA over three generations by showing identical HHV-6 integration sites among a patient with acute lymphoblastic leukemia, his son, and his granddaughter, who were otherwise healthy.[12] One year later, the same group described a patient with Burkitt's lymphoma and HHV-6 integration at 22q13,[13] whose asymptomatic husband had HHV-6 integration at 1q44. By demonstrating that their asymptomatic daughter had HHV-6 integration at 1q44 and at 22q13, the Japanese authors suggested that the daughter's viral genomes were inherited chromosomally by both parents and proposed a model form of HHV-6 latency, in which integrated viral genome can be chromosomally transmitted.[14] These findings have been confirmed by Clark's group from the United Kingdom, who described five different chromosomal integration sites of the HHV-6 genome in nine individuals for whom FISH analysis showed 100% positivity for the HHV-6 signal in the examined cells.[15]

As reviewed by Morissette and Flamand,[16] several integration sites have so far been described, in both patients and healthy subjects, with a higher prevalence of HHV-6B ciHHV-6 cases compared with HHV-6A ciHHV-6 cases. The prevalence of ciHHV-6 varies from less than 1% to up to 5% in different series from Europe and Japan, including adults and children, healthy subjects, and hospital patients.[17] Whether the integrated genome of HHV-6 may persist in an inert state for the lifetime or is capable of either full replication or protein translation has been poorly addressed. In infants with congenital ciHHV-6, the detection of humoral immune responses to HHV-6 at mean titers similar to those present in children without congenital infection, and their persistence even after the expected decline of passive antibody levels, suggested that the integrated genome may produce viral particles to which infants with ciHHV-6 are able to mount a humoral immune response.[18] Maternal and offspring HHV-6 viruses matched, suggesting that integrated latent virus of the mother reactivated and infected the infant *in utero*.[19] Sequence analysis of transplacentally acquired HHV-6 DNA confirmed the occurrence of transmission of a chromosomally integrated reactivated virus.[20]

Consistent with the possibility that ciHHV-6 may reactivate, antibodies to immediate-early and glycoprotein B were detected in a significant proportion of patients and healthy subjects with ciHHV-6 from the series described by Yamanishi's group.[21] *In vitro* induction of HHV-6 immediate-early and late genes has been achieved, following treatment with 12-*O*-tetradecanoyl-phorbol-13-acetate (TPA) and sodium *n*-butyrate, of a Burkitt's lymphoma cell line, the Kakata cell line with HHV-6 integration at 22q13.[13] Moreover, several HHV-6 gene transcripts have been detected in unstimulated PBMCs from a healthy individual with ciHHV-6, and late HHV-6 mRNAs, indicating viral replication, have also been detected in a small proportion of infants with congenital infection due to ciHHV-6.[22] More recently, Arbuckle and colleagues[23] performed *in vitro* experiments demonstrating that HHV-6A integrated in host chromosome may be inducible upon chemical stimuli, such as trichostatin A and TPA, and may produce fully competent virus and infect permissive cells. Furthermore, they suggested that chromosomal integration may represent the sole way through which HHV-6A achieves latency, as they determined that (1) the HHV-6A genome may be found covalently linked to chromosomes, soon after infection; (2) no free circular or linear viral genomes may be detected in permissive latently infected cells; and (3) HHV-6A has been found integrated in three out of three cell clones deriving from infected permissive cells.[23]

CIHHV-6 AND DIAGNOSIS OF HHV-6 ACTIVE INFECTION

The most frequent consequence of ciHHV-6 is the wrongful diagnosis of active HHV-6 infection. Individuals carrying HHV-6 chromosomal integration bear such high levels of HHV-6 DNA in plasma (>3.5 $\log_{10}$ copies per mL) or in peripheral whole blood (>5.5 $\log_{10}$ copies per mL) that persist over time, or in cerebrospinal fluid (4.0 $\log_{10}$ copies per mL), that such findings may be misinterpreted as HHV-6 active infections.[24,25] A single quantitative PCR test on serum or plasma cannot prove whether a patient with ciHHV-6 has an active HHV-6 infection, due to activation of ciHHV-6 or horizontally acquired.[24] When plasma, serum, or cerebrospinal fluid HHV-6 PCR levels are unexpectedly high, the most practical way to confirm that a patient has ciHHV-6 is by quantitative PCR, using whole blood or isolated PBMCs.[24] Individuals with ciHHV-6 have significantly higher viral DNA loads in PBMCs and whole blood than do non-ciHHV-6 individuals, even those immunocompetent individuals with primary HHV-6 infection or immunosuppressed subjects with HHV-6 reactivations.[24]

The search for HHV-6 DNA in hair follicles, nails, or tissue specimens can be performed to confirm the occurrence of ciHHV-6.[24] It has been recommended by Ljungman et al.,[26] for the European Conference on Infections in Leukemia

(ECIL), that ciHHV-6 should be excluded before the diagnosis of HHV-6 active disease can be made, irrespective of the immune status of the patients. Furthermore, it has been stated that ciHHV-6 may only mislead the laboratory diagnosis of HHV-6 active disease, inducing the administration of an undesirable and inappropriate antiviral treatment, especially in transplant patients, where ciHHV-6 may solely reflect either the rate of donor/recipient engraftment or the donor/recipient origin of the phenomenon.[26] Although the education effort made by the ECIL's authors is to be highly appreciated, we would like to introduce a note of caution concerning this statement. Relevant to this, we reported the case of a patient with ciHHV-6A undergoing an allogeneic hematopoietic stem cell transplantation (HSCT) from an unrelated donor for high-risk myelodysplastic syndrome.[27] At day 275, he underwent a bone marrow biopsy because of severe leukopenia and increasing HHV-6 DNA plasma values. GVHD manifestations, which would have been the most plausible cause of viral shedding from the recipient's cells, were absent. Peripheral blood and bone marrow donor chimerism was almost 100%, and a reduction of HHV-6 loads would have been expected with improving donor chimerism in allogeneic HSCT with such a ciHHV-6 donor/recipient combination. The biopsy showed bone marrow aplasia; other viruses commonly responsible for cytopenias were ruled out. After antiviral treatment, HHV-6 DNA levels reduced concomitant with leukocyte count recovery.[27] These *in vivo* findings suggested that ciHHV-6 reactivation may have contributed to bone marrow suppression.

Long-term bone marrow cultures have been established with (1) the bone marrow from the patient alone, composed of an adherent, stromal component of recipient origin and a nonadherent component, mainly of donor origin; and (2) the bone marrow from the patient, further allogeneically stimulated with bone marrow cells from another histo-incompatible subject. Immunocytochemistry with HHV-6A anti-p41/38 showed cytoplasmic reactivity at day 10 in bone marrow adherent cells from the patient cultured in experiment 1 alone, although in the presence of his HLA-unrelated stem cell transplant (SCT) donor's bone marrow nonadherent cells. Similarly, reactivity was observed at day 32 in bone marrow adherent cells in experiment 2 from the patient, further allogeneically stimulated with bone marrow cells from another histo-incompatible subject.[27] These data suggest that ciHHV-6 may produce proteins, at least in *ex vivo* cultures, as a result of the allogeneic reaction between the patient's cells and those of either his HLA-unrelated SCT donor or the histo-incompatible subject. Similarly to human cytomegalovirus (HCMV), which may reactivate from latency in allogeneically stimulated PBMCs,[28] the possibility is also raised that ciHHV-6 may exert its pathogenic potential through the sole expression of viral proteins such as CMV, which may exert a myelosuppressive effect by viral protein expression in bone marrow adherent cells.

In conclusion, ciHHV-6 may occur with a significant frequency both in SCT and solid organ transplant (SOT) patients. The results of molecular studies should be critically evaluated and distinguishing between HHV-6 active infection and ciHHV-6 should be pursued by the comparison of viral loads on different biological fluids. The determination of viral specie is useless for this scope, as HHV-6A and HHV-6B have been found integrated. Quantitative reverse transcriptase PCR, which monitors the expression of HHV-6 genes associated with productive infection (e.g., structural genes), and a quantitative antigenemia assay, which measures the frequency of peripheral blood leukocytes that express lytic antigens, might be useful but their clinical utility for monitoring HHV-6 activity in individuals with ciHHV-6 has not yet been defined.[24] Diagnosis of active infection may be supported by immunohistochemical detection of HHV-6 antigens in biopsy tissues from patients with ciHHV-6 and clinical symptoms suspected for a viral related disease.[27]

By a more conservative clinical point of view, even if it is clear that ciHHV-6 may confound the diagnosis of HHV-6 active infection and that the comparison of viral loads on multiple body fluids and tissues, including hair follicles, may help to identify subjects carrying ciHHV-6, it would be worth considering antiviral therapy even in patients with ciHHV-6 when they show signs and/or symptoms consistent with viral infection, and other known pathogens have been extensively searched for and ruled out.[27]

POTENTIAL CLINICAL OUTCOMES ASSOCIATED WITH CIHHV-6

The clinical significance of ciHHV-6 in transplant recipients is not defined. Several case reports have described ciHHV-6 patients or recipients of ciHHV-6 donor cells in HSCT without apparent reactivation, in spite of significant immunosuppression. The spectrum of clinical manifestations of ciHHV-6 in SCT setting include: (1) episodes of GVHD that may increase the value of HHV-6 load in plasma, and (2) the persistence of high HHV-6 loads on bone marrow blood and when the HHV-6 DNA in plasma is finally reducing. The first is possibly a consequence of increased virus shedding from the immune destruction of recipient's cells by the host's lymphocytes. This is due to the fact that, in subjects carrying ciHHV-6, viral sequences are found in each cell that carries human chromosomal DNA, and any inflammatory process causing an increased cell turnover may lead to variation of HHV-6 DNA in body fluids. The second possibility perhaps reflects different states of chimerism in different compartments; in patients with ciHHV-6 of recipient origin, HHV-6 loads in plasma and whole blood are expected to drop until they almost completely disappear when the non-integrated donor's white blood cells completely replace the integrated recipient's white blood cells. In bone marrow blood, HHV-6 loads are maintained by the persistence of integrated recipient's adherent cells and/or remnant hematopoietic cells.[29] It is possible that in some cases of ciHHV-6 the virus is defective and unable to reactivate[8] and that the integrated viral genome has suffered a mutation in one of the viral genes essential for lytic replication. In the context of transplantation, 21 cases of ciHHV-6 were reported. ciHHV-6B was reported most commonly among SOT recipients, while ciHHV-6A was mostly seen in allogeneic HSCT recipients.[30] None of the 21 patients developed clinical symptoms related to HHV-6 after transplantation.[31] Further details on ciHHV-6 and SOT can be found in Chapter 12. Lee et al. showed that ciHHV-6 status in liver transplants is associated with an increased rate of allograft rejections and opportunistic infections,[31] and Potenza et al.[27] described the case of a ciHHV-6 patient who received an allogeneic stem-cell transplantation and developed leukopenia and blood marrow failure concomitant with an increased plasma load of HHV-6 DNA, which resolved following antiviral treatment. Finally, a wider pretransplantation screening, including HHV-6 quantification, of organ donors and recipients could prevent misinterpretation of ciHHV-6 in the recipients after transplantation and, if applied in a prospective study from a specific geographical area, may help to definitively estimate the prevalence of the ciHHV-6 phenomenon in transplant patients in a given country.

An evaluation of 194 children and infants for encephalitis found a ciHHV-6 prevalence rate of 3.3%, or four times the rate in normal controls, raising the possibility that reactivation of ciHHV-6 may be the trigger of encephalitis in some patients.[32] The occurrence of reactivation in cerebrospinal fluid has been suggested in children undergoing chemotherapy for acute leukemia.[33] Troy et al.[34] and Montoya et al.[35] reported three ciHHV-6 patients, one with severe encephalomyelitis and two with debilitating central nervous system dysfunctions, in whom clinical and virologic improvement correlated with viral load decreases following antiviral treatments, demonstrating for the first time that ciHHV-6 patients presenting with chronic or acute CNS dysfunction might benefit from antiviral therapy.[34,35] This finding supports the emerging hypothesis that in some cells of ciHHV-6 patients the virus reactivates or can become infected by an exogenous HHV-6 strain. In either case, abundant late proteins associated with viral replication are produced; the reactivated cells either die because of lytic virus replication or are killed by immune responses, resulting in a chronic state of inflammation leading to clinical symptoms.[35]

MECHANISM OF CHROMOSOMAL INTEGRATION

Despite growing interest in the many aspects of this oddity that is ciHHV-6, the mechanisms explaining HHV-6 integration are still mainly speculative; however, a few structural elements within HHV-6A and HHV-6B genomes guide us toward research avenues that could explain HHV-6 integration. A brief description of these is provided below. With the exception of HHV-6, no other human herpesvirus readily integrates the human genome; however, Marek's disease virus (MDV), a T-lymphoma causing alphaherpesvirus of chicken, efficiently integrates its genome into the cellular chromosomes. Marek's disease virus shares genomic similarities with human herpes simplex virus 1 and 2.[36] The integration mechanism is not clearly understood, but its genomic structure reveals important information. In fact, the MDV genome consists of one unique long sequence

(U_L) and one unique short sequence (U_S), both flanked by a terminal repeated sequence (TR) and an internal repeated sequence (IR).[37] What appears to be crucial for MDV integration is the presence, in the TR and at the junction of the IR, of short repeated sequences (CCCTAA) identical to telomeric sequences.[36,38] Moreover, the integration sites of MDV are preferentially in the vicinity of chromosomal extremities of its host, within telomeres.[39] HHV-6 does not share the same genomic structure as MDV. The HHV-6 genome is made up of a unique long sequence (U_L), joined by two directly repeated (DR) sequences, which are themselves comprised of assembling/cleavage sites called Pac1 and Pac2.[40–45] Interestingly, HHV-6 also possesses the CCCTAA telomeric repeated sequences (TRSs) identical to human telomeric repeats followed by a Pac2 site. HHV-6 genomes also contain similar but imperfect TRSs, called het(CCCTAA)n, preceding the Pac1 site.[45,46] To date, all HHV-6 integration sites reported are localized with the human telomeric region, without any apparent preference for a specific chromosome (reviewed in Morissette and Flamand[16]). Thus, homologous recombination between viral TRS and human telomeric sequences appears to be one probable hypothesis to explain ciHHV-6 at this location. Depending on the stage of the infectious cycle, three different forms of the viral genome are present (linear, episomal, or concatemeric), and whether one or all genomic forms can integrate the human chromosomes remains to be fully investigated. However, we anticipate that by homologous recombination all forms of DNA can integrate the telomeric region and be reactivated to form a circular viral episome containing a single DR (Figure 15.1).

In a recent report, Huang et al.[47] provide a detailed analysis of the structure of the integrated virus. The right end of the integrated genome (DR_R), closest to the subtelomere region, lacks the Pac2 sequence (likely due to the recombination event; see Figure 15.1), while the left end of the viral genome (DR_L) is lacking the Pac1 sequence. The loss of Pac1 is likely attributable to telomere erosion up until the viral TRS are reached. The viral TRS can then serve as template for telomere lengthening by either the telomerase or alternative processes. The data reported by Huang et al. also indicate that the telomere on the distal end of the integrated virus is frequently the shortest measured in somatic cells.[47] The excision/reactivation of HHV-6 could be possible via a second homologous recombination or by transcriptional amplification after chromatin manipulation for example.[48,49] Huang et al.[47] reported the presence of extra-chromosomal circular HHV-6 molecules, some made up of the entire HHV-6 genome, with a single DR region flanked with Pac1 and Pac2 sequences, suggesting that, once integrated, the virus can excise itself and reconstitute a full-length viral episome.[47] In support, Arbuckle et al.[23] reported evidence that integrated HHV-6 could reactivate and yield infectious virions. These results were recently confirmed by Endo et al.,[50] who provided a convincing demonstration that PBMC cells from a ciHHV-6 subject afflicted with severe combined immunodeficiency (SCID) could produce infectious HHV-6 virions that appeared identical to the integrated HHV-6 but different from other isolates.

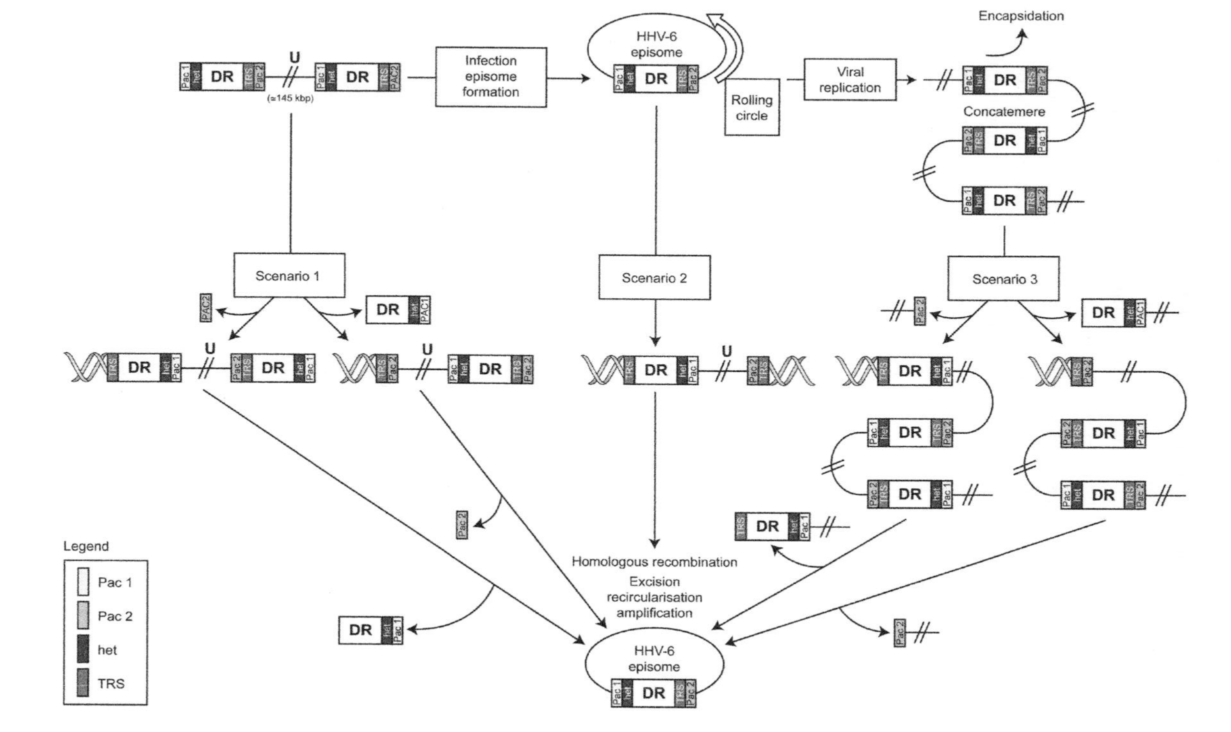

FIGURE 15.1 Proposed models of HHV-6 DNA integration into the telomeric regions of human chromosomes. Three scenarios are proposed depending on the nature of the viral DNA. Scenario 1 describes the integration of a linear viral genome, scenario 2 depicts integration of the viral episome, and scenario 3 portrays the integration of concatemeric viral DNA. Once integrated, the virus could excise itself from the genomic DNA through homologous recombination events between two TRS sequences. The resulting episome could then serve as a replication template for the generation of new infectious virions. DR=Direct repeat; TRS= Telomeric Repeat Sequences; het=heterogeneous TRS.

In their study of infants with transplacentally acquired HHV-6 whose mothers had ciHHV-6, Gravel et al.[20] reached similar conclusions.

It is unlikely that homologous recombination occurs without the participation of cellular factors. Which cellular components could possibly contribute to this integration? Because telomeres are the preferred sites for chromosomal integration, we can speculate that several cellular components related to the biology of telomeres are implicated. For comparison, MDV appears to use some properties of the telomere complexes to develop its oncogenic potential, as MDV oncogenic strains encode for two copies of a protein with 88% homology with the chicken telomerase gene (cTR). This viral telomerase (vTR) can complement the reverse telomeric transcriptase (TERT) in functional assays[51] and potentially plays a role in the elongation and homologous recombination between the viral and the chicken TRS. In the case of HHV-6, Arbuckle et al.[48] have suggested that, because telomeric repeat binding factor 2 (TRF2), which is part of the sheltering complex, can bind the EBV sequences TTAGGGTTA,[52] TRF1/2 could do the same with the TRS of HHV-6 and facilitate viral integration. Many other proteins or cellular complexes could be implicated in ciHHV-6, such as T-circles that are present in the alternative lengthening of telomeres.[53] However, the sole presence of TRS within HHV-6 and the participation of cellular proteins associated with telomeres cannot fully explain HHV-6 integration. Supporting this is the observation that HHV-7, which is very similar to HHV-6 and also possesses the TRS, does not integrate. Hall et al.[54] tested for the presence of HHV-6 or HHV-7 DNA in a large sample of cord blood and found that 1% of samples were positive for ciHHV-6 and found no evidence of ciHHV-7 out of 2129 samples. Thus, HHV-6 possesses at least one attribute, absent from HHV-7, that promotes its integration. The HHV-6 protein that is encoded by open reading frame U94 is the most likely candidate (see next section).

Another important point is to distinguish between *de novo* integration versus inherited ciHHV-6. Inherited ciHHV-6 refers to the transmission from a parent (with ciHHV-6) to an offspring and the integrated HHV-6. Because ciHHV-6 is part of the genome, the transmission occurs according to the law of chromosomal segregation initially postulated by Gregor Mendel such that 50% of descendants have a chance of receiving ciHHV-6. Supporting this is the fact that the same HHV-6 integration sites are observed between parents and children.[14] But, how can HHV-6 colonize the entire body? The simplest explanation is that integration can occur following infection of the gametes prior fecundation. Kaspersen et al.[55] reported that HHV-6 can be intermittently found in the ejaculates of healthy subjects and that HHV-6B could rapidly associate with the spermatozoid acrosome *in vitro*. Several questions, though, remain unanswered, such as whether sperm cell HHV-6 can penetrate the ovule and whether viral integration is possible without altering the first steps of fecundation. For ethical reasons, it will be difficult to answer these questions, but it is more than likely that HHV-6 can infect and integrate its genome into gametes (spermatozoid or ovum). *De novo* HHV-6 integration of somatic diploid cells could be demonstrated by infection of cell lines with HHV-6.[23] As for inherited ciHHV-6, *de novo* integration of HHV-6 has been found to take place within the subtelomeric/telomeric region. It is therefore likely that following infection, a few somatic cells become integrated with HHV-6. Whether HHV-6 genome integration represents the sole form of latency remains to be studied in greater detail.

ROLE FOR HHV-6 U94 IN CHROMOSOMAL INTEGRATION

Human herpesvirus 6 has a distinctive feature not possessed by other human herpesviruses that might favor viral integration: an open reading frame (U94) coding for a protein that shares homologies (24% amino acid identity) with the REP68/78 protein of the parvovirus adeno-associated virus type 2 (AAV-2).[56] The REP68/78 is a nonstructural protein essential for AAV-2 integration within chromosome 19.[57,58] The REP68/78 protein has three basic activities that play a role in AAV-2 integration: DNA-binding, endonuclease, and helicase-ATPase activities. Partial characterization of U94 demonstrated that the protein possesses single-stranded DNA binding activities.[59,60] Mutational analysis of the AAV-2 REP68 has identified several key amino acids that are essential for such functions and that are conserved within U94,[61,62] suggesting that U94 might possess these activities, especially considering the fact that HHV-6 U94 can functionally complement an AAV-2 REP68/78 deletion mutant.[63] Recently, Trempe et al.[64] provided evidence that U94 binds single- and double-stranded DNA containing telomeric motifs. Furthermore, U94 possesses the ability to hydrolyze ATPase into ADP as well as 3′–5′ exonuclease and helicase activities.[64] These facts argue that U94 retains at least some basic functional activities similar to those of REP68/78 that likely play an important role in HHV-6 integration. An ultimate role for U94 in integration will come from work that is in progress that will compare the integration capability of a U94 deletion mutant of HHV-6.

CONCLUSION

Although ciHHV-6 was initially considered an oddity, it is now considered a possible and likely way for HHV-6 to establish latency. Unlike other human herpesviruses that must express at least one viral protein to ensure the maintenance of the viral episome, such as EBNA-1 for EBV, by integrating its DNA within the human genome HHV-6 can remain completely silent and avoid recognition by the immune system. This could explain, at least in part, why the frequency of HHV-6 antigen-reactive T cells in peripheral blood of healthy individuals is much lower than that observed for EBV and CMV. Much remains to be learned about the biological consequences linked with chromosomal integration. The fact that integration takes place within the telomeric regions raises a number of questions pertaining to the long-lasting proliferative potential of cells carrying ciHHV-6. In addition, the fact that HHV-6 does not target a particular chromosome makes it likely that several unrelated pathologies/dysfunctions could be associated with germ-line inheritance of integrated HHV-6 depending on the targeted chromosome. At present, ciHHV-6 raises numerous biological and medical questions. Herpesvirologists must keep an open mind and acknowledge that, although all herpesviruses do share structural similarities, their life cycles can be extremely diverse even in ways initially considered artefactual.

ACKNOWLEDGMENTS

The chapter contribution by Luppi and Potenza is dedicated to retired professor Giuseppe Torelli, our mentor.

REFERENCES

1. Luppi M, Marasca R, Barozzi P, Ferrari S, Ceccherini-Nelli L, Batoni G, et al. Three cases of human herpesvirus-6 latent infection: integration of viral genome in peripheral blood mononuclear cell DNA. *J Med Virol* 1993;**40**(1):44–52.
2. Salahuddin SZ, Ablashi DV, Markham PD, Josephs SF, Sturzenegger S, Kaplan M, et al. Isolation of a new virus, HBLV, in patients with lymphoproliferative disorders. *Science* 1986; **234**(4776):596–601.
3. Jarrett RF, Gledhill S, Qureshi F, Crae SH, Madhok R, Brown I, et al. Identification of human herpesvirus 6-specific DNA sequences in two patients with non-Hodgkin's lymphoma. *Leukemia* 1988;**2**(8):496–502.
4. Torelli G, Marasca R, Luppi M, Selleri L, Ferrari S, Narni F, et al. Human herpesvirus-6 in human lymphomas: identification of specific sequences in Hodgkin's lymphomas by polymerase chain reaction. *Blood* 1991;**77**(10):2251–8.
5. Luppi M, Barozzi P, Garber R, Maiorana A, Bonacorsi G, Artusi T, et al. Expression of human herpesvirus-6 antigens in benign and malignant lymphoproliferative diseases. *Am J Pathol* 1998;**153**(3):815–23.
6. Lacroix A, Collot-Teixeira S, Mardivirin L, Jaccard A, Petit B, Piguet C, et al. Involvement of human herpesvirus-6 variant B in classic Hodgkin's lymphoma via DR7 oncoprotein. *Clin Cancer Res* 2010;**16**(19):4711–21.
7. Torelli G, Barozzi P, Marasca R, Cocconcelli P, Merelli E, Ceccherini-Nelli L, et al. Targeted integration of human herpesvirus 6 in the p arm of chromosome 17 of human peripheral blood mononuclear cells *in vivo*. *J Med Virol* 1995;**46**(3):178–88.
8. Luppi M, Barozzi P, Marasca R, Ceccherini-Nelli L, Torelli G. Characterization of human herpesvirus 6 genomes from cases of latent infection in human lymphomas and immune disorders. *The J Infect Dis* 1993;**168**(4):1074–5.
9. Morris C, Luppi M, McDonald M, Barozzi P, Torelli G. Fine mapping of an apparently targeted latent human herpesvirus type 6 integration site in chromosome band 17p13.3. *J Med Virol* 1999;**58**(1):69–75.
10. Luppi M, Barozzi P, Morris C, Maiorana A, Garber R, Bonacorsi G, et al. Human herpesvirus 6 latently infects early bone marrow progenitors *in vivo*. *J Virol* 1999;**73**(1):754–9.
11. Luppi M, Barozzi P, Bosco R, Vallerini D, Potenza L, Forghieri F, et al. Human herpesvirus 6 latency characterized by high viral load: chromosomal integration in many, but not all, cells. *J Infect Dis* 2006;**194**(7):1020–1. author reply 1–3.

12. Daibata M, Taguchi T, Sawada T, Taguchi H, Miyoshi I. Chromosomal transmission of human herpesvirus 6 DNA in acute lymphoblastic leukaemia. *Lancet* 1998; **352**(9127):543–4.
13. Daibata M, Taguchi T, Taguchi H, Miyoshi I. Integration of human herpesvirus 6 in a Burkitt's lymphoma cell line. *Br J Haematol* 1998;**102**(5):1307–13.
14. Daibata M, Taguchi T, Nemoto Y, Taguchi H, Miyoshi I. Inheritance of chromosomally integrated human herpesvirus 6 DNA. *Blood* 1999;**94**(5):1545–9.
15. Nacheva EP, Ward KN, Brazma D, Virgili A, Howard J, Leong HN, et al. Human herpesvirus 6 integrates within telomeric regions as evidenced by five different chromosomal sites. *J Med Virol* 2008;**80**(11):1952–8.
16. Morissette G, Flamand L. Herpesviruses and chromosomal integration. *J Virol* 2010; **84**(23):12100–9.
17. Leong HN, Tuke PW, Tedder RS, Khanom AB, Eglin RP, Atkinson CE, et al. The prevalence of chromosomally integrated human herpesvirus 6 genomes in the blood of UK blood donors. *J Med Virol* 2007;**79**(1):45–51.
18. Hall CB, Caserta MT, Schnabel K, Shelley LM, Marino AS, Carnahan JA, et al. Chromosomal integration of human herpesvirus 6 is the major mode of congenital human herpesvirus 6 infection. *Pediatrics* 2008;**122**(3):513–20.
19. Hall CB, Caserta MT, Schnabel KC, Shelley LM, Carnahan JA, Marino AS, et al. Transplacental congenital human herpesvirus 6 infection caused by maternal chromosomally integrated virus. *J Infect Dis* 2010;**201**(4):505–7.
20. Gravel A, Hall CB, Flamand L. Sequence analysis of transplacentally acquired human herpesvirus 6 DNA is consistent with transmission of a chromosomally integrated reactivated virus. *J Infect Dis* 2013;**207**(10):1585–9.
21. Tanaka-Taya K, Sashihara J, Kurahashi H, Amo K, Miyagawa H, Kondo K, et al. Human herpesvirus 6 (HHV-6) is transmitted from parent to child in an integrated form and characterization of cases with chromosomally integrated HHV-6 DNA. *J Med Virol* 2004;**73**(3):465–73.
22. Clark DA, Tsao EHF, Leong HN, Ward KN, Nacheva EP, Griffiths PD. Reply to Boutolleau et al. and Luppi et al. *J Infect Dis* 2006;**194**:1022–3.
23. Arbuckle JH, Medveczky MM, Luka J, Hadley SH, Luegmayr A, Ablashi D, et al. The latent human herpesvirus-6A genome specifically integrates in telomeres of human chromosomes *in vivo* and *in vitro*. *Proc Natl Acad Sci USA* 2010;**107**:5563–8.
24. Pellett PE, Ablashi DV, Ambros PF, Agut H, Caserta MT, Descamps V, et al. Chromosomally integrated human herpesvirus 6: questions and answers. *Rev Med Virol* 2012;**22**(3):144–55.
25. Ward KN, Leong HN, Nacheva EP, Howard J, Atkinson CE, Davies NW, et al. Human herpesvirus 6 chromosomal integration in immunocompetent patients results in high levels of viral DNA in blood, sera, and hair follicles. *J Clin Microbiol* 2006;**44**(4):1571–4.
26. Ljungman P, de la Camara R, Cordonnier C, Einsele H, Engelhard D, Reusser P, et al. Management of CMV, HHV-6, HHV-7 and Kaposi's sarcoma herpesvirus (HHV-8) infections in patients with hematological malignancies and after SCT. *Bone Marrow Transplant* 2008; **42**(4):227–40.
27. Potenza L, Barozzi P, Rossi G, Riva G, Vallerini D, Zanetti E, et al. May the indirect effects of CIHHV-6 in transplant patients be exerted through the reactivation of the viral replicative machinery?. *Transplantation* 2011;**92**(9):e49–51. Author reply e2.
28. Soderberg-Naucler C, Fish KN, Nelson JA. Reactivation of latent human cytomegalovirus by allogeneic stimulation of blood cells from healthy donors. *Cell* 1997;**91**(1):119–26.
29. Potenza L, Barozzi P, Masetti M, Pecorari M, Bresciani P, Gautheret-Dejean A, et al. Prevalence of human herpesvirus-6 chromosomal integration (CIHHV-6) in Italian solid organ and allogeneic stem cell transplant patients. *Am J Transplant* 2009;**9**(7):1690–7.

30. Lee SO, Brown RA, Razonable RR. Chromosomally integrated human herpesvirus-6 in transplant recipients. *Transplant Infect Dis* 2012;**14**(4):346–54.
31. Lee SO, Brown RA, Razonable RR. Clinical significance of pretransplant chromosomally integrated human herpesvirus-6 in liver transplant recipients. *Transplantation* 2011;**92**(2):224–9.
32. Ward KN, Thiruchelvam AD, Couto-Parada X. Unexpected occasional persistence of high levels of HHV-6 DNA in sera: detection of variants A and B. *J Med Virol* 2005;**76**(4):563–70.
33. Wittekindt B, Berger A, Porto L, Vlaho S, Gruttner HP, Becker M, et al. Human herpes virus-6 DNA in cerebrospinal fluid of children undergoing therapy for acute leukaemia. *Br J Haematol* 2009;**145**(4):542–5.
34. Troy SB, Blackburn BG, Yeom K, Caulfield AK, Bhangoo MS, Montoya JG. Severe encephalomyelitis in an immunocompetent adult with chromosomally integrated human herpesvirus 6 and clinical response to treatment with foscarnet plus ganciclovir. *Clin Infect Dis* 2008;**47**(12):e93–6.
35. Montoya JG, Neely MN, Gupta S, Lunn MR, Loomis KS, Pritchett JC, et al. Antiviral therapy of two patients with chromosomally-integrated human herpesvirus-6A presenting with cognitive dysfunction. *J Clin Virol* 2012;**55**(1):40–5.
36. Kishi M, Bradley G, Jessip J, Tanaka A, Nonoyama M. Inverted repeat regions of Marek's disease virus DNA possess a structure similar to that of the a sequence of herpes simplex virus DNA and contain host cell telomere sequences. *J Virol* 1991;**65**(6):2791–7.
37. Cebrian J, Kaschka-Dierich C, Berthelot N, Sheldrick P. Inverted repeat nucleotide sequences in the genomes of Marek disease virus and the herpesvirus of the turkey. *Proc Natl Acad Sci USA* 1982;**79**(2):555–8.
38. Kaufer BB, Jarosinski KW, Osterrieder N. Herpesvirus telomeric repeats facilitate genomic integration into host telomeres and mobilization of viral DNA during reactivation. *J Exp Med* 2011;**208**(3):605–15.
39. Robinson CM, Hunt HD, Cheng HH, Delany ME. Chromosomal integration of an avian oncogenic herpesvirus reveals telomeric preferences and evidence for lymphoma clonality. *Herpesviridae* 2010;**1**(1):5.
40. Deng H, Dewhurst S. Functional identification and analysis of *cis*-acting sequences which mediate genome cleavage and packaging in human herpesvirus 6. *J Virol* 1998;**72**(1):320–9.
41. Dominguez G, Dambaugh TR, Stamey FR, Dewhurst S, Inoue N, Pellett PE. Human herpesvirus 6B genome sequence: coding content and comparison with human herpesvirus 6A. *J Virol* 1999;**73**(10):8040–52.
42. Gompels UA, Nicholas J, Lawrence G, Jones M, Thomson BJ, Martin ME, et al. The DNA sequence of human herpesvirus-6: structure, coding content, and genome evolution. *Virology* 1995;**209**(1):29–51.
43. Isegawa Y, Mukai T, Nakano K, Kagawa M, Chen J, Mori Y, et al. Comparison of the complete DNA sequences of human herpesvirus 6 variants A and B. *J Virol* 1999;**73**(10):8053–63.
44. Martin ME, Thomson BJ, Honess RW, Craxton MA, Gompels UA, Liu MY, et al. The genome of human herpesvirus 6: maps of unit-length and concatemeric genomes for nine restriction endonucleases. *J Gen Virol* 1991;**72**(Pt 1):157–68.
45. Thomson BJ, Dewhurst S, Gray D. Structure and heterogeneity of the a sequences of human herpesvirus 6 strain variants U1102 and Z29 and identification of human telomeric repeat sequences at the genomic termini. *J Virol* 1994;**68**(5):3007–14.
46. Gompels UA, Macaulay HA. Characterization of human telomeric repeat sequences from human herpesvirus 6 and relationship to replication. *J Gen Virol* 1995;**76**(Pt 2):451–8.
47. Huang Y, Hidalgo-Bravo A, Zhang E, Cotton VE, Mendez-Bermudez A, Wig G, et al. Human telomeres that carry an integrated copy of human herpesvirus 6 are often short and unstable, facilitating release of the viral genome from the chromosome. *Nucleic Acids Res* 2013.

48. Arbuckle JH, Medveczky PG. The molecular biology of human herpesvirus-6 latency and telomere integration. *Microbes Infect* 2011;**13**(8–9):731–41.
49. Trempe F, Mosrissette G, Gravel A, Flamand L. L'herpesvirus humain de type 6 et l'intégraton chromosomique: conséquences biologiques et médicales associées. *Virologie* 2011;**15**(6):381–93.
50. Endo A, Imai K., Katano H, Inoue N, Ohye T, Kurahashi H, et al. Chromosomally integrated human herpesvirus-6 was activated in a patient with X-linked severe combined immunodeficiency. Presented at 8th International Conference on HHV-6 and -7, Paris, France, April 8–10, 2013; Abstract #4-8.
51. Fragnet L, Blasco MA, Klapper W, Rasschaert D. The RNA subunit of telomerase is encoded by Marek's disease virus. *J Virol* 2003;**77**(10):5985–96.
52. Deng Z, Lezina L, Chen CJ, Shtivelband S, So W, Lieberman PM. Telomeric proteins regulate episomal maintenance of Epstein–Barr virus origin of plasmid replication. *Mol Cell* 2002;**9**(3):493–503.
53. Arbuckle JH, Pantry S, Medveczky PG. The mechanism and significance of integration and vertical transmission of human herpesvirus 6 genome. In: Berencsi G, editor. *Maternal Fetal Transmission of Human Viruses and Their Influence on Tumorigenesis*. Amsterdam: Springer; 2012. pp. 171–94.
54. Hall CB, Caserta MT, Schnabel KC, Boettrich C, McDermott MP, Lofthus GK, et al. Congenital infections with human herpesvirus 6 (HHV6) and human herpesvirus 7 (HHV7). *J Pediatr* 2004;**145**(4):472–7.
55. Kaspersen MD, Larsen PB, Kofod-Olsen E, Fedder J, Bonde J, Hollsberg P. Human Herpesvirus-6A/B binds to spermatozoa acrosome and is the most prevalent herpesvirus in semen from sperm donors. *PLoS One* 2012;**7**(11):e48810.
56. Thomson BJ, Efstathiou S, Honess RW. Acquisition of the human adeno-associated virus type-2 rep gene by human herpesvirus type-6. *Nature* 1991;**351**(6321):78–80.
57. Linden RM, Ward P, Giraud C, Winocour E, Berns KI. Site-specific integration by adeno-associated virus. *Proc Natl Acad Sci USA* 1996;**93**(21):11288–94.
58. Linden RM, Winocour E, Berns KI. The recombination signals for adeno-associated virus site-specific integration. *Proc Natl Acad Sci USA* 1996;**93**(15):7966–72.
59. Dhepakson P, Mori Y, Jiang YB, Huang HL, Akkapaiboon P, Okuno T, et al. Human herpesvirus-6 rep/U94 gene product has single-stranded DNA-binding activity. *J Gen Virol* 2002;**83**(Pt 4):847–54.
60. Mori Y, Dhepakson P, Shimamoto T, Ueda K, Gomi Y, Tani H, et al. Expression of human herpesvirus 6B rep within infected cells and binding of its gene product to the TATA-binding protein *in vitro* and *in vivo*. *J Virol* 2000;**74**(13):6096–104.
61. Walker SL, Wonderling RS, Owens RA. Mutational analysis of the adeno-associated virus type 2 Rep68 protein helicase motifs. *J Virol* 1997;**71**(9):6996–7004.
62. Walker SL, Wonderling RS, Owens RA. Mutational analysis of the adeno-associated virus Rep68 protein: identification of critical residues necessary for site-specific endonuclease activity. *J Virol* 1997;**71**(4):2722–30.
63. Thomson BJ, Weindler FW, Gray D, Schwaab V, Heilbronn R. Human herpesvirus 6 (HHV-6) is a helper virus for adeno-associated virus type 2 (AAV-2) and the AAV-2 rep gene homologue in HHV-6 can mediate AAV-2 DNA replication and regulate gene expression. *Virology* 1994;**204**(1):304–11.
64. Trempe F, Morissette G, Gravel A, Flamand L Enzymatic and biological properties of HHV-6 U94/rep protein. Presented at 8th International Conference on HHV-6 and -7, Paris, France, April 8–10, 2013; Abstract #2–1.

Chapter 16

HHV-6 and HHV-7 in Cardiovascular Diseases and Cardiomyopathies

Dirk Lassner[a], Gerhard R.F. Krueger[b], L. Maximilian Buja[c], and Uwe Kuehl[d]

[a]Institute Cardiac Diagnostics and Therapy, Berlin, Germany, [b]The University of Texas Medical School at Houston, Houston, Texas, [c]The University of Texas Health Science Center at Houston and Texas Heart Institute, St Luke's Episcopal Hospital, Texas Medical Center, Houston, Texas, [d]Charité—Universitätsmedizin Berlin, Campus Benjamin Franklin, Berlin, Germany

INTRODUCTION

Human herpesviruses 6 (HHV-6) and HHV-7 belong to the family of nine different herpesviruses that infect 95 to 100% of humans with a life-long persistence after primary infection during childhood.[1–5] HHV-6 exists as distinct species HHV-6A and HHV-6B, which are genetically related with an overall nucleotide identity of 90% but show consistent differences in their biological, immunological, and molecular properties.[5,6] The complete genomes of HHV-6A and HHV-6B have been sequenced. HHV-6B contains 97 unique genes.[5,7]

Human herpesvirus 7, first reported in 1990, is another lymphotropic member of the betaherpesvirus subfamily of herpesviruses. Comparison of the two fully sequenced HHV-7 isolates confirmed that intrastrain variation is low, although more apparent, in repeat regions close to either ends of the genome.[8–10]

Human herpesviruses 6 is a lymphotropic virus with tropism mainly for CD4+ and CD8+ T cells, B cells, and natural killer cells. Although HHV-6 is a lymphotropic virus, it also infects the vascular endothelial cells. Several studies have identified HHV-6-specific DNA in the vascular endothelium *in vivo* and suggest endothelial cell damage by the virus.[11–13] It has been suggested that endothelial cells and cardiac myocytes might be an important reservoir for viral latency and reactivation.[12]

CD46 is one cell receptor for HHV-6.[14] A recent report also indicates that CD134 is a receptor for HHV-6B.[15] The ubiquitous distribution of CD46 on various cells explains the broad tissue tropism and distribution of HHV-6 in individuals in which infected blood cells are in direct contact with the vascular system.

Human Herpesviruses HHV-6A, HHV-6B & HHV-7.

Because CD46 is expressed on distinct types of immune cells,[7,16] HHV-6A/B and HHV-7 infect, either productively or nonproductively, lymphocytes, monocytes, NK cells, T and B cell subsets, dendritic cells, and stem cells.[16,17] The primary target for viral replication of all three viruses is the CD4+ T lymphocyte. HHV-7 utilizes the CD4 receptor as main receptor to enter target cells[18] and therefore can interfere with human immunodeficiency virus type 1 (HIV-1) infection of CD4+ T lymphocytes.[19] In contrast to HHV-6B, HHV-6A can also replicate in various cytotoxic effector cells such as CD8+ T cells and natural killer cells.[17]

These herpesviruses have developed distinct strategies to downregulate the host immune response by, for example, molecular mimicry caused by functional chemokines and chemokine receptors.[7] The nonproductive infection of macrophages and dendritic cells by HHV-6 thus induces effective suppression of Th1-polarized antiviral immune response.[17] On the other hand, HHV-7 infection causes the elevation of the complement regulation proteins CD46 and CD59, which may be a possible mechanism for HHV-7 to evade humoral immunity by activation of the complement cascade.[20] Deficiency in the expression of these proteins may be associated with a lower level of protection of healthy cells against complement-mediated lysis and may also be responsible for the accumulation of immune complexes in tissues.[21] Immunosuppression in HHV-6-infected patients is enhanced by depletion of CD4+ T cells via direct infection of intrathymic progenitors through induction of apoptosis.[7] One reason for clinical symptoms and diseases could thus be the immunosuppressive herpesvirus infection with consecutively reduced effector function of immune cells.

Regulation of the entry receptor CD46 seems to be a unique feature of HHV-6 and HHV-7. Whereas HHV-7 is upregulating the expression of CD46,[20] HHV-6 infection results in a generalized loss of CD46 expression in lymphoid tissues.[17] Because HHV-6 is a suspected agent in the development of systemic lupus erythematosus (SLE), the reduced expression of CD46 in lymphoid cells of SLE patients might be explained by such virus–receptor interactions.[22]

In common with all human herpesviruses, HHV-6A, HHV-6B, and HHV-7 establish lifelong infection following initial exposure and seroconversion.[2,23,24] Based on their biological properties and cellular tropism, HHV-6 and HHV-7 were individually detected in different cells or tissues accompanying various diseases.[14,25,26] Both viruses were also detected in immunocompromised AIDS patients, suggesting a co-infection or their role as co-factors for different infectious diseases in humans.[28]

Unique among human herpesviruses, the HHV-6 genomes can become integrated into the human genome and be transmitted via germline chromosomal integration.[29,30] In patients with chromosomally integrated human herpesvirus 6 (ciHHV-6), every cell harbors one viral genome. These patients are identified by persistent high viral DNA loads in peripheral blood cells.[30–32] HHV-6 reactivation is possible in every cell type.[10,33] The prevalence of ciHHV6 is approximately 0.2 to 0.8% of human populations.[32] Direct confirmation of ciHHV6 was done by fluorescence *in situ* hybridization (FISH) in various tissues.[10,29,30,33]

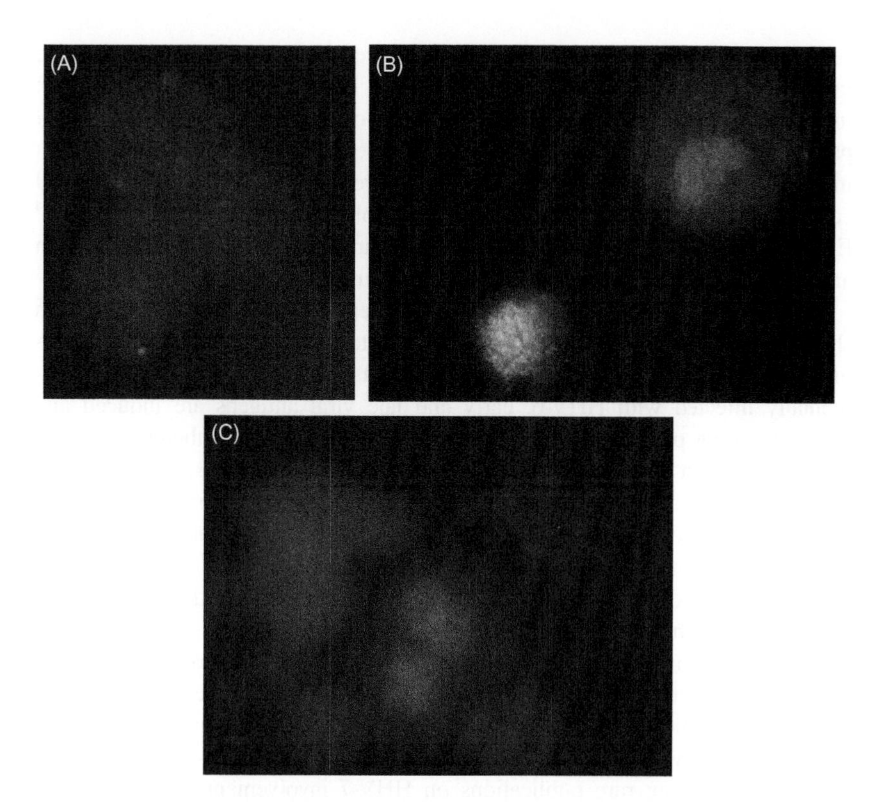

FIGURE 16.1 HHV-6 and HHV-7 infection of lymphocytes and myocardium. FISH analysis of interphase nuclei show (A) a chromosomal integration of HHV-6A genome in lymphocytes of a cardiac patient; (B) lytic HHV-6B infection in Jihan T-cell line; and (C) lytic HHV-7 infection in Jihan T-cell line. HHV-6A and HHV-6B DNA probes were BAC clones, HHV-7 DNA probe was a generated PCR amplicon. *(Photographs courtesy of Benedikt Kaufer and Nina Walleschek, Berlin).*

In FISH analysis, a single chromosomal integration site is demonstrated by one fluorescent dot in the nucleus (Figure 16.1A), whereas lytic infection of HHV-6 or HHV-7 is characterized by massive accumulation of fluorescent foci in the interphase nuclei of analyzed cells (Figure 16.1B and C).

DISEASES OF THE VASCULAR SYSTEM

Endothelial cells are one primary target of herpesvirus infections, which are followed by subsequent entry into the adjacent tissues. The transmission of HHV-6 or HHV7 from infected blood cells to vascular endothelium and the bordering tissues requires an appropriate cellular receptor system. CD46 is strongly expressed in salivary gland epithelial cells and in renal tubular cells; moderately well expressed in lymphoid cells, monocytes, and vascular endothelial cells; and only weakly expressed in muscle cells.[25,34]

Despite the fact that viral DNA was not recovered from peripheral blood mononuclear cells of all patients, HHV-6 has been identified by nested PCR in endothelial cells of the aorta and cardiac microvessels of immunocompetent patients with aortic insufficiency or aneurysm.[12] Because no early antigen but immediate-early and late gene transcripts were detected in aortic endothelial cells, the authors concluded a low-level viral replication occurred at these sites.[34] Endothelial cells may thus represent an important reservoir for viral reactivation and be an important origin for virus-induced injury.[12]

Replicative and latent infection of vascular endothelium by HHV-6 has been shown *in vivo* and *in vitro* using immunohistochemical and molecular techniques.[11–13] If human umbilical vein endothelium cells (HUVECs) are experimentally infected with HHV-6, early and late viral antigens are induced and persist for long periods on cell surfaces. HHV-6-infected endothelial cells thus may well support the attraction of further immunocompetent cells, thereby initiating an inflammatory reaction in the underlying tissue. HHV-6 furthermore upregulates monocyte chemoattractant protein-1 (MCP-1) and interleukin-8 (IL-8) and induces the *de novo* synthesis of RANTES CC chemokine.[13,34–36] The chemokine RANTES is involved in the pathogenesis of cardiovascular disease in mice. A similar effect, however, has not yet been shown for humans.[37]

In contrast to the frequent detection of HHV-6 DNA and of viral antigens in vascular endothelial cells, an association of HHV-6 with vascular disease has been reported only sporadically. Increased levels of HHV-6A antigens and DNA have been observed in endothelial cells of patients with HHV-6 reactivation.[34] There are, however, rare publications on HHV-7 involvement in vascular diseases. Both HHV-6 and HHV-7 DNA has been reported in patients with corneal endothelitis, but its influence on the clinical course of these patients is still uncertain.[38,39] Kawasaki disease has been thought to be related to events associated with a previous HHV-6 infection,[40] but the etiologic role of HHV-6 for Kawasaki syndrome could not be confirmed in another study.[41] A main limitation of both studies is the use of only serologic investigations for determining any possible association of the virus with the vasculitis. Similarly, the only study on Kawasaki disease and HHV-7 was also performed using serological markers.[42]

Infectious agents, including herpes viruses and Chlamydia pneumoniae, have been identified with some frequency in atherosclerotic lesions, but it is still not confirmed that they have a primary role in the pathogenesis of atherosclerosis.[43,44] As discussed above, HHV-6 infection of endothelial cells is able to induce an inflammatory tissue reaction. Kaklikkaya and coworkers[45] found HHV-6 DNA in the carotid artery of 1 of 28 subjects with atherosclerosis, as well as in the iliac artery in 1 of 25 of controls. HHV-7 was identified in neither the atherosclerosis patients nor control cases. The prevalence of both herpesviruses in atherosclerotic lesions is not known, and until now there is no evidence that supports a role for them in atherogenesis.

Although adenovirus, enterovirus, and parvovirus B19V have been implicated in coronary vasculopathy, heart failure, chronic graft failure, and acute rejection, such data for HHV-6 and HHV-7 do not exist.[46,47] Active HHV-6 and HHV-7 infections are not associated with coronary instability induced by infection-driven inflammatory burden;[48] however, endothelial function of the coronary microcirculation is significantly impaired in patients with myocardial virus persistence compared to patients without virus.[49–51]

No association of HHV-6 and giant cell arteritis (GCA) was found in three studies using polymerase chain reaction (PCR) and real-time PCR on vascular biopsies.[52,53] Interestingly, Alvarez-Lafuente and coworkers[54] found no evidence of HHV-6 but did find B19V in temporal artery biopsies of patients with GCA. The prevalence of infections by B19V (54%) and HHV-6 (35%) in GCA is comparable to the prevalence in endomyocardial biopsies (EMBs) of patients with dilated cardiomyopathy.[55] In about 20% of all cardiac biopsies from cardiomyopathic patients, a double infection by HHV-6 and B19V is detectable.[55,56] Similar to HHV-6, parvovirus B19V infects the vascular endothelium. It has been shown that interferon-beta reduces endothelial damage and improves endothelial dysfunction in B19V-positive patients with symptomatic endothelial dysfunction.[49] Treatment of virally induced GCA could further improve endothelial function by reducing parvoviral activity, which might be a cofactor of possible herpesvirus-induced arthritis.

HHV-6 AND HHV-7 AS CARDIAC PATHOGENS

Viral infections of the heart develop within pathologically distinct phases.[57] Most information on this issue has been gained from enteroviral or adenoviral infections that enter the host through the gastrointestinal or respiratory tract and can reside in the reticuloendothelial system as an extracardiac reservoir. They may enter the heart as a secondary target organ and infect cardiomyocytes after binding to a tight junctional protein, the coxsackie and adenovirus receptor (CAR), and the decay-accelerating factor (DAF, CD55). The colocalization of CAR with the coreceptors for adenovirus internalization avβ3 and avβ5 at the myocyte surface suggests that it is an important molecular determinant for the cardiotropism of both viruses in viral heart disease.[58,59] B19V and HHV-6, on the other hand, infect vascular endothelial cells (ECs) or other tissue cells, including myocytes.[49] These different infection sites in cardiac tissue, in addition to different virus variants and virus loads, may explain the heterogenicity of viral heart disease with respect to expression of its phenotype, clinical presentation, and prognosis.

Patients presenting a viral heart disease can be categorized, based on left ventricular function at the time of presentation, into subjects with preserved systolic function, impaired diastolic function, segmental or compensated left ventricular dysfunction, or acute systolic left ventricular compromise.[60] The kind and extent of myocardial compromise depend on the affected cardiac structures and resulting myocardial lesions. Systolic dysfunction following cardiomyocyte infection (e.g., entero- and adenoviruses) and degeneration may result from disturbance of the elements that are responsible for force generation and force transmission (partly reversible), or they may be caused by an irreversible loss of contractile tissue.[55–57] Infection of endothelial cells (e.g., by B19V or herpesviruses) is more often associated with endothelial or diastolic dysfunction but preserved systolic function.[49–51] Under certain circumstances, ongoing myocardial injury may result from chronic immune stimulation induced by incompletely cleared viruses or post-infectious autoimmune processes.[57,61]

If herpesvirus-infected endothelium is not associated with reactive inflammation, the vascular cell layer may be the origin for infection of the underlying tissues. Recently, HHV-6 genomes have been detected in explanted hearts, postmortem,[62] and in the EMBs of patients with clinically suspected myocarditis and dilated cardiomyopathy (DCM).[55,56,63] Short-term follow-up of HHV-6-positive patients revealed a possible association between the clinical course of the disease and the spontaneous course of the viral infection, suggesting an involvement of HHV-6 in acquired cardiomyopathies.[63]

The distribution of viral receptors in the myocardium for well-documented cardiotropic viruses (coxsackievirus, adenovirus) is highly variable, with a significant increase in patients with dilated cardiomyopathy.[59] CAR is a key determinant for the cellular uptake of both viruses and for the molecular pathogenesis of coxsackievirus and adenovirus-associated diseases.[58] With respect to herpesvirus receptors, differentiation of porcine mesenchymal stem cells into cardiomyocytes revealed the expression of CD46.[64] The dynamics of viral receptors in myocardium have been shown for both CAR and CD46. In DCM, CAR and CD46 are also expressed by endothelial cells, suggesting that viruses that target these receptors could more easily gain entry into heart cells after intravascular administration.[65]

Viral infection is a leading cause of myocarditis. In a large series, viral genome was amplified by polymerase chain reaction from 239 (38%) of 624 patients with clinical myocarditis and 30 (20%) of 149 patients with dilated cardiomyopathy.[46] The most frequently detected viruses were adenovirus and enterovirus. Enteroviruses and adenoviruses are common causes of viral myocarditis.[61,63] In the last decade, Epstein–Barr virus (EBV), parvovirus B19, and HHV-6 as new pathogens have been frequently detected in EMBs of patients with clinically suspected myocarditis and DCM in Germany and elsewhere. The clinical presentation and course of the disease are dependent on the type and spontaneous course of virus. Myocardial persistence of both B19V and HHV-6 correlate with a decline in the systolic ejection fraction within 6 months. If the virus is eliminated by the immune response or antiviral treatment, myocardial function improves.[55,56,61,63] Administration of interferon-beta (IFN-β) therapy for chronic enterovirus infection favors virus clearance and reduces virus-induced myocardial injury, improving the long-term survival rate of treated patients.[65]

In a recent series of biopsies from 245 consecutive patients with dilated cardiomyopathy found dominant, 51.4% of the biopsies tested positive for B19V, 21.6% for HHV-6, 9.4% positive for enterovirus, and 1.6% positive for adenovirus. Interestingly, 27.3% had evidence of multiple infections.[55] Active or borderline myocarditis according to the Dallas criteria was not present in any case. Also, lymphocyte and macrophage infiltrates were not significantly different in virus-positive and virus-negative patients. Similar recovery of virus has not been reported in normal hearts of multiorgan donors or patients with valvular heart disease.[67] The findings in the patients with idiopathic left ventricular dysfunction suggest that viral persistence, often presenting as multiple infection, may play a role in the pathogenesis of dilated cardiomyopathy.[55] Mahrholdt and colleagues[56] reported similar frequencies of parvovirus B19 (56%), HHV-6 (18%), and double infections by B19V and HHV-6 (17%) in patients with myocarditis diagnosed by magnetic resonance imaging and EMBs.

Virus-induced microcirculatory dysfunction with impaired coronary blood flow can be important in the pathogenesis of cardiomyopathy. A group of 71 patients with nonischemic cardiomyopathy were evaluated for coronary microcirculatory dysfunction in relationship to viral persistence and myocardial inflammation with endothelial activation.[49] Polymerase chain reaction showed evidence of viral persistence in 43 patients: parvovirus B19 (46%) and HHV-6 (17%). Myocardial inflammation was measured by increased numbers of lymphocytes and expression of endothelial adhesion molecules.[68,69] Endothelial dysfunction occurred in patients with viral persistence independently of myocardial inflammation and endothelial activation but was more pronounced in patients with concurrent inflammation.

In another study, 37 of 70 patients who presented with exertional dyspnea and/or reduced exercise tolerance were confirmed to have isolated diastolic left ventricular dysfunction by echocardiography and cardiac catheterization. In 37 of 70 patients (53%), isolated diastolic dysfunction was confirmed as the cause of their clinical symptoms. In 35 of these patients, cardiotropic viral genomes were detected in EMBs.[51] Ten of these patients had evidence of both coronary endothelial impairment and diastolic dysfunction. All 10 patients were positive for parvovirus B19 genomes, with 7 of the 10 having a mono-infection and 3 of the 10 a co-infection with HHV-6. These data suggest that endothelial and microcirculatory dysfunction could be induced by B19V, supported by co-infection with other cardiotropic viruses (e.g., HHV-6), and as such may be a possible pathomechanism underlying diastolic dysfunction.

Similar to the other herpes viruses, HHV-6 becomes frequently reactivated by infections or drugs with subacute clinical presentations, especially in acquired or drug-induced immunodeficiencies (e.g., transplant recipients) or in patients with autoimmune disorders. It has been suggested that HHV-6 enhances the pathogenicity of other viruses rather than being a pathogen itself.[32,70,71] HHV-6 may directly damage cells, particularly endothelial cells, or induce immune or autoimmune reactions. It furthermore can transactivate other viral infections (e.g., EBV, B19V), which suggests that the co-infection may enhance the pathogenicity of the other viruses.[72] In this respect, even a transient viral infection may be sufficient to cause a chronic disease due to direct cytopathic effects or induction of low-grade inflammation with release of cytokines that alter cell signaling pathways, or may affect extracellular matrix components.[34,35,49,61] These mechanisms may be particularly important in HHV-6-induced myocardial diseases. Reports of HHV-6-associated myocarditis are rather rare despite its frequent detection in DCM.[54,73] Specific conditions may favor either HHV6-associated cardiac diseases, as two cases of herpesvirus-induced myocarditis were correlated with immunosuppressive therapy,[70] or general immunosuppression by double infection with parvovirus B19V.[71]

The high pattern of co-infection by B19V and HHV-6 warrants new attention to the process of viral reactivation. About 50% of all analyzed EMBs were positive for B19V, whereas 10% of them have shown an expression of viral RNA.[74,75] Transactivation was associated with a higher percentage of chest pain and angina while administration of a replication inhibitor was associated with clinical improvement. Inhibition of viral RNA by HHV-6-specific drugs could thus be a therapeutic option for HHV-6-positive patients with unexplained cardiac symptoms.[29]

Although sensitive PCR has detected HHV-6 genomes in distinct organs, the direct visualization of HHV-6 or HHV-7 viral particles or proteins in injured or inflamed vessels has not yet been demonstrated. Recently, however, Krueger and colleagues[76] presented unequivocal evidence for the deposition of HHV-6 antigens in endothelial cells (Figure 16.2) and degenerating cardiomyocytes using immunopathology, *in situ* hybridization, and electron microscopy, with suspected viral particles being found in a cardiomyocyte. We have observed a similar distribution of HHV-6 antigen myocardium and in the perivascular region of vascular endothelium (Figure 16.2). Until now, ultrastructural studies on HHV-7 have only been done in cell cultures.[77]

Based on the overall prevalence of ciHHV-6, patients with unexplained left ventricular dysfunction will also harbor chromosomally integrated HHV-6 at a comparable frequency of approximately 1%.[32,33] Elimination of chromosomally integrated virus is impossible, but patients with symptomatic heart failure get benefit from an antiviral therapy causing a reduction of the transcriptional activity of ciHHV-6.[31,32]

Today only a few reports describe HHV-7 as a cardiac pathogen; in a larger retrospective study with archival cardiac tissue from children with idiopathic dilated cardiomyopathy or congenital heart disease, HHV-6 DNA but no HHV-7 or HHV-8 DNA was detected.[72] We have analyzed EMBs that were primarily checked for HHV-6 DNA (50% were positive by PCR) for the prevalence of HHV-7. HHV-7 DNA was not present in any of the 40 tissue samples but was detected in 2 out of 10 parallel examined peripheral blood cells (PBMCs), and HHV-6 DNA was identified in three blood samples (Table 16.1).[78] With respect to our current knowledge, HHV-7 is not an important cardiac pathogen. Introduction of HHV-7 detection in the routine diagnosis of autopsy material or endomyocardial biopsies may, however, reveal a higher prevalence of this virus in myocardial tissue.[31,55,57,62]

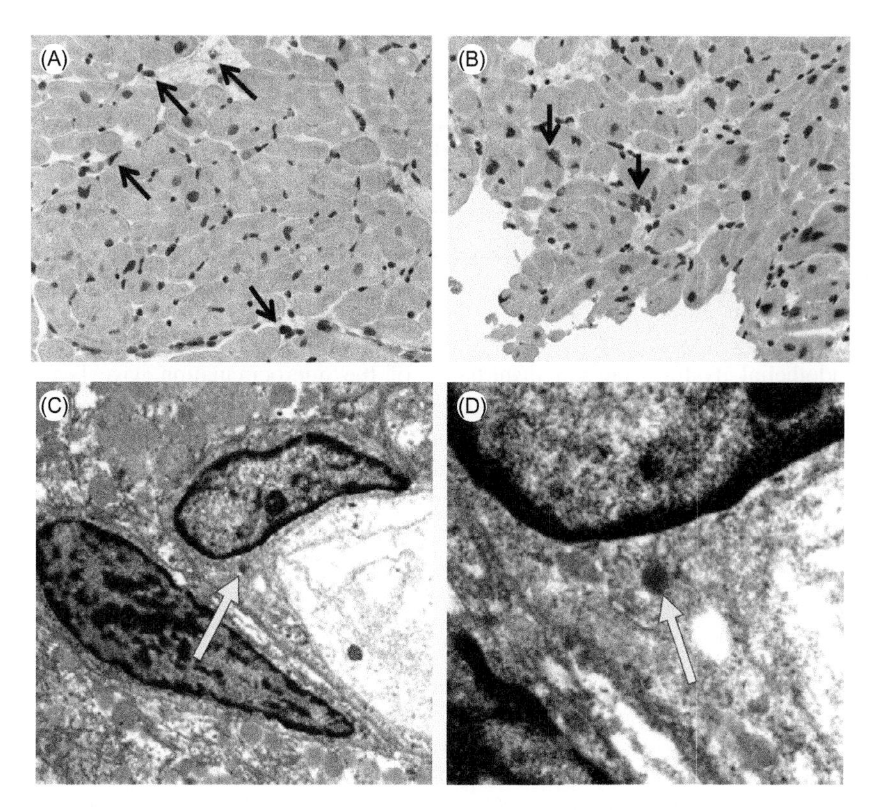

FIGURE 16.2 HHV-6 infection of myocardial tissue. (A, B) Immunohistochemical detection of HHV-6B infection in interstitial region of myocardial tissue. (Specific mAbs directed against (A) p101 of HHV-6 B and (B) envelope glycoprotein 60/110 of HHV-6A and HHV-6B were purchased from Chemicon.) (C, D) Electron microscopy showing herpesvirus particle in vascular endothelium of HHV-6-positive endomyocardial biopsy as well as in AIDS patient (see arrow). *From Kühl U, et al. Herz 2012;* ***37****(6): 637–43.*

CONCLUSIONS

The cardiovascular system may be affected during HHV-6 and HHV-7 infections. Endothelial cells that express the CD46 receptor on their cell surfaces are one of the primary targets of HHV-6. Although sporadic cases of vascular disease and myocarditis have been reported in relation to HHV-6 infection, a high rate of HHV-6 viral genomes has recently been demonstrated in the myocardium of patients with cardiomyopathy and unexplained left ventricular dysfunction. In view of the high frequency of HHV-6 and HHV-7 infections or reactivations in the general population, cardiac involvement by herpesviruses may thus be of clinical relevance even in patients without known preexisting cardiac diseases. These infections should therefore be taken into consideration in patients with unexpected deterioration of heart failure or cardiac symptoms.

TABLE 16.1 Prevalence of HHV-7 in Endomyocardial Biopsies of Cardiac Patients (*n* = 40) (76).

	HHV-6 positive	HHV-7 positive
Heart muscle biopsies (*n* = 40)	20 (50%)	0 (0%)
PBMCs (*n* = 10)	3 (30%)	2 (20%)

Endothelial dysfunction and disturbance of the microcirculation have been associated with B19V, HHV-6, and other viruses even in the absence of myocardial inflammation. An important feature of persisting HHV-6 infection is the transactivation of other latent viruses, which may also contribute to the pathogenesis of cardiomyopathies. However, further systematic studies are necessary to identify the exact role of HHV-6 in the pathogenesis of cardiovascular and myocardial diseases.

REFERENCES

1. De Bolle L, Naesens L, De Clercq E. Update on human herpesvirus 6 biology, clinical features, and therapy. *Clin Microbiol Rev* 2005;**18**(1):217–45.
2. Hall CB, Caserta MT, Schnabel KC, et al. Persistence of human herpesvirus 6 according to site and variant: possible greater neurotropism of variant A. *Clin Infect Dis* 1998;**26**:132–7.
3. Clark DA, Freeland JML, Markie PLK, Jarrett RF, Onions DE. Prevalence of antibody to human herpesvirus 7 by age. *J Infect Dis* 1993;**168**:251–2.
4. Wilborn F, Schmidt CA, Lorenz F, Peng R, Gelderblom H, Huhn D, Siegert W. Human herpesvirus type 7 in blood donors: detection by the polymerase chain reaction. *J Med Virol* 1995; **47**:65–9.
5. Ablashi DV, Balachandran N, Josephs SF, Hung CL, Krueger GR, Kramarsky B, Salahuddin SZ, Gallo RC. Genomic polymorphism, growth properties, and immunologic variations in human herpesvirus-6 isolates. *Virology* 1991;**184**:545–52.
6. Di Luca D, Mirandola P, Ravaioli T, Bigoni B, Cassai E. Distribution of HHV-6 variants in human tissues. *Infect Agents Dis* 1996;**5**:203–14.
7. Dockrell DH. Human herpesvirus 6: molecular biology and clinical features. *J Med Microbiol* 2003;**52**(1):5–18.
8. Nicholas J. Determination and analysis of the complete nucleotide sequence of human herpesvirus. *J Virol* 1996;**70**(9):5975–89.
9. Megaw AG, Rapaport D, Avidor B, Frenkel N, Davison AJ. The DNA sequence of the RK strain of human herpesvirus 7. *Virology* 1998;**244**:119–32.
10. Emery VC, DA. Clark. HHV-6A, 6B, and 7: persistence in the population, epidemiology and transmission. In: Arvin A, Campadelli-Fiume G, Mocarski E, Moore PS, Roizman B, Whitley R, Yamanishi K, editors. *Human Herpesviruses: Biology, Therapy, and Immunoprophylaxis*. Cambridge: Cambridge University Press; 2007. Chapter 49.
11. Wu CA, Shanley JD. Chronic infection of human umbilical vein endothelial cells by human herpesvirus-6. *J Gen Virol* 1998;**79**(5):1247–56.

12. Rotola A, Di Luca D, Cassai E, Ricotta D, et al. Human herpesvirus 6 infects and replicates in aortic endothelium. *J Clin Microbiol* 2000;**38**:3135–6.
13. Caruso A, Favill F, Rotola A, Comar M, et al. Human herpesvirus-6 modulates RANTES production in primary human endothelial cell cultures. *J Med Virol* 2003;**70**:451–8.
14. Santoro F, Kennedy PE, Locate G, Malnati MS, Berger EA, Lusso P. CD46 is a cellular receptor for human herpesvirus 6. *Cell* 1999;**99**:817–27.
15. Tang H, Serada S, Kawabata A, Ota M, Hayashi E, Naka T, Yamanishi K, Mori Y. CD134 is a cellular receptor specific for human herpesvirus-6B entry. *Proc Natl Acad Sci USA* 2013;**110**(22):9096–9.
16. Chevallier P, Robillard N, Illiaquer M, et al. HHV-6 cell receptor CD46 expression on various cell subsets of six blood and graft sources: a prospective series. *J Clin Virol* 2013;**56**(4): 331–5.
17. Lusso P. HHV-6 and the immune system: mechanisms of immunomodulation and viral escape. *J Clin Virol* 2006;**37**(Suppl. 1):S4–10.
18. Lusso P, Secchiero P, Crowley RW, Garzino-Demo A, Berneman ZN, Gallo RC. CD4 is a critical component of the receptor for human herpesvirus 7: interference with human immunodeficiency virus. *Proc Natl Acad Sci USA* 1994;**91**(9):3872–6.
19. Zhang Y, Hatse S, De Clercq E, Schols D. CXC-chemokine receptor 4 is not a coreceptor for human herpesvirus 7 entry into CD4(+) T cells. *J Virol* 2000;**74**(4):2011–6.
20. Takemoto M, Yamanishi K, Mori Y. Human herpesvirus 7 infection increases the expression levels of CD46 and CD59 in target cells. *J Gen Virol* 2007;**88**(5):1415–22.
21. Alegretti AP, Schneider L, Piccoli AK, Xavier RM. The role of complement regulatory proteins in peripheral blood cells of patients with systemic lupus erythematosus. review. *Cell Immunol* 2012;**277**(1–2):1–7.
22. Alegretti AP, Schneider L, Piccoli AK, Monticielo OA, Lora PS, Brenol JC, Xavier RM. Diminished expression of complement regulatory proteins on peripheral blood cells from systemic lupus erythematosus patients. *Clin Dev Immunol* 2012;**2012**:725684.
23. De Bolle L, Van Loon J, De Clercq E, Naesens L. Quantitative analysis of human herpesvirus 6 cell tropism. *J Med Virol* 2005;**75**(**1**):76–85.
24. Pantry SN, Medveczky MM, Arbuckle JH, Luka J, Montoya JG, Hu J, Renne R, Peterson D, Pritchett JC, Ablashi DV, Medveczky PG. Persistent human herpesvirus-6 infection in patients with an inherited form of the virus. *J Med Virol* 2013;**85**(11):1940–6.
25. Ablashi DV, Krueger GRF. The human herpesviruses HHV-6, HHV-7, and HHV-8. In: Ruebsamen-Waigmann H, Deres K, Hewlett G, Welker R, editors. *Viral Infections and Treatment*. New York: Marcel Dekker; 2005. p. 555–89.
26. Dewhurst S. Human herpesvirus type 6 and human herpesvirus type 7 infections of the central nervous system. *Herpes* 2004;**11**(Suppl. 2):105A–11A.
27. Okuno T, Oishi H, Hayashi K, Nonogaki M, Tanaka K, Yamanishi K. Human herpesviruses 6 and 7 in cervixes of pregnant women. *J Clin Microbiol* 1995;**33**:1968–70.
28. Kempf W, Adams V, Hassam S, Schmid M, Moos R, Briner J, Pfaltz M. Detection of human herpesvirus type 6, human herpesvirus type 7, cytomegalovirus and human papillomavirus in cutaneous AIDS-associated Kaposi's sarcoma. *Verh Dtsch Ges Pathol* 1994;**78**:260–4.
29. Tanaka-Taya K, Sashihara J, Kurahashi H, et al. Human herpesvirus 6 (HHV-6) is transmitted from parent to child in an integrated form and characterization of cases with chromosomally integrated HHV-6 DNA. *J Med Virol* 2004;**73**:465–73.
30. Ward KN, Leong HN, Nacheva EP, et al. Human herpesvirus 6 chromosomal integration in immunocompetent patients results in high levels of viral DNA in blood, sera, and hair follicles. *J Clin Microbiol* 2006;**44**:1571–4.

31. Lassner D, Siegismund CS, Stehr J, Rohde M, et al. Recent advances in molecular diagnostics and treatment of heart muscle diseases. *JASMI* 2013;**3**(2):98–109.
32. Pellett PE, Ablashi D, Ambros P, Agut H, Caserta MT, Descamps V, et al. Chromosomally integrated human herpesvirus 6: questions and answers. *Rev Med Virol* 2012;**22**(3):144–55.
33. Strenger V, Aberle SW, Wendelin G, Pfurtscheller K, Nacheva EP, Zobel G, Nagel B. Chromosomal integration of the HHV-6 genome as a possible cause of HHV-6 detection in cardiac tissues. *J Clin Pathol* 2010;**63**(12):1129–30.
34. Buja LM. HHV-6 in cardiovascular pathology. In: Krueger GRF, Ablashi DV, editors. *Human Herpesvirus-6*. Amsterdam: Elsevier; 2005. Chapter 18.
35. Caruso A, Rotola A, Comar M, et al. HHV-6 infects human aortic and heart microvascular endothelial cells, increasing their ability to secrete proinflammatory chemokines. *J Med Virol* 2002;**67**:528–33.
36. Grivel JC, Santoro F, Chen S, Fagá G, et al. Pathogenic effects of human herpesvirus 6 in human lymphoid tissue *ex vivo*. *J Virol* 2003;**77**(15):8280–9.
37. Herder C, Peeters W, Illig T, Baumert J, et al. RANTES/CCL5 and risk for coronary events: results from the MONICA/KORA Augsburg case-cohort, Athero-Express and CARDIoGRAM studies. *PLoS One* 2011;**6**(12):e25734.
38. Yokogawa H, Kobayashi A, Yamazaki N, Sugiyama K. Identification of cytomegalovirus and human herpesvirus-6 DNA in a patient with corneal endotheliitis. *Jpn J Ophthalmol* 2013;**57**(2):185–90.
39. Inoue T, Kandori M, Takamatsu F, Hori Y, Maeda N. Corneal endotheliitis with quantitative polymerase chain reaction positive for human herpesvirus 7. *Arch Ophthalmol* 2010;**128**(4):502–3.
40. Okano M, Luka J, Thiele GM, Sakiyama Y, Matsumoto S, Purtilo DT. Human herpesvirus 6 infection and Kawasaki disease. *J Clin Microbiol* 1989;**27**:2379–80.
41. Marchette NJ, Melish ME, Hicks R, Kihara S, Sam E, Ching D. *J Infect Dis* 1990;**161**:680–4.
42. Burns JC, Newburger JW, Sundell R, Wyatt LS, Frenkel N. Seroprevalence of human herpesvirus 7 in patients with Kawasaki disease. *Pediatr Infect Dis J* 1994;**13**(2):168–9.
43. Kol A, Libby P. The mechanisms by which infectious agents may contribute to atherosclerosis and its clinical manifestations. *Trends Cardiovasc Med* 1998;**8**:191–9.
44. Anderson JL. Infection, antibiotics, and atherothrombosis: end of the road or new beginnings? *N Engl J Med* 2005;**352**:1706–9.
45. Kaklikkaya I, Kaklikkaya N, Birincioglu I, Buruk K, Turan N. Detection of human herpesvirus 6 DNA but not human herpesvirus 7 or 8 DNA in atherosclerotic and nonatherosclerotic vascular tissues. *Heart Surg Forum* 2010;**13**(5):E345–9.
46. Bowles NE, Ni J, Kearney DL, Pauschinger M, Schultheiss HP, McCarthy R, Hare J, Bricker JT, Bowles KR, Towbin JA. Detection of viruses in myocardial tissues by polymerase chain reaction. evidence of adenovirus as a common cause of myocarditis in children and adults. *J Am Coll Cardiol* 2003;**42**(3):466–72.
47. Shirali GS, Ni J, Chinnock RE, Johnston JK, Rosenthal GL, Bowles NE, Towbin JA. Association of viral genome with graft loss in children after cardiac transplantation. *N Engl J Med* 2001;**344**:1498–503.
48. Magnoni M, Malnati M, Cristell N, Coli S, Russo D, Ruotolo G, Cianflone D, Alfieri O, Lusso P, Maseri A. Molecular study of human herpesvirus 6 and 8 involvement in coronary atherosclerosis and coronary instability. *J Med Virol* 2012;**84**(12):1961–6.
49. Vallbracht KB, Schwimmbeck PL, Kühl U, Rauch U, Seeberg B, Schultheiss HP. Differential aspects of endothelial function of the coronary microcirculation considering

myocardial virus persistence, endothelial activation, and myocardial leukocyte infiltrates. *Circulation* 2005;**111**:1784–91.

50. Schmidt-Lucke C, Spillmann F, Bock T, Kühl U, Van Linthout S, Schultheiss HP, Tschöpe C. Interferon beta modulates endothelial damage in patients with cardiac persistence of human parvovirus b19 infection. *J Infect Dis* 2010;**201**(6):936–45.
51. Tschöpe C, Bock CT, Kasner M, et al. High prevalence of cardiac parvovirus B19 infection in patients with isolated left ventricular diastolic dysfunction. *Circulation* 2005;**111**(7):879–86.
52. Helweg-Larsen J, Tarp B, Obel N, Baslund B. No evidence of parvovirus B19, *Chlamydia pneumoniae* or human herpes virus infection in temporal artery biopsies in patients with giant cell arteritis. *Rheumatology (Oxford)* 2002;**41**(4):445–9.
53. Rodriguez-Pla A, Bosch-Gil JA, Echevarria-Mayo JE, et al. No detection of parvovirus B19 or herpesvirus DNA in giant cell arteritis. *Clin Virol* 2004;**31**:11–15.
54. Alvarez-Lafuente R, Fernández-Gutiérrez B, Jover JA, Júdez E, Loza E, Clemente D, García-Asenjo JA, Lamas JR. Human parvovirus B19, varicella zoster virus, and human herpes virus 6 in temporal artery biopsy specimens of patients with giant cell arteritis: analysis with quantitative real time polymerase chain reaction. *Ann Rheum Dis* 2005;**64**(5):780–2.
55. Kühl U, Pauschinger M, Noutsias M, et al. High prevalence of viral genomes and multiple viral infections in the myocardium of adults with "idiopathic" left ventricular dysfunction. *Circulation* 2005;**111**:887–93.
56. Mahrholdt H, Wagner A, Deluigi CC, Kispert E, Hager S, Meinhardt G, Vogelsberg H, Fritz P, Dippon J, Bock CT, Klingel K, Kandolf R, Sechtem U. Presentation, patterns of myocardial damage, and clinical course of viral myocarditis. *Circulation* 2006;**114**(15):1581–90.
57. Schultheiss HP, Kühl U, Cooper LT. The management of myocarditis. *Eur Heart J* 2011;**32**(21):2616–25.
58. Poller W, Fechner H, Noutsias M, Tschoepe C, Schultheiss HP. Highly variable expression of virus receptors in the human cardiovascular system. Implications for cardiotropic viral infections and gene therapy. *Z Kardiol* 2002;**91**(12):978–91.
59. Noutsias M, Fechner H, de Jonge H, et al. Human coxsackie-adenovirus receptor is colocalized with integrins alpha(v)beta(3) and alpha(v)beta(5) on the cardiomyocyte sarcolemma and upregulated in dilated cardiomyopathy: implications for cardiotropic viral infections. *Circulation* 2001;**104**(3):275–80.
60. Pauschinger M, Noutsias M, Lassner D, Schultheiss H-P, Kuehl U. Inflammation, ECG changes and pericardial effusion: Whom to biopsy in suspected myocarditis? *Clin Res Cardiol* 2006;**95**(11):569–83.
61. Kühl U, Pauschinger M, Seeberg B, et al. Viral persistence in the myocardium is associated with progressive cardiac dysfunction. *Circulation* 2005;**112**:1965–70.
62. Yoshikawa T, Ihira M, Suzuki K, Suga S, Kito H, Iwasaki T, Kurata T, Tanaka T, Saito Y, Asano Y. Fatal acute myocarditis in an infant with human herpesvirus 6 infection. *J Clin Pathol* 2001;**54**(10):792–5.
63. Pankuweit S, Maisch B. The heart in viral infections. *Internist (Berl)* 2010;**51**(7):836–43.
64. Moscoso I, Centeno A, López E, et al. Differentiation "*in vitro*" of primary and immortalized porcine mesenchymal stem cells into cardiomyocytes for cell transplantation. *Transplant Proc* 2005;**37**(1):481–2.
65. Toivonen R, Mäyränpää MI, Kovanen PT, Savontaus M. Dilated cardiomyopathy alters the expression patterns of CAR and other adenoviral receptors in human heart. *Histochem Cell Biol* 2010;**133**(3):349–57.

66. Kühl U, Lassner D, von Schlippenbach J, Poller W, Schultheiss HP. Interferon-beta improves survival in enterovirus-associated cardiomyopathy. *J Am Coll Cardiol* 2012;**60**(14):1295–6.
67. Donoso Mantke O, Meyer R, Prösch S, Nitsche A, Leitmeyer K, Kallies R, Niedrig M. High prevalence of cardiotropic viruses in myocardial tissue from explanted hearts of heart transplant recipients and heart donors: a 3-year retrospective study from a German patients' pool. *J Heart Lung Transplant* 2005;**24**(10):1632–8.
68. Noutsias M, Seeberg B, Schultheiss HP, Kühl U. Expression of cell adhesion molecules in dilated cardiomyopathy: evidence for endothelial activation in inflammatory cardiomyopathy. *Circulation* 1999;**99**(16):2124–31.
69. Kühl U, Noutsias M, Schultheiss HP. Immunohistochemistry in dilated cardiomyopathy. *Eur Heart J* 1995;**16**(Suppl):100–6.
70. Fukae S, Ashizawa N, Morikawa S, Yano K. A fatal case of fulminant myocarditis with human herpesvirus-6 infection. *Intern Med* 2000;**39**(8):632–6.
71. Rohayem J, Dinger J, Fischer R, Klingel K, Kandolf R, Rethwilm A. Fatal myocarditis associated with acute parvovirus B19 and human herpesvirus 6 coinfection. *J Clin Microbiol* 2001;**39**(12):4585–7.
72. Krueger GRF, Ablashi DV. Human herpesvirus-6: a short review of its biological behavior. *Intervirology* 2003;**46**:257–69.
73. Comar M, D'Agaro P, Campello C, Poli A, Breinholt 3rd JP, Towbin JA, Vatta M. Human herpes virus 6 in archival cardiac tissues from children with idiopathic dilated cardiomyopathy or congenital heart disease. *J Clin Pathol* 2009;**62**(1):80–3.
74. Kühl U, Rohde M, Lassner D, Gross UM, Escher F, Schultheiss HP. miRNA as activity markers in parvo B19 associated heart disease. *Herz* 2012;**37**(6):637–43.
75. Kuhl U., Lassner D., Dörner A., et al. Transcriptionally active erythrovirus induces reversible cardiac gene expression in human cardiomyopathy. *Basic Res Cardiol* 2013 (in press).
76. Krueger GRF, Rojo J, Lassner D, Kuhl U. Human herpesvirus-6 (HHV-6) is a possible cardiac pathogen: an immunopathological and ultrastructural study. *Rev Med Hosp Gen Mex* 2008;**71**(4):187–91.
77. Klussmann JP, Krueger E, Sloots T, Berneman Z, Arnold G, Krueger GRF. Ultrastructural study of human herpesvirus-7 replication in tissue culture. *Virchows Arch* 1997;**430**(5):417–26.
78. Lassner D. unpublished data.

Chapter 17

HHV-6 in Lymphohematopoietic Diseases

Sylvie Ranger-Rogez
Dupuytren University Teaching Hospital and Faculty of Pharmacy, Limoges, France

INTRODUCTION

Human herpesvirus type 6A (HHV-6A) and HHV-6B are lymphotropic viruses possibly associated with hematologic malignancies, but their implication in these pathologies is difficult to demonstrate because they are ubiquitous and also unable to transform cells *in vitro*, in contrast to Epstein–Barr virus (EBV), another herpesvirus. HHV-6A was initially discovered by Salahuddin et al.[1] in 1986. The virus was isolated from fresh peripheral blood mononuclear cells (PBMCs) from six patients with various lymphoproliferative disorders (LDs): angioimmunoblastic lymphadenopathy, cutaneous T-cell lymphoma, immunoblastic lymphoma, and acute lymphocytic leukemia. The other two patients were suffering from an acquired immunodeficiency syndrome (AIDS)-related lymphoma or from a dermatopathic lymphadenopathy. This new virus was initially reported to selectively infect human B cells, but further studies revealed that it grew preferentially in CD4+ T lymphocytes. Subsequently two variants of HHV-6 could be distinguished: HHV-6A and HHV-6B. They are closely related, and nucleotide sequence variations are 1 to 5% for the conserved genes and 10 to 25% for immediate-early genes. HHV-6A and HHV-6B are now recognized as distinct viral species.[2]

As specific diagnostic tests became available to distinguish between active and latent infections as well as primary and nonprimary virus infections, interpretation of disease associations with HHV-6 became more complicated. In addition, as with many other viral infections, the sole presence of the virus itself does not yet prove that it actually causes disease. As a consequence of this dilemma, the medical literature contains quite controversial opinions about HHV-6 and diseases, sometimes even published by the same authors. This review attempts to give a state-of-the-art description of HHV-6 and lymphohematopoietic diseases, yet it cannot solve many of the problems.

HHV-6 CELLULAR TROPISM

Human herpesvirus 6A and some but not all HHV-6B strains (such as HST) enter cells by use of the CD46 receptor, which is a member of the complement regulatory protein family present on each cell.[3] Though HHV-6 is not found to enter and/or to replicate in every cell type, it is now thought that the virus uses an additional coreceptor or alternatively another receptor. A recent study indicates that CD134 is a receptor for several HHV-6B strains, including HST.[4] Both HHV-6A and HHV-6B readily infect CD34+ hematopoietic stem cells and cells differentiating toward the T-cell lineage, including CD34+ or CD38+ cells, natural killer (NK) cells, γ/δ T lymphocytes, and CD4+ or CD8+ cells. Among the differentiated T cells, nonnaive CD4+ are equally infected by both HHV-6A and HHV-6B, while CD8+ cells are preferentially infected by HHV-6A.[5] In addition, HHV-6 can apparently also infect cells differentiating toward the myelomonocytic lineage, including such highly specialized cellular end stages as dendritic reticular cells and vascular endothelial cells. Nevertheless, complete replication occurs only in T cells.

The reaction of lymphoid tissue components to HHV-6 infection was studied by *ex vivo* techniques. Upregulation in infected cells of certain cell membrane markers (e.g., CD4+, CD21+), which was in part due to virus-induced rigidification of the cell membrane thereby providing one possible basis for superinfection of HHV-6 infected cells by other viruses such as EBV and human immunodeficiency virus (HIV), and downregulation of other markers (CD3, CD46), was demonstrated.[6] It may be added, though, that even under standardized tissue culture conditions, the biological effects of HHV-6 may vary somewhat. This applies not only to the different HHV-6 species but also to different isolates of the same species.

TECHNICAL ADVANCES AND EVOLVING RESULTS

Given the circumstances of the first isolation of HHV-6, its lymphotropism and ability to infect and replicate in human hematopoietic cell lines, and its ability to remain in a latent state, this ubiquitous pathogen was thought to be implicated in LDs. The first studies consisted of serological screening of patients' sera, and the main technique used was the immunofluorescent assay (IFA). Several studies[7,8] noticed a significantly higher seroprevalence of anti-HHV-6 antibodies in sera from patients suffering from Hodgkin's lymphoma (HL) compared to controls (95% vs 76%) in India.[8] Similar results were reported in other LDs, such as low-grade non-Hodgkin's lymphoma (NHL) and acute myeloid leukemia. A statistically significant increase in the levels of anti-HHV-6 antibodies in HL patients compared to healthy controls was also reported,[9] although not found by others.[8] Such results could be interpreted prudently as an immune activation that is not related to the presence of HHV-6, or, inversely, they could be in favor of a role for HHV-6 in such diseases. Levine et al.[10]

TABLE 17.1 Results of Some of the Main Studies Conducted on HHV-6 Prevalence in Lymph Nodes From Hodgkin's Lymphoma Patients

N° samples	N° positive (%)	Subtype of positive HL	Assays	Reference
25	3 (12%)	NSHL	Southern-blot	Torelli et al. *Blood* 1991: **77**: 2251–8
35	0	ND	PCR	Gledhill et al.[21]
39	3 (8%)	NSHL	PCR	Torelli et al.[7]
14	9 (64%)	ND	PCR	Sumiyoshi et al. *Am J Clin Pathol* 1993: **99**: 609–14
45	13 (29%)	most = NSHL	nested-PCR	Di Luca et al. *J Infect Dis* 1994: **170**: 211–5
52	38 (73%)	NSHL: 28 (77.7%), MCHL: 7 (70%)	PCR	Valente et al.[12]
88	11 (13%)	NSHL: 5 (62.5%), M-HL: 4 (66.6%)	nested-PCR	Schmidt et al. *Leuk Res* 2000: **24**: 865–70
47	0	ND	PCR	Shiramizu et al.[22]
37	13 (35.1%)	NSHL: 13 (41.9%), MCHL: 0	Real-time PCR	Collot et al.[15]
39	39%	NSHL: 5 (45.4%), MCHL: 7 (30.4%)	PCR	Tailor et al. *Ann N Y Acad Sci* 2004: **1022**: 282–5
86	68 (79.1%)	NSHL: 61 (83.6%), MCHL: 5 (50%)	Real-time PCR	Lacroix et al.[23]
31	27 (87%)	NSHL: 18 (86%)	PCR	Siddon et al.[26]

ND: not determined; NSHL: nodular sclerosis HL; MCHL: mixed cellularity HL.

noticed a significant increase in titers during the follow-up of 37 patients with HL who relapsed compared to the decrease observed in 39 controls who did not relapse, suggesting that HHV-6 serology may be of prognostic value in this context. Using an enzyme-linked immunosorbent assay (ELISA), Iyengar et al.[11] detected a low percentage of sera positive for anti-p41/38 antibodies among different serum donor categories and a high percentage in the case of patients with African Burkitt's lymphoma (ABL) or HL, indicating the presence of active HHV-6 replication in these patients.

Further studies then employed molecular methods including *in situ* hybridization (ISH) or polymerase chain reaction (PCR) to detect or amplify viral sequences in tissues from patients. Numerous studies were published, with variable and sometimes discordant results; some of them are mentioned in Tables 17.1 and 17.2. HHV-6 DNA has been found in the lymph nodes of

TABLE 17.2 Results of Some of the Main Studies Conducted on Prevalence of HHV-6 in Lymph Node Biopsies of Patients Suffering From non-Hodgkin's Lymphomas

Pathology	N° samples	N° positive (%)	Assays	Reference
NHL	16	2 (12.5%)	ISH and PCR	Borisch et al.[27]
NHL	45	8 (17.8%)	ISH	Yin et al. *Arch Pathol Lab Med* 1993: **117**: 502–6
NHL	113	3 (2.6%)	PCR	Dolcetti et al. *J Med Virol* 1996: **48**: 344–53
NHL	59	34%	PCR	Tailor et al. *Ann N Y Acad Sci* 2004: **1022**: 282–5
NHL	63	14(22%)	PCR	Hernandez-Losa et al., *Cancer*, 2005: **103**: 293–8
NHL (2Bcell-SS)	157	5 (3.2%)	Blot hybridization	Jarrett et al. *Leukemia* 1988: **2**: 496–502
SS lymphomas	14	1 (7.1%)	Nucleic acid hybridization	Fox et al.[32]
NHL B lineage	106	2 (1.9%)	PCR	Torelli et al.[7]
DLBCL	7	2 (28.6%)		Allory et al. *Am J Surg Pathol* 2001: **25**: 865–74
NHL (T cell origin)	41	24 (58.5%)	PCR	Sumiyoshi et al. *J Clin Pathol* 1993: **46**: 1137–38
NHL (B cell origin)	29	18 (62.1%)		
NHL (B cell origin)	36	8 (22.2%)	qRT-PCR	Collot et al. *J Clin Microbiol* 2002: **40**: 2445–51
NHL (T or NK cell origin)	13	3 (23.1%)		
AITL	7	0 (0%)	Southern blot	Josephs et al. *Leukemia* 1988: **2**: 132–5
AITL	8	4 (50%)	PCR	Luppi et al. *Leuk Res* 1993: **17**: 1003–11
AITL-L	4	3 (75%)		
AITL	8	5 (62.5%)	PCR	Sumiyoshi et al. *J Clin Pathol* 1993: **46**: 1137–38
AITL	18	4 (22.2%)	PCR	Vrsalovic et al. *Hematol Oncol* 2004: **22**: 169–77
AITL	49	21 (42.8%)	PCR	Zhou et al.[30]

DLBCL: diffuse large B-cell lymphomas; AITL: angioimmunoblastic T-cell lymphomas; AITL-L: angioimmunoblastic T-cell-like lymphomas; NHL: non-Hodgkin's lymphomas; SS: Sjögren's syndrome.

0 to 85% of patients suffering from LDs, with these figures varying according to the number of patients studied, the variety of pathologies explored, and the assay techniques used. Because of differences in the assays employed and in the sensitivity of the different assays, results reported by the numerous papers are difficult to analyze. Furthermore, the percentage of tumor cells in the tissues examined was not evaluated in most studies. However, amelioration of the employed assays, with increased sensitivity and specificity, improved detection of HHV-6 sequences in biopsies studied.

In order to determine the role of HHV-6 in LDs, various teams attempted to quantify viral genomes in lymph nodes from LD patient samples. Some of them used two techniques of different known sensitivities—Southern-blot analysis and PCR[12]—providing an assay discriminating between low and high levels of viral DNA. More recently, quantitative PCR (qPCR)[13,14] and quantitative real-time PCR (qRTPCR)[15] have been developed.

ATYPICAL LYMPHOPROLIFERATIONS INDUCED BY HHV-6 INFECTION

Atypical lymphoproliferations (APLs) constitute a heterogeneous group of lesions that clinically mimic malignant lymphomas but lack the criteria of monoclonality and malignant transformation. They can be defined as premalignant lymphoproliferations and may finally transform to malignant lymphomas. The incidence of APLs seems to be increased in patients with persistently active infection by lymphotropic viruses. The majority of reported cases are associated with herpesvirus infections—namely, EBV or HHV-6.

Berthold et al.[16] described a Canale–Smith syndrome occurring in two brothers. The patients exhibited lymphoblasts, and peripheral blood examination demonstrated acute lymphocytic leukemia (ALL). Nevertheless, immunophenotyping of PBMCs revealed polyclonality, and no atypical blasts were seen in the bone marrow. Biopsy of an enlarged lymph node from one boy showed an overall intact structure with paracortical expansion of a polymorphic CD30-negative lymphoid population containing many blasts and rare giant cells. ISH revealed focal accumulation of HHV-6 positive cells; HHV-6 was in a productive infectious form as demonstrated by coculture. The boys' mother, who had a past history of "acute lymphocytic leukemia successfully treated" at the age of 3 years, still exhibited persistent lymphocytosis. One of the boys and the mother had an IgG_2 deficiency as well. Interpretation of these data concluded an unusual lymphostimulation by persistently active HHV-6 infection.

Krueger et al.[17] conducted a large study focusing on LDs and APLs associated with persistent infection by EBV and/or HHV-6. They highlighted the improved detection of herpesvirus-associated APLs in immunocompromised subjects such as AIDS patients or allograft recipients. APLs were reported to be histological evidence of excessive infectious mononucleosis or mononucleosis syndrome. Indeed, HHV-6 infections may be accompanied by extensive lymphoproliferative reactions, and it is recognized that HHV-6 primary infection may lead to a mononucleosis-like syndrome in some immunocompetent patients. Because the diagnosis of HHV-6 is not always undertaken in cases of APLs, its association with this pathology is probably underestimated.

Interestingly, a survey of seven children who developed a primary HHV-6 infection described modifications appearing in blood cells leading to a myelodysplasic syndrome that disappeared spontaneously after 1 to 2 months.[18] One can think that, at this stage, the virus can integrate cells or affect the bone marrow precursor response to growth factors, leading to the development of malignant cells.

The role of HHV-6 in the onset of Kikuchi–Fujimoto disease, initially postulated, seems now to be ruled out.[19] Rosai–Dorfman disease, or sinus histiocytosis with massive lymphadenopathy, is a rare and benign lymphadenopathy characterized by nonmalignant proliferation of histiocytic cells within lymph-node sinuses and lymphatics in extranodal sites. Detection of late HHV-6 antigens has been reported in several patients.[20] The infection appears to be restricted to abnormal histiocytes that constitute the proliferating elements.

HODGKIN'S LYMPHOMA AS THE MAIN LYMPHOPROLIFERATIVE DISORDER ASSOCIATED WITH HHV-6 IN IMMUNOCOMPETENT SUBJECTS

Hodgkin's lymphoma (HL) is characterized by the expansion of Reed–Sternberg (RS) cells, which are now postulated to be of B-cell lineage. Numerous teams have worked on the association of HHV-6 with HL, testing first the seroprevalence and then the prevalence of viral DNA in lymph nodes from patients. Results reported are variable and sometimes contradictory, probably due to the difference in assays employed and also in patients tested as discussed previously. Table 17.1 summarizes some results obtained by molecular methods. Some research teams detected no HHV-6 in lymph nodes from HL patients;[21,22] in contrast, a prevalence as high as 82% was found by others. In our experience, the same lymph node population examined by two different assays revealed a positivity of 35.1% or 78%. Although it is difficult to reach definitive conclusions regarding such findings, it seems that HHV-6 can be detected in more than one-third of HL biopsies, suggesting that it might be associated with at least a subset of this disorder.

Characteristics of HHV-6 Found in Lymph Nodes

Viral Quantification

As mentioned previously, a few papers have attempted to quantify viral genomes in samples from HL patients. Ohyashiki et al.[14] observed significantly higher levels of HHV-6 DNA by qPCR–ELISA for all LD patients combined (21 blood specimens and 19 lymph nodes) compared to controls (PBMCs from 23 healthy volunteers, 4 lymph nodes with reactive hyperplasia). Valente et al.[12] obtained <10,000 HHV-6 copies per µg DNA for positive lymph node samples, while Secchiero et al.[13] reported <10 to 1000 HHV-6 genome equivalents per µg DNA among AIDS–HL samples. A qRT-PCR developed in our laboratory revealed from 100 to 864,640 copies per µg DNA.[15] All the reports underline the fact that HHV-6 may be present in variable quantities in lymph nodes according to patients studied, although in higher amounts than those found in healthy subjects.

HHV-6B Most Often Detected

The detection of HHV-6 types in HL samples has been reported by several authors, and in most cases HHV-6B was found almost exclusively.[12,15] This is not surprising considering the greater lymphotropism of HHV-6B compared to HHV-6A.

HHV-6 and Nodular Sclerosis HL (NSHL) in Young Adults

Epstein–Barr virus, another herpesvirus frequently associated with HL, is not distributed equally among the different subtypes of positive HL: 70% of mixed cellularity, greater than 95% of lymphocyte depleted, 10 to 40% of nodular sclerosis, and almost absent from lymphocyte-predominant HL subtypes. In addition, HL cases associated with EBV occurred most frequently in children or in the elderly. Previous epidemiological studies suggested multiple etiologies for HL and led to the hypothesis of an infectious viral non-EBV etiology for cases occurring in young adults. It was therefore interesting to look for the presence of HHV-6 sequences according to HL subtype.

Torelli et al.[7] described three HHV-6 positive cases of HL, belonging to the NS lymphocyte-depletion subgroup, which occurred in young women (27, 28, and 31 years old). Similarly, different authors obtained the highest prevalence and highest mean copy for NSHL (Table 17.1). In a large study conducted on 86 adult HL patients, we obtained a high prevalence (83.6%) of HHV-6 in NSHL.[23] However, the NSHL subtype was the most frequent in HL patients examined. In the same study, the mean age of NSHL patients positive for HHV-6 and negative for EBV was 29.5 years, whereas it was 45 years for patients who were HHV-6 negative and EBV positive. Therefore, HHV-6 seems to be more predominant in NSHL and much more so in young adults, arguing for a potential role for this virus in the etiology of HL in this context. Clark et al.[9] found an association of increased HHV-6 seropositivity and geometric mean titer ratio with HL among young adults lacking social contact in the family group, suggesting that those patients were exposed late to HHV-6.

HHV-6 Association with EBV

Some cases of HL are associated with EBV, especially cases of mixed cellularity occurring in children and in the elderly. Although HHV-6 and EBV infect different cells, they can be present in the same HL tissue. Using two different real-time PCRs (one for HHV-6 and one for EBV), we found that 47.7% of HL patients were positive for both viruses in their lymph nodes. In this case, EBV was often weakly present compared to HHV-6.[23] Identical findings have been reported by others.[7,12] We demonstrated recently by immunohistochemistry (IHC) that in some cases HHV-6 and EBV can colocalize in the same cell.[25]

Presence of HHV-6 in Various Cell Types and Reed–Sternberg Cells

Cells harboring HHV-6 were studied in positive lymphoid lesions by ISH and/or by IHC.[12,24] An elevated number of cells carrying HHV-6 was found in cases of HL not yet in reactive lymphoid hyperplasia. Virus-infected cells were identified primarily as lymphohistiocytic cells, less frequently as Hodgkin's and Reed–Sternberg cells. In some studies, HHV-6-positive cells appeared small and lymphoid in appearance.[17] In other studies, RS or HL cells expressing HHV-6 nonstructural and/or structural antigens were detected with monoclonal antibodies (MAbs) in up to 20% of HL patients. In our experience, 39.5% of lymph nodes from HHV-6-positive HL exhibited RS cells expressing HHV-6 structural antigens (HHV-6A or HHV-6B gp116/54/64) as determined by the use of MAbs.[25] In these specimens and also in other HHV-6-positive lymph nodes, HHV-6 was also detected in lymphocytes, plasmocytes, and histiocytic cells. This phenomenon has also been described for EBV, which, when present, may be localized in RS cells and also in lymphocytes. Furthermore, using a polyclonal antibody synthesized in our laboratory, we examined the expression of the oncoprotein DR7 of HHV-6B (detailed later) in these tissues and detected the protein in 28 out of 38 lymph nodes tested (73.7%); DR7B was principally found in RS cells and to a lesser extent in other lymphoid cells, signaling the expression of this oncogenic protein in transformed cells. Similarly, Siddon et al.[26] identified HHV-6 DNA in 86% of tissue samples from 31 lymph node cases of NSHL and identified HHV-6 via IHC in malignant RS cells in nearly half of the cases. Studies regarding HHV-6 positivity in HL have long led to conflicting results. It seems more and more that HHV-6 is involved in the outcome and/or maintenance of NSHL.

DETECTION OF HHV-6 IN OTHER LYMPHOPROLIFERATIVE DISEASES

Non-Hodgkin's Lymphomas in Immunocompetent Subjects

Some teams studied the presence of HHV-6 in lymph nodes from NHL patients; as for HL (and probably more so), the results reported vary largely.

This could be due to the great diversity of LD examined in each study, with often just a few cases representing each hematological entity. More recently, the widespread use of PCR has led to more homogeneous results, and finally it can be concluded that HHV-6 is found in a quarter of NHL cases and more in cases of angioimmunoblastic T-cell lymphoma (AITL) (Table 17.2). Histologically, HHV-6 positive LD cases were not distinguishable from negative ones.

B-Cell Lymphomas

Based on the detection of active replication, HHV-6 was suspected to be associated with some cases of African Burkitt's lymphoma.[11] HHV-6 sequences were also occasionally detected in lymph nodes from European Burkitt's lymphoma cases and at a high copy number. Other B-cell lymphomas are not frequently reported to be associated with HHV-6, and the virus probably has no direct role in the development of B-cell NHL. Two positive B-cell lymphomas of high-grade malignancy with clonal proliferations, originating from elderly patients, were reported.[27] Using qPCR–ELISA, Ohyashiki et al.[14] obtained a low amount of HHV-6 (6.4 copies per μg DNA) for a B-cell lymphoma, although in other B-cell lymphomas they studied, the number of HHV-6 genomes was less than 5 copies per μg of DNA in blood and lymph node specimens. Four HHV-6-positive cases were found among 14 low-grade primary ocular lymphomas arising from mucosa-associated lymphoid tissue (MALT) (28.5%).[28] Finally, a primary mediastinal large B-cell lymphoma associated with HHV-6 was discovered among 24 cases studied.[29] HHV-6 DNA was rarely detected in tumor cells and mostly in healthy tissue.

T-Cell Lymphomas

One of the initial isolates of HHV-6[1] originated from a patient suffering from angioimmunoblastic T-cell lymphoma. After HL, AITL probably constitutes the most frequent hemopathy associated with HHV-6. In this case, also, findings reported by various authors vary considerably (Table 17.2). Nevertheless, it can be admitted today that HHV-6 is present in a large number of cases if AITL. AITL is a peripheral T-cell lymphoma characterized by the obliteration of lymph node architecture and replacement by an arborescence of proliferating small vessels with a hypocellular polymorphous infiltrate. Progression to lymphoma occurs when lymphoid cells are prominent and form clusters. Expanded B-cell clones are sometimes present beside the T-cell clones and are mostly CD10 positive, which accounts for only a small proportion of the total population. The majority of cells are comprised of reactive immunoblastic B and T cells, plasma cells, macrophages, and eosinophils localized in a substantial vascular network. A fatal outcome is rapidly observed in most cases. The disease is thought to be due to excessive immune activity triggered by an antigen (a viral antigen and HHV-6 could represent a good candidate), which leads to a clonal expansion of B and T lymphocytes. Zhou et al.[30] did not find HHV-6 in neoplastic T cells themselves, ruling out a direct role for the virus. They suggested that HHV-6 may induce microenvironmental conditions that promote recruitment of the inflammatory cells, contributing to the lymphoma progression. Other T-cell lymphomas seem to be rarely associated with HHV-6. Because HHV-6 was rarely found in cutaneous T-cell lymphomas or in mycosis fungoide tissues (1.1 to 3.3%), it seems improbable that it could play a role in these pathologies.

Cerebral Lymphomas and Lymphomas with a Predilection for the Nervous System

A few papers have examined the presence of HHV-6 in cerebral lymphomas and reported an absence or a very low number of positive samples. Viali et al.[31] reported the case of a 53-year-old man who developed an angiotropic large-cell lymphoma that manifested as neurological symptoms resulting from intravascular lymphomatosis and related to HHV-6.

Lymphomas in Sjögren's Syndrome Patients

A few cases of patients suffering from Sjögren's syndrome followed by non-Hodgkin's lymphoma associated with HHV-6 during the multistep course of the disease have been described.[17,32]

ACUTE LYMPHOBLASTIC LEUKEMIA

Few studies have been conducted on acute leukemias. A high copy number of the HHV-6 genome has been demonstrated in one case of a Philadelphia chromosome-positive early pre-B-cell acute lymphoblastic leukemia (ALL). A case of ALL occurring two months after a mononucleosis syndrome related to HHV-6 was also reported.[33] HHV-6 was not detected by PCR on DNA extracted from Guthrie cards for Swedish children who later developed ALL or for controls.[34]

HHV-6-ASSOCIATED LYMPHOPROLIFERATIVE DISORDERS IN IMMUNOCOMPROMISED INDIVIDUALS

In contrast to EBV, HHV-6 seems to be more rarely found in lymphoproliferative tissues in a immunosuppressive context.

Posttransplant LDs

A γ/δ-T-cell lymphoma associated with HHV-6, occurring in a 48-year old man who was kidney transplanted 4 years previously, was reported.[35] A case of large-cell immunoblastic lymphoma following HHV-6 viremia and bone marrow infection, after allogeneic bone marrow transplant, has also been described. Both cases were EBV negative. A pediatric ALL was also reported after stem cell transplantation.[36]

AIDS-Related LDs

Some studies, conducted in either pediatric or adult patients, have demonstrated no positivity for HHV-6 in AIDS-related LDs, while others have reported about 30 to 40% AIDS lymphomas harboring HHV-6 sequences.[37,38] HHV-6B was largely predominant. Although it is quite difficult to reach a general consensus on this question, most authors, citing the low number of DNA copies in AIDS-associated LDs, have argued against a strong association between this virus and the genesis of AIDS lymphomas.

PROPOSED MECHANISMS FOR HHV-6 IN LYMPHOPROLIFERATIVE DISORDERS

Human herpesvirus 6 can be implicated in LDs directly or it can be a cofactor for cell transformation; therefore, different mechanisms have been put forth.

Transforming Capacities of HHV-6

The 3.9-kbp *SalI*-L fragment located within the direct repeat region of HHV-6A was shown to transform the murine NIH3T3 cell line, human epidermal keratinocytes RHEK-1, and both primary and established rodent cells.[39,40] This transforming activity was localized to the DR7 gene and cells expressing DR7 protein-induced tumors when injected into immunodeficient nude mice, while cells expressing truncated DR7 protein did not.[41] Moreover, DR7 protein from HHV-6A strain U1102 binds to the human tumor suppressor protein p53, which is a major control point of the cell cycle, involving inhibition of p53-activated transcription.[42]

Because DR7A and DR7B share only 42.2% homology, and because HHV-6B is found much more often in hematological malignancies than HHV-6A, it seemed interesting to examine the oncogenic properties of DR7B. Using a two-hybrid system, we demonstrated that DR7B was also able to bind to human p53.[25] This interaction may induce or participate in lymphomagenesis, considering the fact that HHV-6 is able to retain p53 within the cytoplasm, leading to p53 accumulation and thus protecting infected cells from apoptosis.[43–45]

HHV-6 Transactivation of Genes

Transactivation of Human Genes

An interesting feature associated with HL is the abundant constitutive activation of the p50/p65 nuclear factor kappa B (NF-κB) complex in cultured RS cells, this factor being normally observed only for limited time intervals after stimulation with diverse inducers. Moreover, constitutive NF-κB activation has been shown to be required for survival and proliferation of RS cells, preventing them from undergoing apoptosis under stress conditions.[46] We showed that DR7B was able to transactivate p105/p50 and p65 NF-κB subunits in HEp-2 cells and to activate the NF-κB complex.[25] HHV-6B could therefore induce or reinforce NF-κB constitutive transactivation in H/RS cells. We also demonstrated that DR7B promotes overexpression of Id2, which is an inhibitor of the E2A transcription factor.[25] E2A plays important roles in promoting cell differentiation and suppressing cell growth. Id2 was strongly and uniformly expressed in RS cells. Moreover, DR7B overexpression led to marked upregulation of CD30 and CD83 and downregulation of CD99 molecules. The modified expression of these markers was linked to the ability of DR7B to change proliferative and apoptotic features (unpublished data).

Transactivation of Viral Genes

EBV Transactivation

Epstein–Barr virus, considered to be an oncogenic virus, is found in up to 60% of HHV-6-positive LD tissues. The intensity of signals obtained for EBV is weaker than for HHV-6, suggesting that EBV sequences are markedly less represented.[7] We demonstrated that both viruses can be present in the same cells. The infection of EBV latently infected cells with HHV-6A results in activation of EBV replication, mediated via a cyclic AMP-responsive element located within the EBV ZEBRA promoter. This promoter controls the EBV gene product ZEBRA, which is responsible for disrupting EBV latency and initiating the lytic replication cascade.[47] Activation of LMP1 and EBNA2 expression after HHV-6A infection of EBV-positive Burkitt's lymphoma cell lines has also been described. The presence of positive and negative regulatory elements responsive to HHV-6A infection in LMP1 regulatory sequences have also been demonstrated.[48]

HHV-8 Transactivation

Human herpesvirus 6, which is occasionally present in HHV-8-related LDs,[38,49] is able to induce HHV-8 replication as demonstrated by coculture of HHV-6-infected T cells with HHV-8 latently infected cells. HHV-8 replication results from activation of the first promoter activated during HHV-8 replication (ORF-50 promoter).[50]

Human Endogenous Retrovirus (HERV) Transactivation

Interestingly, the infection of a human T-leukemia cell line with HHV-6 was shown to produce a molecule consisting of a partial HERV polymerase gene.[51]

Ability of HHV-6 to Integrate into Host Cell Chromosomal DNA

Today, there is no evidence that chromosomally integrated HHV-6 (ciHHV-6) is able to induce or to participate in lymphoproliferation. We found that ciHHV-6 had the same range of incidence among a population of patients with hematological malignancies and a control group. Nevertheless, clinical cases of patients suffering from HL or other hematological malignancies are often reported.

Role of HHV-6 in Cell Proliferation by Dysregulation of the Cytokine Network

The hypothesis that HHV-6 may contribute to the development of LDs by deregulation of cytokine control rather than by direct oncogenic involvement has been suggested. HHV-6 has been demonstrated to contain chemokine (i.e., U83) and chemokine receptor (i.e., U12) genes. Also, HHV6 can modulate the production of chemokines and cytokines, which can promote the survival and proliferation of cells. It can also influence the response to these molecules.

Takaku et al.[52] studied the network of dynamic gene and protein interactions occurring during infection of an adult T-leukemia cell line by HHV-6B using a microarray, and they analyzed the data using a Bayesian statistical framework. They reported the possible association between chemokine genes regulating the Th1/Th2 balance and genes regulating T-cell proliferation.

An interesting development is the design of a computer model to simulate cell changes happening in LDs and disturbances of the T-cell immune system.[53] The model uses the concept that these disturbances, identified as proliferation, differentiation, or inhibition factors, may lead to hyperplastic, aplastic, or neoplastic diseases. This computer model simulated acute and chronic persistent HHV-6 infections to study the influence of cytokines or chemokines in the Canale–Smith syndrome.

CONCLUSION

Human herpes 6 is most certainly involved in the outcome of nodular sclerosis Hodgkin's lymphoma cases, particularly in young adults. It may also be implicated in some cases of angioimmunoblastic T-cell lymphoma, although mechanisms involved in these two pathologies may be different. Most rarely, it may be implicated in the outcome of other hematological malignancies such as lymphomas.

REFERENCES

1. Salahuddin SZ, Ablashi DV, Markham PD, et al. Isolation of a new virus, HBLV, in patients with lymphoproliferative disorders. *Science* 1986;**234**:596–601.

2. Adams MJ, Cartsens EB. Ratification vote on taxonomic proposals to the International Committee on Taxonomy of Viruses. *Arch Virol* 2012;**157**:1411–22.
3. Santoro F, Kennedy PE, Locatelli G, et al. CD46 is a cellular receptor for human herpesvirus 6. *Cell* 1999;**99**:817–27.
4. Tang H, Serada S, Kawabata A, Ota M, Hayashi E, Naka T, Yamanishi K, Mori Y. CD134 is a cellular receptor specific for human herpesvirus-6B entry. *Proc Natl Acad Sci USA* 2013;**110**(22):9096–9.
5. Grivel JC, Santoro F, Chen S, et al. Pathogenic effects of human herpesvirus-6 in human lymphoid tissue *ex vivo*. *J Virol* 2003;**77**:8280–9.
6. Schonnebeck M, Krueger GF, Braun M, et al. Human herpesvirus-6 infection may predispose cells to superinfection by other viruses. *In vivo* 1991;**5**:255–63.
7. Torelli G, Marasca R, Montorsi M, et al. Human herpesvirus-6 in non-AIDS related Hodgkin's and non-Hodgkin's lymphomas. *Leukemia* 1992;**6**(Suppl. 3):46S–8S.
8. Shanavas KR, Kala V, Vasudevan DM, et al. Anti-HHV-6 antibodies in normal population and in cancer patients in India. *J Exp Pathol* 1992;**6**:95–105.
9. Clark DA, Alexander FE, McKinney PA, et al. The seroepidemiology of human herpesvirus-6 (HHV-6) from a case-control study of leukaemia and lymphoma. *Int J Cancer* 1990;**45**:829–33.
10. Levine PH, Ebbesen P, Ablashi DV, et al. Antibodies to human herpes virus-6 and clinical course in patients with Hodgkin's disease. *Int J Cancer* 1992;**51**:53–7.
11. Iyengar S, Levine PH, Ablashi D, et al. Sero-epidemiological investigations on human herpesvirus 6 (HHV-6) infections using a newly developed early antigen assay. *Int J Cancer* 1991;**49**:551–7.
12. Valente G, Secchiero P, Lusso P, et al. Human herpesvirus 6 and Epstein–Barr virus in Hodgkin's disease: a controlled study by polymerase chain reaction and *in situ* hybridization. *Am J Pathol* 1996;**149**:1501–10.
13. Secchiero P, Zella D, Crowley RW, et al. Quantitative PCR for human herpesviruses 6 and 7. *J Clin Microbiol* 1995;**33**:2124–30.
14. Ohyashiki JH, Abe K, Ojima T, et al. Quantification of human herpesvirus 6 in healthy volunteers and patients with lymphoproliferative disorders by PCR-ELISA. *Leuk Res* 1999;**23**:625–30.
15. Collot S, Petit B, Bordessoule D, et al. Real-time PCR for quantification of human herpesvirus 6 DNA from lymph nodes and saliva. *J Clin Microbiol* 2002;**40**:2445–51.
16. Berthold F, Krueger GR, Tesch H, et al. Monoclonal B cell proliferation in lymphoproliferative disease associated with herpes virus type 6 infection. *Anticancer Res* 1989;**9**:1511–8.
17. Krueger GR, Manak M, Bourgeois N, et al. Persistent active herpes virus infection associated with atypical polyclonal lymphoproliferation (APL) and malignant lymphoma. *Anticancer Res* 1989;**9**:1457–76.
18. Kagialis-Girard S, Durand B, Mialou V, et al. Human herpesvirus-6 infection and transient acquired myelodysplasia in children. *Pediatr Blood Cancer* 2006;**47**:543–8.
19. Rosado FG, Tang YW, Hasserjian RP, et al. Kikuchi–Fujimoto lymphadenitis: role of parvovirus B-19, Epstein–Barr virus, human herpesvirus 6, and human herpesvirus-8. *Hum Pathol* 2013;**44**:255–9.
20. Luppi M, Barozzi P, Garber R, et al. Expression of human herpesvirus-6 antigens in benign and malignant lymphoproliferative diseases. *Am J Pathol* 1998;**153**:615–23.
21. Gledhill S, Gallagher A, Jones DB, et al. Viral involvement in Hodgkin's disease: detection of clonal type A Epstein–Barr virus genomes in tumour samples. *Br J Cancer* 1991;**64**:227–32.

22. Shiramizu B, Chang CW, Cairo MS. Absence of human herpesvirus-6 genome by polymerase chain reaction in children with Hodgkin disease: a Children's Cancer Group Lymphoma Biology Study. *J Pediatr Hematol Oncol* 2001;**23**:282–5.
23. Lacroix A, Jaccard A, Rouzioux C, et al. HHV-6 and EBV DNA quantitation in lymph nodes of 86 patients with Hodgkin's lymphoma. *J Med Virol* 2007;**79**:349–56.
24. Luppi M, Barozzi P, Garber R, et al. Expression of human herpesvirus-6 antigens in benign and malignant lymphoproliferative diseases. *Am J Pathol* 1998;**153**:815–23.
25. Lacroix A, Collot-Teixeira S, Mardivirin L, et al. Involvement of human herpesvirus-6 variant B in classic Hodgkin's lymphoma via DR7 oncoprotein. *Clin Cancer Res* 2010;**16**:4711–21.
26. Siddan A, Lozovatsky L, Mohamed A, et al. Human herpesvirus 6 positive Reed–Sternberg cells in nodular sclerosis Hodgkin lymphoma. *Br J Haematol* 2012;**158**:635–43.
27. Borisch B, Ellinger K, Neipel F, et al. Lymphadenitis and lymphoproliferative lesions associated with the human herpes virus-6 (HHV-6). *Virchows Arch B Cell Pathol Incl Mol Pathol* 1991;**61**:179–87.
28. Daibata M, Komatsu T, Taguchi H. Human herpesviruses in primary ocular lymphoma. *Leuk Lymphoma* 2000;**37**:361–5.
29. Kolonic SO, Dzebro S, Kusec R, et al. Primary mediastinal large B-cell lymphoma: a single-center study of clinicopathologic characteristics. *Int J Hematol* 2006;**83**:331–6.
30. Zhou Y, Attygalle AD, Chuang SS, et al. Angioimmunoblastic T-cell lymphoma: histological progression associates with EBV and HHV-6B viral load. *Br J Haematol* 2007;**138**:44–53.
31. Viali S, Hutchinson DO, Hawkins TE, et al. Presentation of intravascular lymphomatosis as lumbosacral polyradiculopathy. *Muscle Nerve* 2000;**23**:1295–300.
32. Fox RI, Saito I, Chan EK, et al. Viral genomes in lymphomas of patients with Sjögren's syndrome. *J Autoimmun* 1989;**2**:449–55.
33. Seror E, DeVillartay P, Leverger G, et al. HHV-6 infection and acute lymphoblastic leukemia in a child. *Arch Pediatr* 2008;**15**:37–40.
34. Gustafsson B, Bogdanovic G, et al. Specific viruses were not detected in Guthrie cards from children who later developed leukemia. *Pediatr Hematol Oncol* 2007;**24**:607–13.
35. Lin WC, Moore JO, Mann KP, et al. Post transplant CD8+ gammadelta T-cell lymphoma associated with human herpes virus-6 infection. *Leuk Lymphoma* 1999;**33**:377–84.
36. Vila L, Moreno L, Andrés MM, et al. Could other viruses cause pediatric posttransplant lymphoproliferative disorder? *Clin Transl Oncol* 2008;**10**:422–5.
37. Fillet AM, Raphael M, Visse B, et al. Controlled study of human herpesvirus 6 detection in acquired immunodeficiency syndrome-associated non-Hodgkin's lymphoma. *J Med Virol* 1995;**45**:106–12.
38. Asou H, Tasaka T, Said JW, et al. Co-infection of HHV-6 and HHV-8 is rare in primary effusion lymphoma. *Leuk Res* 2000;**24**:59–61.
39. Razzaque A, Williams O, Wang J, et al. Neoplastic transformation of immortalized human epidermal keratinocytes by two HHV-6 DNA clones. *Virology* 1993;**195**:113–20.
40. Thompson J, Choudhury S, Kashanchi F, et al. A transforming fragment within the direct repeat region of human herpesvirus type 6 that transactivates HIV-1. *Oncogene* 1994;**9**:1167–75.
41. Kashanchi F, Araujo J, Doniger J, et al. Human herpesvirus 6 (HHV-6) ORF-1 transactivating gene exhibits malignant transforming activity and its protein binds to p53. *Oncogene* 1997;**14**:359–67.
42. Collot-Teixeira S, Bass J, Denis F, et al. Human tumor suppressor p53 and DNA viruses. *Rev Med Virol* 2004;**14**:301–19.
43. Takemoto M, Mori Y, Ueda K, et al. Productive human herpesvirus 6 infection causes aberrant accumulation of p53 and prevents apoptosis. *J Gen Virol* 2004;**85**:869–79.

44. Oster B, Kofod-Olsen E, Bundgaard B, et al. Restriction of human herpesvirus 6B replication by p53. *J Gen Virol* 2008;**89**:1106–13.
45. Oster B, Bundgaard B, Hupp TR, et al. Human herpesvirus 6B induces phosphorylation of p53 in its regulatory domain by a CK2- and p38-independent pathway. *J Gen Virol* 2008;**89**:87–96.
46. Bargou RC, Emmerich F, Krappmann D, et al. Constitutive nuclear factor-κB-RelA activation is required for proliferation and survival of Hodgkin's disease tumor cells. *J Clin Invest* 1997;**100**:2961–9.
47. Flamand L, Menezes J. Cyclic AMP-responsive element-dependent activation of Epstein–Barr virus ZEBRA promoter by human herpesvirus 6. *J Virol* 1996;**70**:1784–91.
48. Cuomo L, Trivedi P, de Grazia U, et al. Upregulation of Epstein–Barr virus-encoded latent membrane protein by human herpesvirus 6 superinfection of EBV-carrying Burkitt lymphoma cells. *J Med Virol* 1998;**55**:219–26.
49. Nakayama-Ichiyama S, Yokote T, Iwaki K, et al. Co-infection of human herpesvirus-6 and human herpesvirus-8 in primary cutaneous diffuse large B-cell lymphoma, leg type. *Br J Haematol* 2011;**155**:514–5.
50. Lu C, Zeng Y, Huang Z, et al. Human herpesvirus 6 activates lytic cycle replication of Kaposi's sarcoma-associated herpesvirus. *Am J Pathol* 2005;**166**:173–83.
51. Prusty BK, zur Hausen H, Schmidt R, et al. Transcription of HERV-E and HERV-E-related sequences in malignant and non-malignant human haematopoietic cells. *Virology* 2008;**382**:37–45.
52. Takaku T, Ohyashiki JH, Zhang Y, et al. Estimating immunoregulatory gene networks in human herpesvirus type 6–infected T cells. *Biochem Biophys Res Commun* 2005;**336**:469–77.
53. Krueger GR, Brandt ME, Wang G, et al. TCM-1: a nonlinear dynamical computational model to simulate cellular changes in the T cell system; conceptional design and validation. *Anticancer Res* 2003;**23**:123–35.

Chapter 18

HHV-6A and HHV-6B in Cancer

John R. Crawford
University of California and Rady Children's Hospital, San Diego, California

INTRODUCTION

Human herpesvirus 6 (HHV-6) was first isolated in 1986 in patients with acquired immune deficiency syndrome (AIDS) and lymphoproliferative disorders.[1] HHV-6 is a member of the Betaherpesvirinae subfamily of DNA viruses, along with human cytomegalovirus (CMV) and HHV-7. HHV-6 exists as two distinct species, HHV-6A and HHV-6B, which share greater than 90% sequence homology.[2–5] HHV-6A and HHV-6B have been implicated in a wide variety of human diseases, including encephalitis, multiple sclerosis, seizures, drug-induced hypersensitivity, organ disease, neuroinflammation, chronic fatigue syndrome, and cancer.[6–8] Exanthem subitum is the only known human disease directly caused by HHV-6B infection.[9] HHV-6A is thought to exhibit more neurotropism than HHV-6B.[10] Although HHV-6A and HHV-6B share considerable sequence homology, they have been recently formally classified as two distinct viruses.[11] More recently, HHV-6 has been shown to exhibit chromosomal integration that may provide further insight into the potential biologic mechanisms of HHV-6 infection.[12–15]

Early studies into the role of HHV-6 and cancer were performed by Puri et al.,[16] who transfected NIH3T3 fibroblasts with genomic HHV-6; upon injection into nude mice, they formed undifferentiated fibrosarcomas.[16] Subsequent work demonstrated the ability of HHV-6 DNA to transform epidermal keratinocytes *in vitro*.[17,18] Although there has been no *in vivo* association between HHV-6 and cancer, viral infections have been well established in human malignancies and include lymphoma, nasopharyngeal carcinoma, post-transplant lymphoproliferative disease, leiomyosarcoma (Epstein–Barr virus), cervical carcinoma (human papillomavirus), Kaposi's sarcoma (human herpesvirus 8), T-cell leukemia (human T-cell lymphotropic virus type 1, or HTLV-1), and hepatocellular carcinoma (hepatitis B and C).[19] In this review, the most well-studied associations between HHV-6A and HHV-6B and cancer—namely,

Human Herpesviruses HHV-6A, HHV-6B & HHV-7.

hematologic, cervical, and brain malignancies—will be discussed with regard to potential epigenetic and oncomodulatory disease mechanisms.

HHV-6A AND HHV-6B IN LYMPHOMA

Lymphoma is a hematologic malignancy that can broadly be separated into two varieties: Hodgkin's lymphoma (HL) and non-Hodgkin's lymphoma (NHL) based on World Health Organization criteria.[20] HL has been the model system for viral-induced cancer, as up to 40% of patients with HL have associated EBV infection.[21–23] Both HL and NHL are among the most well studied of diseases in association with HHV-6A and HHV-6B.[24–38] A summary of the major studies regarding the detection of HHV-6A and HHV-6B are summarized in Table 18.1. Torelli et al.[24] first proposed an association between HHV-6 and HL in 1991 as demonstrated by an increase in anti-HHV-6 titers in HL compared to controls. While polymerase chain reaction (PCR) analysis was positive for HHV-6 in only 3 of 25 HL patients, 0 of 41 NHL were positive for HHV-6. Sumiyoshi et al.[25] reported much higher rates of PCR detection of HHV-6 in HL (64%), as well as high percentages in B-cell lymphoma (62%), T-cell lymphoma (59%), and anaplastic large-cell lymphoma (50%).[25] Overall, the percentage of HHV-6 variants in HL ranged from 7 to 80% among studies when utilizing conventional or real-time PCR.[24–38] Similar variances have also been observed in the percentage of HHV-6A versus HHV-6B variants among investigators. For example, Schmidt et al.[31] and Collot et al.[32] reported a majority of HHV-6A (73 to 92%) in lymphoma samples versus the 43 to 93% HHV-6B reported by Zhou et al.[36] and Lacroix et al.[37] More recent work by Lacroix et al.[37] and Siddon et al.[38] has demonstrated a high percentage of HHV-6 in Reed–Sternberg cells, a hallmark histologic feature of HL.

TABLE 18.1 Detection of Human Herpesvirus 6 in Lymphoma

Ref.	Study Material	Detection Method	Major Findings
Torelli et al.[21]	Serum and lymph nodes (NHL, HL)	IFA, PCR	Increased anti-HHV-6 titers in HL compared to controls; 3/25 HL HHV-6+ DNA, 0/41 NHL HHV-6+ DNA by PCR
Sumiyoshi et al.[25]	Lymph nodes (HL, BCL, TCL)	PCR	64% HL, 62% BCL, 59% TCL, 50% anaplastic large-cell lymphoma HHV-6+ DNA by PCR
Luppi et al.[26]	Lymph nodes (AILD, AILD-L)	PCR	7/12 HHV-6+ DNA by PCR2 HHV-6A, 2 HHV-6B, 1 both
Trovato et al.[27]	Lymph nodes (HL)	PCR	7% pediatric HL HHV-6+ and EBV DNA by PCR

(Continued)

TABLE 18.1 (Continued)

Ref.	Study Material	Detection Method	Major Findings
Valente et al.[28]	Lymph nodes (HL)	PCR, ISH, IHC	73% HHV-6+ by DNA PCR, 82% HHV-6+ by ISH, 8% HHV-6+ by IHC, 55% EBV+ HHV-6 + ; no detection in RS cells
Luppi et al.[29]	Lymph nodes (NHL, HL, AILD, other)	IHC	Detection of p41 early antigen in 2 RS cells in HL
Ohyashiki et al.[30]	Lymph nodes (BCL, TCL, other)	PCR, PCR–ELISA	20% BCL, 50% TCL HHV-6+ DNA by PCR; only HHV-6B detected
Schmidt et al.[31]	Lymph nodes (HL, NHL)	PCR, nPCR	13% HL HHV-6+ DNA by nPCR, 73% HHV-6A, 50% HHV-6+ HHV-7+
Collot et al.[32]	Lymph nodes, saliva (HL, BCL, TCL, other)	qPCR	35% HD, 22% BCL, 23% TCL HHV-6+ by qPCR, 92% HHV-6A, 100% nodular sclerosis HHV-6+
Vrsalovic et al.[33]	Lymph nodes (AITL)	PCR	22% AITL HHV-6+ DNA by PCR
Hernández-Losa et al.[34]	HL, NHL	PCR	80% HL, 59% NHL HHV-6+ DNA by PCR, >60% normal splenic lymphocytes HHV-6+
Lacroix et al.[35]	Lymph nodes (HL)	qPCR	79% HL HHV-6+ DNA by qPCR, 93% HHV-6B, 47% co-detection HHV-6 and EBV, 84% nodular sclerosis HHV-6+
Zhou et al.[36]	Lymph nodes (AITL)	PCR	21/49 HHV-6B+ by PCR; co-detection HHV-6B and EBV in 17/49
Lacroix et al.[37]	Lymph nodes (HL)	IHC, qPCR	HHV-6 DR7B 74% HHV-6+, EBV LMP1 negative samples; gp116/54/64 late antigen in 40% of RS cells
Siddon et al.[38]	Lymph nodes (NSHL)	PCR, IHC, FISH	85% HHV-6+ DNA by PCR, 48% RS cells HHV-6+ compared to 24% EBV+ RS cells

Abbreviations:*AILD, angioimmunoblastic lymphoma with dysproteinemia; AILD-L, AILD-like lymphoma; AITL, Angioimmunoblastic T-cell lymphoma; BCL, B-cell lymphoma; ELISA, enzyme-linked immunosorbent assay; FISH, fluorescence in situ hybridization; HL, Hodgkin's lymphoma; IHC, immunohistochemistry; LMP1, latent membrane protein 1; NHL, non-Hodgkin's lymphoma; nPCR, nested PCR; NSHL, nodular sclerosis Hodgkin's lymphoma; PCR, nested polymerase chain reaction; qPRC, real-time quantitative polymerase chain reaction; RS cells, Reed–Sternberg cells; TCL, T-cell lymphoma.*

Among the NHL groups, angioimmunoblastic T-cell lymphoma in particular, HHV-6 DNA detection by PCR ranges from 22 to 63%, yet no study has detected viral antigen by immunohistochemistry (IHC) that is indicative of a latent infection.[25,26,29,30,32,33,36] One common finding that has been reported is the detection of HHV-6 in the nodular sclerosis form of HL.[32,37] In terms of the prognostic value of HHV-6 detection in HL, Lacroix et al.[37] reported more favorable outcomes in those patients who were HHV-6 positive, who also happened to be a younger set of patients. Perhaps the more interesting results have come from investigations of the incidence of co-infection of HHV-6 and EBV ranging from 7 to 55% by PCR or IHC.[27,28,35,36] Future studies will be investigating possible synergistic effects between HHV-6 and EBV in lymphomagenesis.

HHV-6A AND HHV-6B IN LEUKEMIA

The association between HHV-6 and acute leukemia was first proposed by Ablashi et al.[39] who discovered that a small group of children with acute lymphoblastic leukemia (ALL) had elevated HHV-6 antibody titers. The same group in a larger series of 50 controls and 50 patients with ALL failed to find a significant difference.[40] Gentile et al.[41] confirmed these negative findings but reported a small but significant association between HHV-6 antibody titers in patients with acute myeloid leukemia (AML). Salonen et al.[42] studied 40 children with leukemia (33 ALL, 7 AML) for the incidence of HHV-6 immunoglobulin M (IgM) and IgG antibodies and found that 40% of patients were positive for HHV-6 IgM compared to 8% of matched controls. Work by Bogdanovic et al.[43] studying Guthrie cards at birth from 54 patients with a subsequent diagnosis of leukemia found no detectable HHV-6 DNA by PCR, thus excluding congenital HHV-6 as a potential etiology of leukemia in this group. One understudied area regarding the association of HHV-6A and HHV-6B with leukemia is the potential role of chromosomal integration. Chromosomally integrated HHV-6 (ciHHV-6) has been shown to be present in up to 2% of the population of the United Kingdom.[12] HHV-6 was first demonstrated to exhibit chromosomal integration in a Burkitt's lymphoma cell line and subsequent transmission of ciHHV-6 in ALL.[44,45] However, one of the largest studies performed by Hobacek et al.[46] showed only 5 of 339 patients with leukemia to have ciHHV-6, making it unlikely to be a major oncogenic mechanism of leukemia. In summary, current evidence does not support a direct role of HHV-6A and HHV-6B in the oncogenesis of leukemia.

HHV-6A AND HHV-6B IN CERVICAL CARCINOMA

Cervical carcinoma is the quintessential example of viral oncogenesis and represents the second most common cancer in females.[47] Caused by the chromosomal integration of HPV into the host genome, viral proteins E6 and E7 inactivate both p53 and pRB tumor suppressor proteins that significantly increase the risk of activating mutations that lead to cancer.[48,49] The detection of HHV-6A and

HHV-6B in cervical carcinoma was first reported by Chen et al.,[50] who reported 6/72 squamous cervical carcinoma and cervical intraepithelial neoplasm positive for HHV-6 by PCR (4/6 HHV-6B) compared to 0/30 normal cervical specimens. It was also the first study to show co-localization of HHV-6 with HPV-16 by ISH. As summarized in Table 18.2, HHV-6 variants have been detected by DNA PCR in 1 to 25% of cases of cervical carcinoma.[50–56] Arguably the more interesting data are with regard to the association of HPV with HHV-6 and other viruses in cases of cervical carcinoma. Tran-Thanh et al.[54] reported a nearly fivefold increase in HPV-16 positivity by PCR when HHV-6 was also positive. Broccolo et al.[55] reported dual positivity of CMV and HHV-6 by PCR in 25% of squamous intraepithelial lesions versus 8% dual positivity in normal controls. Most recently, Szostek et al.[56] reported lower rates of HPV co-infected with HHV-6 (10%) compared to CMV (52%) or EBV (22%); however, HHV-6 and CMV were co-detected with HPV by PCR in 20% of cases. Taken together, current data suggest that cervical carcinoma involves co-infection with HPV subtypes and other herpesvirus family members, including HHV-6, whose potential role in malignancy is not well characterized at this time.

HHV-6A AND HHV-6B IN BRAIN TUMORS

Central nervous system tumors occur at an incidence of 16.5/100,000 person-years and are a significant cause of morbidity and mortality in both children and adults.[57] Brain tumors represent a histologically and molecularly diverse group of neoplasms of varied anatomical location, growth characteristics, and survival. The potential role of viruses in the oncogenesis of brain tumors has been most extensively studied in association with polyomavirus and CMV and has been the subject of a prior review.[58] In brief, human polyomaviruses (JC virus, BK virus, SV40 virus) have been detected in a variety of brain tumor subtypes with varying frequencies by PCR, Southern blot, and immunohistochemical analysis without a clear tropism.[58] CMV, on the other hand, has been detected predominantly in glioblastoma multiforme (GBM) in adults as assayed by PCR, ISH, and IHC, the detection of which has been validated by several groups.[59–63] Furthermore, CMV has been detected in both matched serum and tumor samples of varied frequency in GBM patients, suggestive of a possible hematogeneous spread of CMV into the central nervous system.[63] The association with CMV and GBM prompted a clinical trial to assess the efficacy of valganciclovir in combination with standard radiation and chemotherapy. While overall survival was unchanged, explorative analysis showed trends for improved survival in those patients on antiviral therapy greater than 6 months.[64] Given the genomic similarities between CMV and HHV-6,[65] a similar screening of HHV-6A and HHV-6B in pediatric and adult brain tumors has been performed by several laboratories whose major findings are summarized in Table 18.3.[66–73] Luppi et al. first reported detection of HHV-6 in 37% of glial tumors by nPCR.[66] A follow-up study by Chan et al. in 1999 reported 6% GBM, 14% meningioma, and 60% ependymoma HHV-6 positivity

TABLE 18.2 Detection of Human Herpesvirus 6 in Cervical Carcinoma

Ref.	Study Material	Detection Method	Major Findings
Chen et al.[50]	Cervical biopsy (*n* = 72)	PCR, ISH	6/72 SCC and CIN HHV-6+ DNA by PCR (4/6 HHV-6B), 0/30 normal cervical specimens HHV-6+; first report of HPV-16 co-localizing with HHV-6 by ISH
Wang et al.[51]	Cervical biopsy (*n* = 8)	PCR	2/8 samples HHV-6+ DNA by PCR
Yadav et al.[52]	Cervical biopsy (*n* = 34; 8 normal)	IHC, ISH	54% carcinoma HHV-6+ by IHC versus 63% control; 39% carcinoma = HHV-6 by ISH versus 13% controls
Romano et al.[53]	Cervical cytology (*n* = 109; 85 abnormal)	PCR	1/109 HHV-6B+ DNA by PCR
Tran-Thanh et al.[54]	Cervical biopsy (*n* = 300; 50 cervical carcinoma, 65 HSIL, 80 LSIL, 125 normal)	PCR	2% HHV-6+ DNA by PCR; 5% HPV-16+ were HHV-6+ versus 1.3% HHV-6+ HPV-16–; HSIL 6.2% HHV-6 DNA+ versus 0.8% normal controls
Broccolo et al.[55]	Cervical biopsy (*n* = 208)	qPCR	25% HHV-6+, 66% CMV+, 6% HHV-7+ by DNA PCR; 40% HSIL HHV-6+ versus 14% normal controls, CMV+ HHV-6+ in 25% SIL versus 8% normal controls
Szostek et al.[56]	Cervical biopsy (60 HPV+)	qPCR	72% HPV+ co-infected with CMV (52%), EBV (22%), HHV-7 (14.5%), HHV-6 (10%), HSV1/2(9%); HHV-6 and CMV co-infected with HPV-16 in 19%

Abbreviations: *CIN, cervical intraepithelial neoplasm; EPV, Epstein–Barr virus; HPV, human papillomavirus; HSIL, high grade squamous intraepithelial lesion; IHC, immunohistochemistry; ISH, in situ hybridization; LSIL, low-grade squamous intraepithelial lesion; PCR, nested polymerase chain reaction; qPRC, real-time quantitative polymerase chain reaction; SCC, squamous cervical carcinoma; SIL, squamous intraepithelial lesion.*

TABLE 18.3 Detection of Human Herpesvirus 6 in Primary Central Nervous System Tumors

Ref.	Sample Size	Detection Method	HHV-6 DNA Detection Rate by Tumor Subtype	HHV-6 Protein Detection by Tumor Subtype	HHV-6 Distinction
Chan et al.[67]	98	nPCR	14% meningioma 6% GBM 60% ependymoma	NP	3 HHV-6A 6 HHV-6B
Luppi et al.[66]	37	nPCR	16% neuroglial	NP	NP
Cuomo et al.[68]	118	nPCR	45% GBM 26% meningioma 50% ependymoma 33% astrocytoma 38% other	40%	33 HHV-6A 10 HHV-6B
Neves et al.[69]	35	qPCR	ND	NP	NP
Crawford et al.[70]	150	nPCR, IHC, ISH	66–71% glioma 56–72% JPA 64–71% GBM 25–70% meningioma 20–58% medulloblastoma	47% glioma 44% JPA 29% GBM 0% meningioma 20% medulloblastoma	43 HHV-6A 17 HHV-6B
Crawford et al.[71]	282	nPCR, IHC, ISH	32–48% glioma 20–60% JPA 30–47% GBM 32% meningioma 50–62% oligodendroglioma	32–38% glioma 16–24% JPA 28–32% GBM 14–16% meningioma 19–24% oligodendroglioma	6 HHV-6A 8 HHV-6B
Duncan et al.[72]	40	DK	ND	NP	NP
Chi et al.[73]	40	nPCR, IHC, Cx	43% gliomoa 41% astrocytoma 50% GBM 33% oligodendroglioma		1 HHV-6A (tumor cyst)

Abbreviations: *Cx, culture; DK, digital karyotyping; GBM, glioblastoma multiforme; IHC, immunohistochemistry; ISH, in situ hybridization; JPA, juvenile pilocytic astrocytoma; ND, not detected; NP, not performed; nPCR, nested polymerase chain reaction; qPCR, real-time quantitative polymerase chain reaction.*

by PCR with a predominance of HHV-6B.[67] Cuomo et al. found a higher percentage of GBM HHV-6 positivity by PCR (45%) and was the first to demonstrate HHV-6 viral antigen (40%) in malignant brain tumors.[68] Crawford et al.[70,71] reported the most extensive adult and pediatric series to date utilizing tissue microarrays for IHC, nested PCR, and ISH. In pediatric tumors, there was a glial tropism for HHV-6 DNA, as well as viral antigen detection, with a predominance of HHV-6A.[70] Similar glial tropisms for antigen detection were found in the adult studies utilizing two separate antibodies;[71] however, in both studies, there was no significant effect of HHV-6 detection and survival.[70,71] Chi et al.[73] in 2012 confirmed the presence of HHV-6 DNA and antigen in a large series of adult gliomas with similar results. More importantly, for the first time they isolated HHV-6A from a glioma cyst drained during surgery, as evidenced by inoculation of cord blood mononuclear cells, and confirmed by PCR and IHC.[73] Overall, the constellation of reports supports the presences of HHV-6A and HHV-6B in pediatric and adult brain tumors; however, a role for HHV-6 in neuro-oncogenesis or oncomodulation remains to be determined.

MECHANISMS OF HHV-6A AND HHV-6B IN CANCER

Human herpesvirus 6A and HHV-6B have been detected in a variety of cancers, either in isolation or combination with other known oncogenic viruses. Although these association studies do not serve as direct evidence for a role of HHV-6 in cancer, *in vitro* studies using a variety of cell culture systems may provide important insight into potential oncologic disease mechanisms, as depicted in Figure 18.1. The term *oncomodulation* was first used to describe the potential role of CMV in malignant gliomagenesis. It refers to the effects of viral proteins and noncoding nucleic acids through disruption of intracellular signaling pathways in the absence of direct transformation.[74–76] One potential mechanism of oncomodulation is through the inflammatory properties of HHV-6A and HHV-6B infection.[8] Both HHV-6A and HHV-6B have been shown to induce a Th1–Th2 shift in T cells through induction of interleukin-10 (IL-10) production[77] and IL-12 inhibition by dendritic cells and macrophages.[78,79] HHV-6 has been to shown to upregulate expression of a number of inflammatory cytokines in peripheral blood mononuclear cells, including IL-1β, tumor necrosis factor (TNF), and interferon (IFN).[80,81] In cultured astrocytes, latent HHV-6 infection has also been associated with altered cytokine expression.[82] HHV-6A and HHV-6B infection of both lymphoid tissue and cultured astrocytes can induce a variety of chemokines, including CCL-5, CCL-3, CCL-2, CXCL-2, and CXCL-8.[83–87] In fact, HHV-6 open reading frame U83 encodes a chemokine that has been demonstrated to be functional by Zou et al.[88] This constellation of inflammatory signals mediated by HHV-6A and HHV-6B could have potential epigenetic effects on tumor growth, invasion, and metastasis.

A second major pathway for HHV-6A and HHV-6B oncomodulation is through direct interaction of HHV-6 proteins with key oncogenic cell signaling

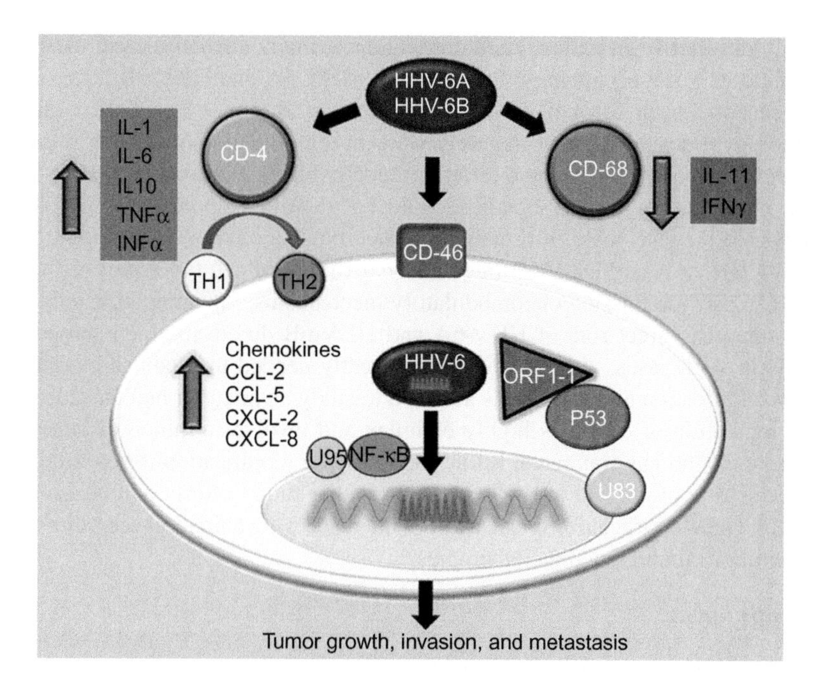

FIGURE 18.1 HHV-6Aand HHV-6B in cancer: potential oncomodulatory mechanisms.

pathways. One example is the HHV-6 ORF-1 *trans*-activating gene that has been shown to bind to p53, a ubiquitous tumor suppressor gene.[89] A second potential candidate is the HHV-6 immediate-early gene U95, which has been shown to bind nuclear factor B (NFB), another key transcription factor in oncogenesis.[90]

A third potential oncomodulatory pathway for HHV-6A and HHV-6B is through chromosomal integration mechanisms.[91,92] Although much is still unknown about the chromosomal mechanisms of integration, diversity of integration sites, and consequences of gene disruption, there is a lot of enthusiasm for the role of ciHHV-6 in cancer, given the established history of a number of viruses and oncogenesis.

A final potential mechanism for HHV-6A and HHV-6B in cancer, and the most plausible, is through synergistic interactions with other co-infected viruses.[93] As has been shown in hematologic and cervical carcinoma, there is a high prevalence of co-infection with HHV-6A and HHV-6B as well as with other viruses. Future studies will be aimed at identifying a potential synergistic association between HHV-6 and known transforming viruses.

DISCUSSION

While there is mounting evidence that HHV-6A and HHV-6B DNA and antigen are associated with a number of human cancers, most of the work is highly descriptive in nature. As a consequence, many challenges still exist in the field of HHV-6 oncology. For example, HHV-6A and HHV-6B have only just recently been recognized as two distinct viruses in spite of their close homology. Because most of the early work in identification of HHV-6 in cancer did not specifically involve differentiation between HHV-6A and HHV-6B, it is possible that new associations may be identified upon more complete analysis. Likewise, more information on specific gene expression profiling following HHV-6A and HHV-6B infection in a variety of cell types will be more likely to provide missing oncomodulatory mechanisms. Epidemiologic data do not support a direct role of HHV-6A and HHV-6B due to the high seropositivity. In such cases, animal models are greatly needed to establish causality between infection and disease, as are used to study a variety of human cancers. The regulation of cancer is highly complex and involves a milieu of genetic, epigenetic, and environmental influences. Disease modification is best studied in the setting of clinical trials. It is the hope that future efforts will be geared toward HHV-6-targeted clinical trials with respect to human cancer through antiviral or immune-mediated therapies.

REFERENCES

1. Salahuddin SZ, Ablashi DV, Markham PD, et al. Isolation of a new virus, HBLV, in patients with lymphoproliferative disorders. *Science* 1986;**234**:596–601.
2. Flamand L, Komaroff AL, Arbuckle JH, et al. Review, part 1: Human herpesvirus-6—basic biology, diagnostic testing, and antiviral efficacy. *J Med Virol* 2010;**82**:1560–8.
3. Isegawa Y, Mukai T, Nakano K, et al. Comparison of the complete DNA sequences of human herpesvirus 6 variants A and B. *J Virol* 1999;**73**:8053–63.
4. Dominguez G, Dambaugh TR, Stamey FR, et al. Human herpesvirus 6B genome sequence: coding content and comparison with human herpesvirus 6A. *J Virol* 1999;**73**:8040–52.
5. Gravel A, Ablashi D, Flamand L. Complete genome sequence of early passaged human herpesvirus 6a (GS strain) isolated from North America. *Genome Announc* 2013;**1**:e00012–13.
6. Yao K, Crawford JR, Komaroff AL, et al. Review, part 2: Human herpesvirus-6 in central nervous system diseases. *J Med Virol* 2010;**82**:1669–78.
7. Ablashi DV, Devin CL, Yoshikawa T, et al. Review, part 3: Human herpesvirus-6 in multiple non-neurological diseases. *J Med Virol* 2010;**82**:1903–10.
8. Reynaud JM, Horvat B. Human herpesvirus 6 and neuroinflammation. *ISRN Virol* 2013;**2013**:1–11.
9. Yamanishi K, Okuno T, Shiraki K, et al. Identification of human herpesvirus-6 as a causal agent for exanthem subitum. *Lancet* 1988;**1**:1065–7.
10. Hall CB, Caserta MT, Schnabel KC, et al. Persistence of human herpesvirus 6 according to site and variant: possible greater neurotropism of variant A. *Clin Infect Dis* 1998;**26**:132–7.
11. Adams MJ, Carstens EB. Ratification vote on taxonomic proposals to the International Committee on Taxonomy of Viruses. *Arch Virol* 2012;**157**:1411–22.
12. Leong HN, Tuke PW, Tedder RS, et al. The prevalence of chromosomally integrated human herpesvirus 6 genomes in the blood of UK blood donors. *J Med Virol* 2007;**79**:45–51.
13. Hall CB, Caserta MT, Schnabel K, et al. Chromosomal integration of human herpesvirus 6 is the major mode of congenital human herpesvirus 6 infection. *Pediatrics* 2008;**122**:513–20.

14. Clark DA, Ward KN. Importance of chromosomally integrated HHV-6A and -6B in the diagnosis of active HHV-6 infection. *Herpes* 2008;**15**:28–32.
15. Pellett PE, Ablashi DV, Ambros PF, et al. Chromosomally integrated human herpesvirus 6: questions and answers. *Rev Med Virol* 2012;**22**:144–55.
16. Puri RK, Leland P, Razzaque A. Antigen(s)-specific tumour-infiltrating lymphocytes from tumour induced by human herpes virus-6 (HHV-6) DNA transfected NIH 3T3 transformants. *Clin Exp Immunol* 1991;**83**:96–101.
17. Razzaque A. Oncogenic potential of human herpesvirus-6 DNA. *Oncogene* 1990;**5**:1365.
18. Razzaque A, Williams O, Wang J, et al. Neoplastic transformation of immortalized human epidermal keratinocytes by two HHV-6 DNA clones. *Virology* 1993;**195**:113–20.
19. Liao JB. Viruses and human cancer. *Yale J Biol Med* 2006;**79**:115–22.
20. Jaffe ES, Harris NL, Stein H, Vardiman JW, editors. *WHO Classification of Tumours of Tumors: Pathology and Genetics of Tumours of Haematopoietic and Lymphoid Tissues.* Lyon, France: IARC Press; 2001.
21. Diepstra A, Niens M, Vellenga E, et al. Association with HLA class I in Epstein–Barr-virus-positive and with HLA class III in Epstein–Barr-virus-negative Hodgkin's lymphoma. *Lancet* 2005;**365**:2216–24.
22. Küppers R. B cells under influence: transformation of B cells by Epstein–Barr virus. *Nat Rev Immunol* 2003;**10**:801–12.
23. Ogata M. Human herpesvirus 6 in hematological malignancies. *J Clin Exp Hematop* 2009;**49**:57–67.
24. Torelli G, Marasca R, Luppi M, et al. Human herpesvirus-6 in human lymphomas: identification of specific sequences in Hodgkin's lymphomas by polymerase chain reaction. *Blood* 1991;**77**:2251–8.
25. Sumiyoshi Y, Kikuchi M, Ohshima K, et al. Analysis of human herpes virus-6 genomes in lymphoid malignancy in Japan. *J Clin Pathol* 1993;**46**:1137–8.
26. Luppi M, Marasca R, Barozzi P, et al. Frequent detection of human herpesvirus-6 sequences by polymerase chain reaction in paraffin-embedded lymph nodes from patients with angioimmunoblastic lymphadenopathy and angioimmunoblastic lymphadenopathy-like lymphoma. *Leuk Res* 1993;**17**:1003–11.
27. Trovato R, Di Lollo S, Calzolari A, et al. Detection of human herpesvirus-6 and Epstein–Barr virus genome in childhood Hodgkin's disease. *Pathologica* 1994;**86**:500–3.
28. Valente G, Secchiero P, Lusso P, et al. Human herpesvirus 6 and Epstein–Barr virus in Hodgkin's disease: a controlled study by polymerase chain reaction and *in situ* hybridization. *Am J Pathol* 1996;**149**:1501–10.
29. Luppi M, Barozzi P, Garber R, et al. Expression of human herpesvirus-6 antigens in benign and malignant lymphoproliferative diseases. *Am J Pathol* 1998;**153**:815–23.
30. Ohyashiki JH, Abe K, Ojima T, et al. Quantification of human herpesvirus 6 in healthy volunteers and patients with lymphoproliferative disorders by PCR-ELISA. *Leuk Res* 1999;**23**:625–30.
31. Schmidt CA, Oettle H, Peng R, et al. Presence of human beta- and gamma-herpes virus DNA in Hodgkin's disease. *Leuk Res* 2000;**24**:865–70.
32. Collot S, Petit B, Bordessoule D, et al. Real-time PCR for quantification of human herpesvirus 6 DNA from lymph nodes and saliva. *J Clin Microbiol* 2002;**40**:2445–51.
33. Vrsalovic MM, Korac P, Dominis M, et al. T- and B-cell clonality and frequency of human herpes viruses-6, -8 and Epstein–Barr virus in angioimmunoblastic T-cell lymphoma. *Hematol Oncol* 2004;**22**:169–77.

34. Hernández-Losa J, Fedele CG, Pozo F, et al. Lack of association of polyomavirus and herpesvirus types 6 and 7 in human lymphomas. *Cancer* 2005;**103**:293–8.
35. Lacroix A, Jaccard A, Rouzioux C, et al. HHV-6 and EBV DNA quantitation in lymph nodes of 86 patients with Hodgkin's lymphoma. *J Med Virol* 2007;**79**:1349–56.
36. Zhou Y, Attygalle AD, Chuang SS, et al. Angioimmunoblastic T-cell lymphoma: histological progression associates with EBV and HHV6B viral load. *Br J Haematol* 2007;**138**:44–53.
37. Lacroix A, Collot-Teixeira S, Mardivirin L, et al. Involvement of human herpesvirus-6 variant B in classic Hodgkin's lymphoma via DR7 oncoprotein. *Clin Cancer Res* 2010;**16**:4711–21.
38. Siddon A, Lozovatsky L, Mohamed A, et al. Human herpesvirus 6 positive Reed–Sternberg cells in nodular sclerosis Hodgkin lymphoma. *Br J Haematol* 2012;**158**:635–43.
39. Ablashi DV, Josephs SF, Buchbinder A, et al. Human B-lymphotropic virus (human herpesvirus-6). *J Virol Methods* 1988;**21**:29–48.
40. Levine PH, Ablashi DV, Saxinger WC, et al. Antibodies to human herpes virus-6 in patients with acute lymphocytic leukemia. *Leukemia* 1992;**6**:1229–31.
41. Gentile G, Mele A, Ragona G, et al. Human herpes virus-6 seroprevalence and leukaemias: a case-control study. GIMEMA (Gruppo Italiano Malattie Ematologiche dell' Adulto). *Br J Cancer* 1999;**80**:1103–6.
42. Salonen MJ, Siimes MA, Salonen EM, et al. Antibody status to HHV-6 in children with leukaemia. *Leukemia* 2002;**16**:716–9.
43. Bogdanovic G, Jernberg AG, Priftakis P, et al. Human herpes virus 6 or Epstein–Barr virus were not detected in Guthrie cards from children who later developed leukaemia. *Br J Cancer* 2004;**91**:913–5.
44. Daibata M, Taguchi T, Taguchi H, et al. Integration of human herpesvirus 6 in a Burkitt's lymphoma cell line. *Br J Haematol* 1998;**102**:1307–13.
45. Daibata M, Taguchi T, Sawada T, et al. Chromosomal transmission of human herpesvirus 6 DNA in acute lymphoblastic leukaemia. *Lancet* 1998;**352**:543–4.
46. Hubacek P, Muzikova K, Hrdlickova A, et al. Prevalence of HHV-6 integrated chromosomally among children treated for acute lymphoblastic or myeloid leukemia in the Czech Republic. *J Med Virol* 2009;**81**:258–63.
47. González Martín A. Molecular biology of cervical cancer. *Clin Transl Oncol* 2007;**9**:347–54.
48. DiMaio D, Liao JB. Human papillomaviruses and cervical cancer. *Adv Virus Res* 2006;**66**:125–59.
49. de Freitas AC, Gurgel AP, Chagas BS, et al. Susceptibility to cervical cancer: an overview. *Gynecol Oncol* 2012;**126**:304–11.
50. Chen M, Wang H, Woodworth CD, et al. Detection of human herpesvirus 6 and human papillomavirus 16 in cervical carcinoma. *Am J Pathol* 1994;**145**:1509–16.
51. Wang H, Chen M, Berneman ZN, et al. Detection of human herpesvirus-6 in paraffin-embedded tissue of cervical cancer by polymerase chain reaction. *J Virol Methods* 1994;**47**:297–305.
52. Yadav M, Arivananthan M, Kumar S. HHV-6 antigen and viral DNA detected in cervical cells from archived tissue using histochemical staining and hybridization. *Clin Diagn Virol* 1996;**7**:23–33.
53. Romano N, Romano FM, Viviano E, et al. Rare association of human herpesvirus 6 DNA with human papillomavirus DNA in cervical smears of women with normal and abnormal cytologies. *J Clin Microbiol* 1996;**34**:1589–91.
54. Tran-Thanh D, Koushik A, Provencher D, et al. Detection of human herpes virus type 6 DNA in precancerous lesions of the uterine cervix. *J Med Virol* 2002;**68**:606–10.

55. Broccolo F, Cassina G, Chiari S, et al. Frequency and clinical significance of human beta-herpesviruses in cervical samples from Italian women. *J Med Virol* 2008;**80**:147–53.
56. Szostek S, Zawilinska B, Kopec J, et al. Herpesviruses as possible cofactors in HPV-16-related oncogenesis. *Acta Biochim Pol* 2009;**56**:337–42.
57. Dolecek TA, Propp JM, Stroup NE, et al. CBTRUS statistical report: primary brain and central nervous system tumors diagnosed in the United States in 2005–2009. *Neuro-Oncol* 2012;**14**(Suppl. 5):v1–49.
58. Saddawi-Konefka R, Crawford JR. Chronic viral infection and primary central nervous system malignancy. *J Neuroimmune Pharmacol* 2010;**5**:387–403.
59. Cobbs CS, Harkins L, Samanta M, et al. Human cytomegalovirus infection and expression in human malignant glioma. *Cancer Res* 2002;**62**:3347–50.
60. Scheurer ME, El-Zein R, Bondy ML, et al. RE: "Lack of association of herpesviruses with brain tumors." *J Neurovirol* 2007;**13**:85. author reply 86–7.
61. Miller G. Brain cancer. A viral link to glioblastoma? *Science* 2009;**323**:30–1.
62. Scheurer ME, Bondy ML, Aldape KD, Albrecht T, El-Zein R. Detection of human cytomegalovirus in different histological types of gliomas. *Acta Neuropathol* 2008;**116**:79–86.
63. Mitchell DA, Xie W, Schmittling R, et al. Sensitive detection of human cytomegalovirus in tumors and peripheral blood of patients diagnosed with glioblastoma. *Neuro Oncol* 2008;**10**:10–18.
64. Stragliotto G, Rahbar A, Solberg NW, et al. Effects of valganciclovir as an add-on therapy in patients with cytomegalovirus-positive glioblastoma: a randomized, double-blind, hypothesis-generating study. *Int J Cancer* 2013;**133**(5):1204–13.
65. Lawrence GL, Chee M, Craxton MA, et al. Human herpesvirus 6 is closely related to human cytomegalovirus. *J Virol* 1990;**64**:287–99.
66. Luppi M, Barozzi P, Maiorana A, et al. Human herpesvirus-6: a survey of presence and distribution of genomic sequences in normal brain and neuroglial tumors. *J Med Virol* 1995;**47**:105–11.
67. Chan PK, Ng HK, Cheng AF. Detection of human herpesviruses 6 and 7 genomic sequences in brain tumours. *J Clin Pathol* 1999;**52**:620–3.
68. Cuomo L, Trivedi P, Cardillo MR, et al. Human herpesvirus 6 infection in neoplastic and normal brain tissue. *J Med Virol* 2001;**63**:45–51.
69. Neves AM, Thompson G, Carvalheira J, et al. Detection and quantitative analysis of human herpesvirus in pilocytic astrocytoma. *Brain Res* 2008;**1221**:108–14.
70. Crawford JR, Santi MR, Thorarinsdottir HK, et al. Detection of human herpesvirus-6 variants in pediatric brain tumors: association of viral antigen in low grade gliomas. *J Clin Virol* 2009;**46**:37–42.
71. Crawford JR, Santi MR, Cornelison R, et al. Detection of human herpesvirus-6 in adult central nervous system tumors: predominance of early and late viral antigens in glial tumors. *J Neurooncol* 2009;**95**:49–60.
72. Duncan CG, Leary RJ, Lin JC, et al. Identification of microbial DNA in human cancer. *BMC Med Genomics* 2009;**2**:22.
73. Chi J, Gu B, Zhang C, et al. Human herpesvirus 6 latent infection in patients with glioma. *J Infect Dis* 2012;**206**:1394–8.
74. Michaelis M, Doerr HW, Cinatl J. The story of human cytomegalovirus and cancer: increasing evidence and open questions. *Neoplasia* 2009;**11**:1–9.
75. Cinatl Jr J, Nevels M, Paulus C, Michaelis M. Activation of telomerase in glioma cells by human cytomegalovirus: another piece of the puzzle. *J Natl Cancer Inst* 2009;**101**:441–3.

76. Kofman A, Marcinkiewicz L, Dupart E, et al. The roles of viruses in brain tumor initiation and oncomodulation. *J Neurooncol* 2011;**105**:451–66.
77. Arena A, Liberto MC, Iannello D, et al. Altered cytokine production after human herpes virus type 6 infection. *New Microbiol* 1999;**22**:293–300.
78. Smith A, Santoro F, Di Lullo G, et al. Selective suppression of IL-12 production by human herpesvirus 6. *Blood* 2003;**102**:2877–84.
79. Smith AP, Paolucci C, Di Lullo G, et al. Viral replication-independent blockade of dendritic cell maturation and interleukin-12 production by human herpesvirus 6. *J Virol* 2005;**79**:2807–13.
80. Flamand L, Gosselin J, D'Addario M, et al. Human herpesvirus 6 induces interleukin-1 beta and tumor necrosis factor alpha, but not interleukin-6, in peripheral blood mononuclear cell cultures. *J Virol* 1991;**65**:5105–10.
81. Kikuta H, Nakane A, Lu H, et al. Interferon induction by human herpesvirus 6 in human mononuclear cells. *J Infect Dis* 1990;**162**:35–8.
82. Yoshikawa T, Asano Y, Akimoto S, et al. Latent infection of human herpesvirus 6 in astrocytoma cell line and alteration of cytokine synthesis. *J Med Virol* 2002;**66**:497–505.
83. Caruso A, Favilli F, Rotola A, et al. Human herpesvirus-6 modulates RANTES production in primary human endothelial cell cultures. *J Med Virol* 2003;**70**:451–8.
84. Caruso A, Rotola A, Comar M, et al. HHV-6 infects human aortic and heart microvascular endothelial cells, increasing their ability to secrete proinflammatory chemokines. *J Med Virol* 2002;**67**:528–33.
85. Grivel JC, Ito Y, Fagà G, et al. Suppression of CCR5- but not CXCR4-tropic HIV-1 in lymphoid tissue by human herpesvirus 6. *Nat Med* 2001;**7**:1232–5.
86. Grivel JC, Santoro F, Chen S, et al. Pathogenic effects of human herpesvirus 6 in human lymphoid tissue *ex vivo*. *J Virol* 2003;**77**:8280–9.
87. Meeuwsen S, Persoon-Deen C, Bsibsi M, et al. Modulation of the cytokine network in human adult astrocytes by human herpesvirus-6A. *J Neuroimmunol* 2005;**164**:37–47.
88. Zou P, Isegawa Y, Nakano K, et al. Human herpesvirus 6 open reading frame U83 encodes a functional chemokine. *J Virol* 1999;**73**:5926–33.
89. Kashanchi F, Araujo J, Doniger J, et al. Human herpesvirus 6 (HHV-6) ORF-1 transactivating gene exhibits malignant transforming activity and its protein binds to p53. *Oncogene* 1997;**14**:359–67.
90. Takemoto M, Shimamoto T, Isegawa Y, et al. The R3 region, one of three major repetitive regions of human herpesvirus 6, is a strong enhancer of immediate-early gene U95. *J Virol* 2001;**75**:10149–60.
91. Razzaque A. Oncogenic potential of human herpesvirus-6 DNA. *Oncogene* 1990;**5**:1365.
92. Amirian ES, Scheurer ME. Chromosomally-integrated human herpesvirus 6 in familial glioma etiology. *Med Hypotheses* 2012;**2**:193–6.
93. Amirian ES, Adler-Storthz K, Scheurer ME. Associations between human herpesvirus-6, human papillomavirus and cervical cancer. *Cancer Lett* 2013;**336**:18–23.

Chapter 19

Treating HHV-6 Infections

The Laboratory Efficacy and Clinical Use of Anti-HHV-6 Agents

Joshua C. Pritchett[a], Lieve Naesens[b], and Jose Montoya[c]
[a]The HHV-6 Foundation, Santa Barbara, California, [b]Rega Institute, KU Leuven, Leuven, Belgium, [c]Stanford University School of Medicine, Stanford, California

INTRODUCTION

HHV-6 Infection

Human herpesvirus-6 is the collective name for HHV-6A and HHV-6B, two closely related herpesviruses[1] that have a combined seroprevalence of over 90% in adults worldwide.[2,3] HHV-6B is typically transmitted via saliva, and primary infection usually occurs between 6 months and 2 to 3 years of age.[2] In approximately 30% of children, primary HHV-6B infection causes roseola infantum (exanthema subitum, or sixth disease) presenting with high-grade fever followed by a characteristic mild rash that is sometimes accompanied by benign febrile convulsions and rarely by status epilepticus.[4] Although little is known about the transmission of and diseases associated with primary HHV-6A infection, one study indicates that about 40% of African children undergo primary infection with HHV-6A—a rate significantly higher than the prevalence of HHV-6A primary infection in many other communities worldwide.[5]

Disease Associations and Need for Treatment of Acute HHV-6 Infections

Severe and sometimes fatal cases of acute HHV-6 infection (AI)—whether congenital or primary, or resulting from reactivation or reinfection—have been documented, warranting pharmacological intervention for the clinical presentation described in these cases.[6] The treatment of HHV-6 reactivation or AI is of particular importance in the posttransplant setting, especially in the context of hematopoietic stem cell transplantation (HSCT).[7] Reactivation of latent HHV-6 has been reported in transplant recipients, resulting in the development

Human Herpesviruses HHV-6A, HHV-6B & HHV-7.

of fatal HHV-6 encephalitis and other neurological sequelae attributable to this reactivation event.[8–11] Cord blood transplant recipients have an especially high rate of associated encephalitis due to posttransplant HHV-6 reactivation.[12–16] Finally, HHV-6B reactivation has been firmly associated with drug-induced hypersensitivity syndrome (DIHS) and the related drug rash with eosinophilia and systemic symptoms (DRESS).[17–19]

Additional symptoms and conditions that have been associated with HHV-6B in immunocompromised patients include exanthematous rash, fever, seizures,[20] encephalopathy,[21] limbic encephalitis,[22] amnesia,[23,24] cognitive dysfunction,[9] lymphadenopathy,[25] colitis,[26–28] and hepatitis.[29,30] Less well-established associations have been reported for renal failure,[31,32] liver failure,[33,34] hemophagocytic syndrome,[35,36] myocarditis,[37–39] pneumonitis,[40–42] hypogammaglobulinemia,[43] and arteriopathies.[44,45]

Human herpesvirus 6A reactivation has been associated with neuroinflammatory diseases such as multiple sclerosis[46–48] and encephalitis.[49] Other clinical manifestations associated with HHV-6A include syncytial giant-cell hepatitis in liver transplant patients[50] and patients with HIV infection and AIDS.[51–53] The virus is also more frequently detected in the plasma of bone marrow transplant patients.[54,55] In addition, HHV-6A has been implicated as a significant factor in Hashimoto's thyroiditis (HT), an autoimmune disorder that is the most common of all thyroid diseases.[56]

Although drug treatment for HHV-6 AI and reactivation is currently practiced only in a limited number of clinical settings (such as HHV-6-associated encephalitis), an increased awareness of HHV-6 diseases associated with acute infection in both immunocompetent and immunocompromised patients—likely due to the increased usage of HHV-6 diagnostic assays such as HHV-6 quantitative polymerase chain reaction (qPCR) in the clinical setting—has resulted in growing interest in evaluating the best treatment options for HHV-6 disease.[6]

Antiviral Treatment Options

Regarding the treatment of HHV-6, no antiviral agents have been officially developed by pharmaceutical companies. In the absence of specific and formally approved anti-HHV-6 therapeutics, this chapter reviews the existing literature on the *in vitro* efficacy of the three available anti-cytomegalovirus (CMV) drugs that clinicians most often turn to for the treatment of HHV-6 AI: (1) the nucleoside analog *(val)ganciclovir*, (2) the nucleotide analog *cidofovir*, and (3) the pyrophosphate analog *foscarnet* (Table 19.1). CMV (HHV-5) is a closely related betaherpesvirus with 67% homology to HHV-6A[57] and similar homology to HHV-6B.[58,59]

IN VITRO EFFICACY OF COMMONLY USED ANTI-HHV-6 AGENTS

The medical literature database PubMed was systematically reviewed for all publications regarding the *in vitro* anti-HHV-6 efficacy of ganciclovir,

TABLE 19.1 Available Antivirals Currently Used in the Treatment of Acute HHV-6 Infections

Drug Name	Brand Name	Structure	Main Use[a]	Target	*In Vitro* Selectivity Against HHV-6	Cross Blood–Brain Barrier?	Clinical Risks
(Val) Ganciclovir	Cytovene, Valcyte	Nucleoside analog	CMV	Viral polymerase	Moderate[b]	Yes[106]	Bone marrow suppression
Cidofovir	Vistide	Nucleotide analog	CMV	Viral polymerase	Good[b]	Minimal[107]	Renal toxicity
Foscarnet	Foscavir	Pyrophosphate analog	CMV	Viral polymerase	Excellent[b]	Yes[91]	Renal toxicity

[a]*As reported by the U.S. Food and Drug Administration.*
[b]*See Table 19.2 for an efficacy comparison of these compounds.*

cidofovir or foscarnet. Data from each of these sources was gathered and sorted according to the assay, measurement, and virus strain and cell line used in each study to determine the average anti-HHV-6 effectiveness of each compound. Information on the various materials, assays, and laboratory parameters used by each study to calculate the anti-HHV-6 efficacy of each antiviral compound (see Table 19.2) is presented below.

Virus Strains[1]

The HHV-6A virus strain was

- GS—Isolated from a U.S. patient with a lymphoproliferative disorder; National Cancer Institute, Washington, D.C.[60,61]

The HHV-6B virus strains were

- HST—Isolated from a Japanese patient with exanthema subitum (ES)[62]
- Z-29—Isolated from a human immunodeficiency virus 1 (HIV-1)-positive AIDS patient from Zaire; Centers for Disease Control and Prevention, Atlanta, GA[63]
- CDVR1—Cidofovir (CDV)-resistant mutant strain obtained from HST by *in vitro* selection[64]

Cell Lines[2]

The cell lines for propagation of HHV-6A were

- HSB-2—T-lymphoblast cell line derived from the peripheral blood buffy coat of a patient with acute lymphoblastic leukemia and propagated as a tumor in newborn Syrian hamsters
- Sup-T1—T-lymphoblast cell line isolated from the plural effusion of an 8-year-old child with lymphoblastic leukemia

[1] Although some studies included data on additional HHV-6 strains, these data could not be used in the efficacy comparison because they were grown in uncommon cell lines, were known to be resistant to a given antiviral compound (CDVR1), or were not common enough to produce a standard data compilation. To avoid skewing the data, only results from the viral strains listed here have been included to produce a standardized selectivity index for each antiviral compound.

[2] Although some studies included data on additional cell models used to determine drug efficacy (such as fresh lymphocytes isolated from peripheral blood or cord blood), these data could not be used in the comparison because these cells are not indicative of normal HHV-6 infection. For example, peripheral blood mononuclear cells (PBMCs) have often been predisposed to HHV-6 prior to *in vitro* infection and therefore may yield a super-infection upon reintroduction of HHV-6 that can disable the cell's innate antiviral mechanisms and produce an infection that is stronger than an ordinary primary infection, skewing efficacy results. To avoid these complications, only data that expresses HHV-6 infection in the T-cell lines listed here have been included to produce a standardized selectivity index for each antiviral compound.

TABLE 19.2 Comparison of the *In Vitro* Anti-HHV-6 Efficacy of GCV, CDV, and PFA

Assay[a]	Ganciclovir (GCV)				Cidofovir (CDV)				Foscarnet (PFA)				Ref.
	Strain/ Cell Line[b]	AVA[c] (μM)	TOX[d] (μM)	SI[f]	Strain/ Cell Line[b]	AVA[c] (μM)	TOX[d] (μM)	SI[f]	Strain/ Cell Line[b]	AVA[c] (μM)	TOX[d] (μM)	SI[f]	
FACS	—	—	—	—	**A** GS/ Sup-T1	11.7	159	13.6	**A** GS/ HSB-2	13.7	166	121.	Williams et al.[65]
FACS	**A** GS/ HSB-2	25	54	2.2	—	—	—	—	**A** GS/ HSB-2	5.8	818	141.0	Reymen et al.[66]
FACS	**A** GS/ HSB-2	41.6[e]	62.7[e]	1.5	**A** GS/ Sup-T1	7.16[e]	209.8[e]	29.3	**A** GS/ HSB-2	49.5[e]	521[e]	10.5	Long et al.[67]
FACS	**A** GS/ Sup-T1	29.8[e]	155[e]	5.2	**A** GS/ HSB-2	16.5[e]	141.8[e]	8.6	**A** GS/ Sup-T1	32.8[e]	444[e]	13.5	Long et al.[67]
FACS	**B** HST/ MT4	6.4	89.6	14.0	**B** HST/ MT4	0.95	50.4	53.1	**B** HST/ MT4	6	348	58.0	Manichanh et al.[68]
FACS	**A** GS/ HSB-2	31.9	50	1.6	**A** GS/ HSB-2	9	25.9	2.9	**A** GS/ HSB-2	16	1250	78.1	De Bolle et al.[69]
FACS	**B** Z-29/ MOLT3	68.6	50	0.7	**B** Z-29/ MOLT3	9.8	55.6	5.7	**B** Z-29/ MOLT3	25.4	1016	40.0	De Bolle et al.[69]
FACS	—	—	—	—	**A** GS/ HSB-2	3.1	150	48.4	—	—	—	—	Kushner et al.[70]
FACS	—	—	—	—	**A** GS/ HSB-2	5.7	ND[g]	—	—	—	—	—	Kern et al.[71]

(*Continued*)

TABLE 19.2 (Continued)

Assay[a]	Ganciclovir (GCV)				Cidofovir (CDV)				Foscarnet (PFA)				Ref.
	Strain/ Cell Line[b]	AVA[c] (μM)	TOX[d] (μM)	SI[f]	Strain/ Cell Line[b]	AVA[c] (μM)	TOX[d] (μM)	SI[f]	Strain/ Cell Line[b]	AVA[c] (μM)	TOX[d] (μM)	SI[f]	
IFA	**A** GS/ HSB-2	59.2[e]	ND[g]	—	**A** GS/ HSB-2	46[e]	ND[g]	—	**A** GS/ HSB-2	45.8[e]	ND[g]	—	Long et al.[67]
IFA	**A** GS/ Sup-T1	14.1[e]	ND[g]	—	**A** GS/ Sup-T1	5.7[e]	ND[g]	—	**A** GS/ Sup-T1	27.6[e]	ND[g]	—	Long et al.[67]
IFA	**A** GS/ HSB-2	25	200	8.0	—	—	—	—	**A** GS/ HSB-2	49	1519	31.0	Akesson-Johansson et al.[72]
qPCR	—	—	—	—	—	—	—	—	**A** GS/ HSB-2	14.2	400	28.2	Naesens et al.[73]
qPCR	—	—	—	—	—	—	—	—	**B** Z-29/ MOLT3	18.1	400	22.1	Naesens et al.[73]
qPCR	**A** GS/ HSB-2	Inactive	100	—	**A** GS/ HSB-2	12.6	60	4.8	**B** HST/ MT4	8.9	759	85.3	Naesens et al.[73]
qPCR	**B** CDVR1/ MT4	12.9	ND[g]	—	—	—	—	—	**A** CDVR1/ MT4	17.8	ND[g]	—	Bonnafous et al.[64]
qPCR	**B** HST/ MT4	3.1	ND[g]	—	**B** HST/ MT4	1.8	ND[g]	—	**B** HST/ MT4	15	ND[g]	—	Bonnafous et al.[64]

qPCR	BHST/ MT4	4.5	43	9.6	**B** HST/ MT4	1.8	37	20.3	**B** HST/ MT4	9.7	85	8.7	Bounaadja et al.[75]
qPCR	—	—	—	—	**B** HST/ MT4	1.7	ND[g]	—	—	—	—	—	Bonnafous et al.[74]
DNA hyb.	—	—	—	—	**A** GS/ HSB-2	0.33	6.8	20.6	—	—	—	—	Prichard et al.[76]
DNA hyb.	—	—	—	—	**B** Z-29/ MOLT3	2.3	8.1	3.5	—	—	—	—	Prichard et al.[76]
DNA hyb.	—	—	—	—	**A** GS/ HSB-2	2.7	ND[g]	—	—	—	—	—	Williams-Aziz et al.[77]
DNA hyb.	—	—	—	—	**B** Z-29/ MOLT3	5.4	ND[g]	—	—	—	—	—	Williams-Aziz et al.[77]
DNA hyb.	**A** GS/ HSB-2	7.2	20	2.8	**A** GS/ HSB-2	2.9	25	8.6	**A** GS/ HSB-2	4.6	500	108.7	De Clercq et al.[78]

[a]*More information on the various assays used to determine the antiviral efficacy can be found in the section of Antiviral Assays this chapter.*
[b]*More information on the viral strains and cell lines used to determine the antiviral efficacy can be found in the section Antiviral Assays of this chapter.*
[c]*AVA is the measurement of antiviral activity against HHV-6, such as the effective concentration 50% (EC_{50}). Low AVA values signify increased anti-HHV-6 activity for a compound. More information on the various methods used to define AVA can be found in the section of Antiviral Assays this chapter.*
[d]*TOX is the measurement of a compound's cellular cytotoxicity, such as the cytotoxicity concentration 50% (CC_{50}). High TOX values signify decreased potential for a compound to be harmful against host cells during treatment. More information on the various methods used to define TOX can be found in the section Antiviral Assays of this chapter.*
[e]*These values were originally reported in μg/mL and have been converted to μM by using average molecular weight values for GCV (255.23 g/mol), CDV (279.18 g/mol), and PFA (191.95 g/mol).*
[f]*The selectivity index (SI) of a compound is calculated by dividing the compound's reported AVA value into its reported TOX value. High SI values are favorable, as they show a drug's potential to have cellular cytotoxicity at high concentrations, while the drug's effectiveness is seen at comparatively low concentrations. This is the widely used parameter for expressing a compound's antiviral efficacy.*
[g]*ND, test not done.*

The cell lines for propagation of HHV-6B were

- MOLT3—T-lymphoblast cells isolated from a patient with acute lymphoblastic leukemia
- MT4—Cells isolated from a patient with adult T-cell leukemia; a cell line transformed by human T-lymphotropic virus-I (HTLV-I), which has a protein that counteracts a certain innate mechanism of the cell to combat viral infection, making these cells more permissive to some virus infections

Antiviral Assays

The studies reviewed in this chapter utilized different methods to determine the anti-HHV-6 efficacy of each antiviral compound under consideration (Table 19.2). A brief description of the methods employed by each group is given here.

Fluorescence Activated Cell Sorting (FACS)

Live cells are incubated with virus and antiviral compounds. After several days, the virus-infected cells are stained with a primary antibody that is specific for an HHV-6 protein, followed by a secondary antibody that is labeled with a fluorescent marker. The fluorescence intensity of the individual cells is measured by FACS and assisted by specialized computer programs. A diminished fluorescence intensity indicates that the test compound is exhibiting antiviral activity.[65–71]

Immunofluorescence Assay (IFA)

This assay is similar to the FACS method, the main difference being that detection is done by fluorescence microscopy. Live cells are incubated with virus and antiviral compounds. After several days, the virus-infected cells are stained with a primary antibody that is specific for an HHV-6 protein, followed by a secondary antibody that is labeled with a fluorescent marker. The fluorescence intensity of the individual cells is estimated by fluorescence microscopy, and the number of fluorescent cells is counted. In case the compound is active, the number of fluorescent cells is lower.[67,72]

Polymerase Chain Reaction (qPCR, RT-PCR)

Two primers (nucleotide sequences) and a fluorescent DNA probe complementary to HHV-6-specific DNA sequences are designed and manufactured. The cells are incubated with virus and test compounds, and after a period of days DNA extracts are prepared and PCR analyzed. During the PCR reaction, the viral DNA is amplified by a polymerase enzyme in about 40 sequential cycles, during which the fluorescence signal increases above a certain threshold value. The more viral DNA copies present in the DNA extract, the faster the threshold value is reached. In samples coming from an active test compound, fewer viral DNA copies are detected. The PCR technique is very sensitive and, hence, care must be taken to ensure that an irrelevant DNA signal is not amplified in the process, giving a false-positive result. In contrast to qualitative PCR analysis (qPCR), in which the result of the PCR reaction is measured after the 40-cycle amplification is completed, real-time PCR (RT-PCR) allows accurate determination of the number of viral DNA copies in the test samples.[64,73–75]

DNA Hybridization

Prior to the wide application of real-time PCR assays, DNA hybridization represented an alternative (yet much less sensitive) method for quantification of DNA. The assay also requires an HHV-6-specific DNA probe that is labeled with a radioisotope (such as ^{32}P) or a marker (such as digoxigenin). The cells are incubated with virus and test compounds, and after a period of days DNA extracts are prepared. The DNA is spotted onto a membrane and then hybridized with the probe. The membrane is subjected to a staining procedure to visualize the DNA spots. The more viral DNA in the original extracts, the more intense the signal will be.[76–78]

Measurements and Data Calculations

The above techniques were employed to determine two parameters that allow determination of a compound's antiviral efficacy: the antiviral activity (AVA) against HHV-6 and the cytotoxicity to the host cell (TOX). The AVA and TOX values for a given compound are dependent on the assay employed, and these values are often combined to produce the selectivity index (SI) for this particular compound. The values (as reported in the literature reviewed in this study) used to illustrate a compound's antiviral activity and cytotoxicity are listed below.

Antiviral Activity (AVA) Values

EC_{50}

This value, used to illustrate AVA by the majority of the publications reviewed in this report,[66,67,70,71,73,76–78] is defined as the *effective concentration 50%* (EC_{50}), or the concentration of a test compound that produces 50% inhibition of virus replication. In the case of antiviral evaluation in cell culture, the EC_{50} is the compound concentration in which the amount of virus is 50% compared to what is detected in the untreated, virus-infected control.

IC_{50}

This value, employed by many groups to illustrate AVA,[64,65,68,69,72] is defined as the *inhibitory concentration 50%* (IC_{50}), or the concentration of antiviral compound necessary to reduce the number of HHV-6-infected cells by 50%.

Cytotoxicity (TOX) Values

CC_{50}

This value, used to express TOX by the majority of the publications reviewed in this report,[65–70,76] is defined as the *cytotoxicity concentration 50%* (CC_{50}), or the concentration of a drug that will kill half the cells in an uninfected cell culture.

CIC_{50}

This value, employed by Akesson-Johansson et al.[72] to illustrate TOX, is defined as the concentration of antiviral compounds sufficient to inhibit cellular proliferation by 50%.

MCC

This value, employed by two groups,[73,78] expresses TOX as the *minimum cytotoxic concentration*, or the concentration of compound necessary to cause minimal alterations in cell morphology as determined by microscopy.

Calculating the Average Selectivity Index (SI) for Each Compound

The selectivity index (SI) is a ratio that measures the window between cytotoxicity and antiviral activity by dividing the given AVA value into the TOX value (AVA/TOX). The higher the SI ratio, the theoretically more effective and safe a drug would be during *in vivo* treatment for a given viral infection. The ideal drug would be cytotoxic only at very high concentrations and have antiviral activity at very low concentrations, thus yielding a high SI value (high AVA/low TOX) and thereby able to eliminate the target virus at concentrations well below its cytotoxic concentration. The selectivity index of a compound is a widely accepted parameter used to express a compound's *in vitro* efficacy in the inhibition of virus replication. We have calculated the selectivity index for each study that reported both HHV-6 AVA and TOX values (Table 19.2).

Results

The data that were collected, sorted, and calculated are displayed in Table 19.2. A two-tailed, unpaired Student's *t*-test was performed to determine the possible difference of each compound's efficacy against HHV-6A versus HHV-6B (not shown). No statistically significant difference in the treatment of HHV-6B versus HHV-6A was observed for GCV ($p = 0.67$), CDV ($p = 0.85$), or PFA ($p = 0.94$). Thus, although HHV-6A and HHV-6B are considered separate virus species,[58,79] the efficacy of the compounds in question appears to be the same for the two HHV-6 viruses. This is not unexpected, as the three agents considered here target the HHV-6 polymerase, a protein that is highly conserved in HHV-6A and HHV-6B.

CLINICAL TREATMENT AND MANAGEMENT OF HHV-6 INFECTIONS

Although no internationally approved guidelines currently exist for the clinical treatment of HHV-6 acute infection (AI), The International Herpesvirus Management Forum and the American Society of Transplantation Infectious Diseases Community of Practice have recommended the initiation of antiviral therapy in cases of HHV-6 encephalitis.[80–83] HHV-6 encephalitis is a condition that occurs in the immunocompromised and immunocompetent.[26,84] It has a high rate of morbidity and thus requires urgent treatment to prevent death or permanent sequelae.[8,9,11,85–87] When possible, the reversal of immunosuppression may play a key role in the management of HHV-6 AI or reactivation; however, this option is not available in many cases, so administration of antiviral medication is a highly necessary element for the effective clinical management of HHV-6 reactivation.[6,88]

Foscarnet is currently considered the preferential treatment option for HHV-6 encephalitis in patients with anemia, as the administration of ganciclovir poses an additional risk of dose-limiting hematological toxicity.[89–91] Risks associated with foscarnet include renal toxicity[91] as well as complications from catheter-related deep vein thrombosis and infection. Unlike cidofovir, foscarnet cannot be administered in a peripheral vein. Of the three compounds, an oral prodrug is currently only available for ganciclovir (Valcyte).[92] The ability of the drug to penetrate the blood–brain barrier (BBB) is of critical importance in cases of HHV-6 central nervous system (CNS) infection. Both foscarnet and ganciclovir have been found to successfully penetrate the BBB.[93] It is assumed that cidofovir does not effectively cross the BBB. GCV, CDV, and PFA inhibit CMV much more selectively than HHV-6—the oral prodrug of GCV (valganciclovir) has significantly enhanced bioavailability—and all three drugs have a considerable risk for serious side effects, in particular when prolonged treatment is required.[65,77] There is a great need for the development of novel antivirals to more effectively treat acute HHV-6 infections and reactivation.

The existing data (Table 19.2) suggest that foscarnet should be considered the most selective *in vitro* inhibitor of HHV-6 among the three drugs currently used to treat acute HHV-6 infections in the clinical setting. Foscarnet has already been suggested as the preferential antiviral therapy to be used in the treatment and prevention of HHV-6 reactivation in HSCT patients.[94,95] However, further investigation on the benefits of foscarnet and other novel pharmacological agents in the specific clinical management of acute HHV-6 infections is warranted.

NEW OPTIONS FOR MANAGEMENT OF HHV-6 INFECTIONS

Several compounds in various stages of clinical development have shown effectiveness against HHV-6 *in vitro* (Table 19.3). Of these, there are three that

TABLE 19.3 Potential New Anti-HHV-6 Agents

Development Stage	Compound Name	Developing Company or Laboratory	Antiviral Spectrum and FDA Approval Status (If Applicable)[a]	*In Vitro* Efficacy Against HHV-6	Cross Blood–Brain Barrier?	Associated Clinical Risk?	Refs.
Experimentally available; some trials completed, seeking full commercial approval	CMX001 (HDP-CDV)	Chimerix, Inc. (USA)	FDA fast-track approval for treatment of CMV, adenovirus, and smallpox	Excellent	Yes	Low clinical risk for toxicity; gastrointestinal tract concerns	Bonnafous et al.,[74] Williams-Aziz,[77] Quenelle et al.,[96] Painter et al.[97]
	Artesunate (ART)	CDC (USA), Guilin Pharmaceutical Co. (China), Saokim, Ltd. (Intl.)	FDA investigational (IND) approval for malaria	Very good	Yes	Low risk, some associated heart complications	Miilbradt et al.,[98] Shapira et al.,[99] Hakacova et al.[100]
	Valomaciclovir (H2G prodrug, EPB-348)	Epiphany Biosciences (USA)	Completed FDA Phase IIB trials for VZV (shingles) and Phase IIA trials for EBV (infectious mononucleosis)	Very good	Unknown	Very low clinical risk for toxicity	De Clercq et al.,[101] Yao et al.,[102] Tyring et al.,[103] Andrei et al.[104]
	Ampligen (rintatolimod)	Hemispherx Biopharma (USA)	FDA CR letter issued following trials for treatment of infectious agents associated with ME/CFS[c]	Good[b]	Yes	Hepatotoxicity, abdominal pain, irregular heartbeat	Ablashi et al.[108]

In active pharmaceutical development	Cyclopropavir (MBX-400)	Microbiotix, Inc. (USA)	CMV	Good	Unknown	Safety trial pending	James et al.[109]
Showed efficacy in laboratory testing	S2242	Hoechst (Germany)	Many herpesviruses	Good	Unknown	Unknown	Neyts et al.,[110] De Clercq et al.[111]
	Arylsulfone derivatives	Rega Institute (Belgium)	Betaherpes-viruses	Good	Unknown	Unknown	Naesens et al.[112]
	Nexavir (formerly Kutapressin)	Nexco Pharma (USA)	EBV, HHV-6, CMV (generally prescribed for CFS patients)	Good[b]	Unknown	Very low	Ablashi et al.,[113] Rosenfeld et al.[114]
	AZT-lipid-PFA	Tumingen University Clinic (Germany)	Broad antiviral (enteroviruses, herpesviruses)	Good	Unknown	Unknown	Yao et al.[102]
	ZSM derivatives	Wayne State University of Medicine (USA)	CMV other herpesviruses	Moderate	Unknown	Unknown	Kern et al.[71]
	BDCRB	University of Michigan–Glaxo Wellcome, Inc. (USA)	CMV	Moderate	Unknown	Unknown	Prichard et al.,[76] Chen et al.,[115] Yoshida et al.[116]
	A-5021 (AV-038)	Ajinomoto Co., Inc. (Japan)	HSV	Moderate	Unknown	Unknown	De Clercq et al.[117]

[a] *FDA approval rating and trial stage reported by the U.S. Food and Drug Administration (FDA) as of March 2013.*
[b] *Only qualitative studies have been conducted on this compound to determine its in vitro efficacy.*
[c] *ME/CFS, myalgic encephalomyelitis/chronic fatigue syndrome.*

show superior antiviral efficacy. CMX001 is an antiviral drug developed by Chimerix, Inc. This agent, a lipophilic prodrug of CDV, is a lipid-ester derivative of CDV showing enhanced cellular permeability compared to underivatized CDV. This derivation increases the *in vitro* activity against HHV-6 by a factor 100-fold (compared to underivatized CDV), allowing for effectiveness at subcytotoxic concentrations.[74] There is also evidence suggesting that CMX001 can successfully penetrate the BBB.[96] As of this writing, CMX001 is entering Phase III clinical trials as a broad-acting anti-DNA virus agent in transplant patients. It is already being utilized via compassionate use in immunocompromised patients suffering from severe adenovirus infections.[97] Artesunate, a drug used globally for the treatment of malaria, has shown excellent efficacy against HHV-6 *in vitro*. It readily crosses the BBB[98] and is being explored in several centers for the compassionate-use treatment of drug-resistant CMV.[99] In addition, artesunate has been recently demonstrated as an effective treatment for antiviral-resistant HHV-6B myocarditis in a pediatric patient.[100] Valomaciclovir, an anti-varicella zoster virus (VZV) drug initially developed by Epiphany Biosciences, Inc., has also shown remarkable efficacy against HHV-6 *in vitro*. Its ability to penetrate the BBB is unknown,[101,102] although it has previously shown *in vitro* efficacy against HHV-6 in neural cells.[102] This compound is currently in Phase III clinical trials for the treatment of shingles (caused by VZV)[103] and in Phase II trials for the treatment of infectious mononucleosis caused by Epstein–Barr virus (EBV).[104] Beyond these four agents in advanced stages of clinical testing, a number of compounds have shown superior anti-HHV-6 efficacy *in vitro* but development of them has been discontinued for reasons unknown—including compounds such as CMV423, which showed unprecedented selectivity against HHV-6 *in vitro* but was eventually deemed unsuitable for clinical applications.[69]

In addition, a novel immunotherapy has been recently developed to limit the effects of multiple viruses known to cause complications following HSCT. The therapy focuses on the rapid generation of polyclonal cytotoxic T lymphocytes (CTLs) that are specific for 15 antigens from seven different viruses: EBV, CMV, adenovirus, polyomavirus BK, HHV-6, respiratory syncytial virus (RSV), and influenza virus.[105,106] Adoptive T-cell immunotherapy offers advantages over conventional antiviral treatment strategies, as there is no risk for drug toxicity and the immunotherapy can be used in patients infected with strains that are resistant to commonly used antivirals. This approach has already been utilized to limit the effects of adenovirus, CMV, and EBV in the clinical setting and has been shown to treat and prevent EBV-related post-transplant lymphoproliferative disease in several studies by the same group.[107] Clinical investigation into the effectiveness of this strategy for the specific treatment of HHV-6 AI and reactivation is now underway.

The establishment of formal treatment guidelines for the effective clinical management of HHV-6 is desperately needed. While the significance of HHV-6 AI is becoming more clearly defined in some clinical settings, including CNS infections, solid organ transplant (SOT) patients, and HSCT, the specific role of HHV-6 AI or reactivation remains to be fully established. As the diagnostic capabilities in this field are improving, standardization of care for HHV-6 AI and reactivation will become more realistic. However, large-scale studies at both the molecular and clinical levels are needed to obtain a more complete understanding of the clinical impact of HHV-6A and HHV-6B and to identify more optimal therapies to manage these emerging viruses.

ACKNOWLEDGMENTS

We would like to thank Kristin Loomis, Nathan Zemke, Louise Chatlynne, and Dharam Ablashi of the HHV-6 Foundation for their editorial and substantive contributions to this chapter.

REFERENCES

1. Salahuddin SZ, Kelley AS, Krueger GR, Josephs SF, Gupta S, Ablashi DV. Human herpes virusHHV-6 (HHV-6) in diseases. *Clin Diagn Virol* 1993;**1**(2):81–100.
2. Hall CB, Caserta MT, Schnabel KC, et al. Characteristics and acquisition of human herpesvirus (HHV) 7 infections in relation to infection with HHV-6. *J Infect Dis* 2006;**193**(8):1063–9.
3. Leach CT, Sumaya CV, Brown NA. Human herpesvirusHHV-6: clinical implications of a recently discovered, ubiquitous agent. *J Pediatr* 1992;**121**(2):173–81.
4. Zerr DM, Meier AS, Selke SS, et al. A population-based study of primary human herpesvirus 6 infection. *N Engl J Med* 2005;**352**(8):768–76.
5. Bates M, Monze M, Bima H, et al. Predominant human herpesvirus 6 variant A infant infections in an HIV-1 endemic region of Sub-Saharan Africa. *J Med Virol* 2009;**81**(5):779–89.
6. Agut H. Deciphering the clinical impact of acute human herpesvirus 6 (HHV-6) infections. *J Clin Virol* 2011;**52**(3):164–71.
7. Akhyani N, Fotheringham J, Yao K, Rashti F, Jacobson S. Efficacy of antiviral compounds in human herpesvirusHHV-6–infected glial cells. *J Neurovirol* 2006;**12**(4):284–93.
8. Zerr DM, Boeckh M, Delaney C, et al. HHV-6 reactivation and associated sequelae after hematopoietic cell transplantation. *Biol Blood Marrow Transplant* 2012;**18**(11):1700–8.
9. Zerr DM, Fann JR, Breiger D, et al. HHV-6 reactivation and its effect on delirium and cognitive functioning in hematopoietic cell transplantation recipients. *Blood* 2011;**117**(19):5243–9.
10. Kawamura Y, Sugata K, Ihira M, et al. Different characteristics of human herpesvirus 6 encephalitis between primary infection and viral reactivation. *J Clin Virol* 2011;**51**(1):12–19.
11. Sakai R, Kanamori H, Motohashi K, et al. Long-term outcome of human herpesvirus-6 encephalitis after allogeneic stem cell transplantation. *Biol Blood Marrow Transplant* 2011;**17**(9):1389–94.
12. Scheurer ME, Pritchett JC, Amirian ES, Zemke NR, Lusso P, Ljungman P. HHV-6 encephalitis in umbilical cord blood transplantation: a systematic review and meta-analysis. *Bone Marrow Transplant* 2013;**48**(4):574–80.

13. Chevallier P, Robillard N, Illiaquer M, et al. HHV-6 cell receptor CD46 expression on various cell subsets of six blood and graft sources: a prospective series. *J Clin Virol* 2013;**56**(4):331–5.
14. Hill JA, Koo S, Guzman Suarez BB, et al. Cord-blood hematopoietic stem cell transplant confers an increased risk for human herpesvirus-6-associated acute limbic encephalitis: a cohort analysis. *Biol Blood Marrow Transplant* 2012;**18**(11):1638–48.
15. Mori Y, Miyamoto T, Nagafuji K, et al. High incidence of human herpes virus 6-associated encephalitis/myelitis following a second unrelated cord blood transplantation. *Biol Blood Marrow Transplant* 2010;**16**(11):1596–602.
16. Mata S, Guidi S, Nozzoli C, et al. Human herpesvirus 6-associated limbic encephalitis in adult recipients of unrelated umbilical cord blood transplantation. *Bone Marrow Transplant* 2008;**42**(10):693–5.
17. Pritchett JC, Nanau RM, Neuman MG. The link between hypersensitivity syndrome reaction development and human herpes virus-6 reactivation. *Int J Hepatol* 2012;**2012**:723062.
18. Gentile I, Talamo M, Borgia G. Is the drug-induced hypersensitivity syndrome (DIHS) due to human herpesvirus 6 infection or to allergy-mediated viral reactivation? Report of a case and literature review. *BMC Infect Dis* 2010;**10**:49.
19. Tohyama M, Hashimoto K. New aspects of drug-induced hypersensitivity syndrome. *J Dermatol* 2011;**38**(3):222–8.
20. Yamashita N, Morishima T. HHV-6 and seizures. *Herpes* 2005;**12**(2):46–9.
21. Ichiyama T, Ito Y, Kubota M, Yamazaki T, Nakamura K, Furukawa S. Serum and cerebrospinal fluid levels of cytokines in acute encephalopathy associated with human herpesvirus-6 infection. *Brain Dev* 2009;**31**(10):731–8.
22. Seeley WW, Marty FM, Holmes TM, et al. Post-transplant acute limbic encephalitis: clinical features and relationship to HHV6. *Neurology* 2007;**69**(2):156–65.
23. Gorniak RJ, Young GS, Wiese DE, Marty FM, Schwartz RB. MR imaging of human herpesvirus-6-associated encephalitis in 4 patients with anterograde amnesia after allogeneic hematopoietic stem-cell transplantation. *AJNR Am J Neuroradiol* 2006;**27**(4):887–91.
24. Bollen AE, Wartan AN, Krikke AP, Haaxma-Reiche H. Amnestic syndrome after lung transplantation by human herpes virus-6 encephalitis. *J Neurol* 2001;**248**(7):619–20.
25. Maric I, Bryant R, Abu-Asab M, et al. Human herpesvirus-6-associated acute lymphadenitis in immunocompetent adults. *Mod Pathol* 2004;**17**(11):1427–33.
26. Revest M, Minjolle S, Veyer D, Lagathu G, Michelet C, Colimon R. Detection of HHV-6 in over a thousand samples: new types of infection revealed by an analysis of positive results. *J Clin Virol* 2011;**51**(1):20–4.
27. Lamoth F, Jayet PY, Aubert JD, et al. Case report: human herpesvirus 6 reactivation associated with colitis in a lung transplant recipient. *J Med Virol* 2008;**80**(10):1804–7.
28. Amo K, Tanaka-Taya K, Inagi R, et al. Human herpesvirus 6B infection of the large intestine of patients with diarrhea. *Clin Infect Dis* 2003;**36**(1):120–3.
29. Chevret L, Boutolleau D, Halimi-Idri N, et al. Human herpesvirus-6 infection: a prospective study evaluating HHV-6 DNA levels in liver from children with acute liver failure. *J Med Virol* 2008;**80**(6):1051–7.
30. Yoshikawa T. Human herpesvirus 6 causes hepatitis in transplant recipients. *Intern Med* 2006;**45**(7):417–8.
31. Chapenko S, Folkmane I, Ziedina I, et al. Association of HHV-6 and -7 reactivation with the development of chronic allograft nephropathy. *J Clin Virol* 2009;**46**(1):29–32.
32. Tohyama M, Hashimoto K, Yasukawa M, et al. Association of human herpesvirus 6 reactivation with the flaring and severity of drug-induced hypersensitivity syndrome. *Br J Dermatol* 2007;**157**(5):934–40.

33. Lautenschlager I, Hockerstedt K, Linnavuori K, Taskinen E. Human herpesvirus-6 infection after liver transplantation. *Clin Infect Dis* 1998;**26**(3):702–7.
34. Harma M, Hockerstedt K, Lautenschlager I. Human herpesvirus-6 and acute liver failure. *Transplantation* 2003;**76**(3):536–9.
35. Marabelle A, Bergeron C, Billaud G, Mekki Y, Girard S. Hemophagocytic syndrome revealing primary HHV-6 infection. *J Pediatr* 2010;**157**(3):511.
36. Dharancy S, Crombe V, Copin MC, et al. Fatal hemophagocytic syndrome related to human herpesvirus-6 reinfection following liver transplantation: a case report. *Transplant Proc* 2008;**40**(10):3791–3.
37. Kuhl U, Pauschinger M, Noutsias M, et al. High prevalence of viral genomes and multiple viral infections in the myocardium of adults with "idiopathic" left ventricular dysfunction. *Circulation* 2005;**111**(7):887–93.
38. Leveque N, Boulagnon C, Brasselet C, et al. A fatal case of human herpesvirus 6 chronic myocarditis in an immunocompetent adult. *J Clin Virol* 2011;**52**(2):142–5.
39. Mahrholdt H, Wagner A, Deluigi CC, et al. Presentation, patterns of myocardial damage, and clinical course of viral myocarditis. *Circulation* 2006;**114**(15):1581–90.
40. Yamaguchi N, Takatsuka H, Wakae T, et al. Idiopathic interstitial pneumonia following stem cell transplantation. *Clin Transplant* 2003;**17**(4):338–46.
41. Cone RW, Hackman RC, Huang ML, et al. Human herpesvirus 6 in lung tissue from patients with pneumonitis after bone marrow transplantation. *N Engl J Med* 1993;**329**(3):156–61.
42. Hammerling JA, Lambrecht RS, Kehl KS, Carrigan DR. Prevalence of human herpesvirus 6 in lung tissue from children with pneumonitis. *J Clin Pathol* 1996;**49**(10):802–4.
43. Kano Y, Inaoka M, Shiohara T. Association between anticonvulsant hypersensitivity syndrome and human herpesvirus 6 reactivation and hypogammaglobulinemia. *Arch Dermatol* 2004;**140**(2):183–8.
44. Takatsuka H, Wakae T, Mori A, et al. Endothelial damage caused by cytomegalovirus and human herpesvirus-6. *Bone Marrow Transplant* 2003;**31**(6):475–9.
45. Matsuda Y, Hara J, Miyoshi H, et al. Thrombotic microangiopathy associated with reactivation of human herpesvirus-6 following high-dose chemotherapy with autologous bone marrow transplantation in young children. *Bone Marrow Transplant* 1999;**24**(8):919–23.
46. Soldan SS, Berti R, Salem N, et al. Association of human herpes virus 6 (HHV-6) with multiple sclerosis: increased IgM response to HHV-6 early antigen and detection of serum HHV-6 DNA. *Nat Med* 1997;**3**(12):1394–7.
47. Martinez A, Alvarez-Lafuente R, Mas A, et al. Environment–gene interaction in multiple sclerosis: human herpesvirus 6 and MHC2TA. *Hum Immunol* 2007;**68**(8):685–9.
48. Garcia-Montojo M, De Las Heras V, Dominguez-Mozo M, et al. Human herpesvirus 6 and effectiveness of interferon beta1b in multiple sclerosis patients. *Eur J Neurol* 2011;**18**(8):1027–35.
49. Crawford JR, Kadom N, Santi MR, Mariani B, Lavenstein BL. Human herpesvirus 6 rhombencephalitis in immunocompetent children. *J Child Neurol* 2007;**22**(11):1260–8.
50. Potenza L, Luppi M, Barozzi P, et al. HHV-6A in syncytial giant-cell hepatitis. *N Engl J Med* 2008;**359**(6):593–602.
51. Lusso P, Crowley RW, Malnati MS, et al. Human herpesvirus 6A accelerates AIDS progression in macaques. *Proc Natl Acad Sci USA* 2007;**104**(12):5067–72.
52. Lusso P, Gallo RC. Human herpesvirus 6 in AIDS. *Immunol Today* 1995;**16**(2):67–71.
53. Emery VC, Atkins MC, Bowen EF, et al. Interactions between beta-herpesviruses and human immunodeficiency virus *in vivo*: evidence for increased human immunodeficiency viral load in the presence of human herpesvirus 6. *J Med Virol* 1999;**57**(3):278–82.

54. Nitsche A, Muller CW, Radonic A, et al. Human herpesvirus 6A DNA Is detected frequently in plasma but rarely in peripheral blood leukocytes of patients after bone marrow transplantation. *J Infect Dis* 2001;**183**(1):130–3.
55. Secchiero P, Carrigan DR, Asano Y, et al. Detection of human herpesvirus 6 in plasma of children with primary infection and immunosuppressed patients by polymerase chain reaction. *J Infect Dis* 1995;**171**(2):273–80.
56. Caselli E, Zatelli MC, Rizzo R, et al. Virologic and immunologic evidence supporting an association between HHV-6 and Hashimoto's thyroiditis. *PLoS Pathog* 2012;**8**(10):e1002951.
57. Gompels UA, Nicholas J, Lawrence G, et al. The DNA sequence of human herpesvirus-6: structure, coding content, and genome evolution. *Virology* 1995;**209**(1):29–51.
58. Dominguez G, Dambaugh TR, Stamey FR, Dewhurst S, Inoue N, Pellett PE. Human herpesvirus 6B genome sequence: coding content and comparison with human herpesvirus 6A. *J Virol* 1999;**73**(10):8040–52.
59. Isegawa Y, Mukai T, Nakano K, et al. Comparison of the complete DNA sequences of human herpesvirus 6 variants A and B. *J Virol* 1999;**73**(10):8053–63.
60. Salahuddin SZ, Ablashi DV, Markham PD, et al. Isolation of a new virus, HBLV, in patients with lymphoproliferative disorders. *Science* 1986;**234**(4776):596–601.
61. Ablashi DV, Balachandran N, Josephs SF, et al. Genomic polymorphism, growth properties, and immunologic variations in human herpesvirus-6 isolates. *Virology* 1991;**184**(2):545–52.
62. Aubin JT, Collandre H, Candotti D, et al. Several groups among human herpesvirus 6 strains can be distinguished by Southern blotting and polymerase chain reaction. *J Clin Microbiol* 1991;**29**(2):367–72.
63. Lopez C, Pellett P, Stewart J, et al. Characteristics of human herpesvirus-6. *J Infect Dis* 1988;**157**(6):1271–3.
64. Bonnafous P, Boutolleau D, Naesens L, Deback C, Gautheret-Dejean A, Agut H. Characterization of a cidofovir-resistant HHV-6 mutant obtained by *in vitro* selection. *Antiviral Res* 2008;**77**(3):237–40.
65. Williams SL, Hartline CB, Kushner NL, et al. *In vitro* activities of benzimidazole D- and L-ribonucleosides against herpesviruses. *Antimicrob Agents Chemother* 2003;**47**(7):2186–92.
66. Reymen D, Naesens L, Balzarini J, Holy A, Dvorakova H, De Clercq E. Antiviral activity of selected acyclic nucleoside analogues against human herpesvirus 6. *Antiviral Res* 1995;**28**(4):343–57.
67. Long MC, Bidanset DJ, Williams SL, Kushner NL, Kern ER. Determination of antiviral efficacy against lymphotropic herpesviruses utilizing flow cytometry. *Antiviral Res* 2003;**58**(2):149–57.
68. Manichanh C, Grenot P, Gautheret-Dejean A, Debre P, Huraux JM, Agut H. Susceptibility of human herpesvirus 6 to antiviral compounds by flow cytometry analysis. *Cytometry* 2000;**40**(2):135–40.
69. De Bolle L, Andrei G, Snoeck R, et al. Potent, selective and cell-mediated inhibition of human herpesvirus 6 at an early stage of viral replication by the non-nucleoside compound CMV423. *Biochem Pharmacol* 2004;**67**(2):325–36.
70. Kushner NL, Williams SL, Hartline CB, et al. Efficacy of methylenecyclopropane analogs of nucleosides against herpesvirus replication *in vitro*. *Nucleosides Nucleotides Nucleic Acids* 2003;**22**(12):2105–19.

71. Kern ER, Kushner NL, Hartline CB, et al. *In vitro* activity and mechanism of action of methylenecyclopropane analogs of nucleosides against herpesvirus replication. *Antimicrob Agents Chemother* 2005;**49**(3):1039–45.
72. Akesson-Johansson A, Harmenberg J, Wahren B, Linde A. Inhibition of human herpesvirus 6 replication by 9-[4-hydroxy-2-(hydroxymethyl)butyl]guanine (2HM-HBG) and other antiviral compounds. *Antimicrob Agents Chemother* 1990;**34**(12):2417–9.
73. Naesens L, Bonnafous P, Agut H, De Clercq E. Antiviral activity of diverse classes of broad-acting agents and natural compounds in HHV-6-infected lymphoblasts. *J Clin Virol* 2006;**37**(Suppl. 1):S69–75.
74. Bonnafous P, Bogaert S, Godet AN, Agut H. HDP-CDV as an alternative for treatment of human herpesvirus-6 infections. *J Clin Virol* 2013;**56**(2):175–6.
75. Bounaadja L, Piret J, Goyette N, Boivin G. Evaluation of Epstein–Barr virus, human herpesvirus 6 (HHV-6), and -8 antiviral drug susceptibilities by use of real-time-PCR-based assays. *J Clin Microbiol* 2013;**51**(4):1244–6.
76. Prichard MN, Frederick SL, Daily S, et al. Benzimidazole analogs inhibit human herpesvirus 6. *Antimicrob Agents Chemother* 2011;**55**(5):2442–5.
77. Williams-Aziz SL, Hartline CB, Harden EA, et al. Comparative activities of lipid esters of cidofovir and cyclic cidofovir against replication of herpesviruses *in vitro*. *Antimicrob Agents Chemother* 2005;**49**(9):3724–33.
78. De Clercq E, Andrei G, Snoeck R, et al. Acyclic/carbocyclic guanosine analogues as antiherpesvirus agents. *Nucleosides Nucleotides Nucleic Acids* 2001;**20**(4–7):271–85.
79. Adams MJ, Carstens EB. Ratification vote on taxonomic proposals to the International Committee on Taxonomy of Viruses (2012). *Arch Virol* 2012;**157**(7):1411–22.
80. Flamand L, Komaroff AL, Arbuckle JH, Medveczky PG, Ablashi DV. Review, part 1: Human herpesvirus-6—basic biology, diagnostic testing, and antiviral efficacy. *J Med Virol* 2010;**82**(9):1560–8.
81. Razonable RR, Lautenschlager I. Impact of human herpes virus 6 in liver transplantation. *World J Hepatol* 2010;**2**(9):345–53.
82. Dewhurst S. Human herpesvirus type 6 and human herpesvirus type 7 infections of the central nervous system. *Herpes* 2004;**11**(Suppl. 2):105A–11A.
83. Razonable RR. Human herpesviruses 6, 7 and 8 in solid organ transplant recipients. *Am J Transplant* 2013;**13**(Suppl. 3):67–77. quiz, 78.
84. Trabue CH, Bloch KC, Myers JW, Moorman JP. Case report and literature review: HHV-6-associated meningoencephalitis in an immunocompetent adult. *Herpes* 2008;**15**(2):33–5.
85. Yao K, Crawford JR, Komaroff AL, Ablashi DV, Jacobson S. Review, part 2: Human herpesvirus-6 in central nervous system diseases. *J Med Virol* 2010;**82**(10):1669–78.
86. Hoshino A, Saitoh M, Oka A, et al. Epidemiology of acute encephalopathy in Japan, with emphasis on the association of viruses and syndromes. *Brain Dev* 2012;**34**(5):337–43.
87. Troy SB, Blackburn BG, Yeom K, Caulfield AK, Bhangoo MS, Montoya JG. Severe encephalomyelitis in an immunocompetent adult with chromosomally integrated human herpesvirus 6 and clinical response to treatment with foscarnet plus ganciclovir. *Clin Infect Dis* 2008;**47**(12):e93–6.
88. Razonable RR. Infections due to human herpesvirus 6 in solid organ transplant recipients. *Curr Opin Organ Transplant* 2010;**15**(6):671–5.
89. Ishiyama K, Katagiri T, Ohata K, et al. Safety of pre-engraftment prophylactic foscarnet administration after allogeneic stem cell transplantation. *Transpl Infect Dis* 2012;**14**(1):33–9.

90. Zerr DM. Human herpesvirus 6 (HHV-6) disease in the setting of transplantation. *Curr Opin Infect Dis* 2012;**25**(4):438–44.
91. Wagstaff AJ, Faulds D, Goa KL. Aciclovir. A reappraisal of its antiviral activity, pharmacokinetic properties and therapeutic efficacy. *Drugs* 1994;**47**(1):153–205.
92. Jacobsen T, Sifontis N. Drug interactions and toxicities associated with the antiviral management of cytomegalovirus infection. *Am J Health Syst Pharm* 2010;**67**(17):1417–25.
93. Sjovall J, Bergdahl S, Movin G, Ogenstad S, Saarimaki M. Pharmacokinetics of foscarnet and distribution to cerebrospinal fluid after intravenous infusion in patients with human immunodeficiency virus infection. *Antimicrob Agents Chemother* 1989;**33**(7):1023–31.
94. Ogata M, Satou T, Kawano R, et al. Plasma HHV-6 viral load-guided preemptive therapy against HHV-6 encephalopathy after allogeneic stem cell transplantation: a prospective evaluation. *Bone Marrow Transplant* 2008;**41**(3):279–85.
95. Vu T, Carrum G, Hutton G, Heslop HE, Brenner MK, Kamble R. Human herpesvirus-6 encephalitis following allogeneic hematopoietic stem cell transplantation. *Bone Marrow Transplant* 2007;**39**(11):705–9.
96. Quenelle DC, Lampert B, Collins DJ, Rice TL, Painter GR, Kern ER. Efficacy of CMX001 against herpes simplex virus infections in mice and correlations with drug distribution studies. *J Infect Dis* 2010;**202**(10):1492–9.
97. Painter W, Robertson A, Trost LC, Godkin S, Lampert B, Painter G. First pharmacokinetic and safety study in humans of the novel lipid antiviral conjugate CMX001, a broad-spectrum oral drug active against double-stranded DNA viruses. *Antimicrob Agents Chemother* 2012;**56**(5):2726–34.
98. Milbradt J, Auerochs S, Korn K, Marschall M. Sensitivity of human herpesvirus 6 and other human herpesviruses to the broad-spectrum antiinfective drug artesunate. *J Clin Virol* 2009;**46**(1):24–8.
99. Shapira MY, Resnick IB, Chou S, et al. Artesunate as a potent antiviral agent in a patient with late drug-resistant cytomegalovirus infection after hematopoietic stem cell transplantation. *Clin Infect Dis* 2008;**46**(9):1455–7.
100. Hakacova N, Klingel K, Kandolf R, Engdahl E, Fogdell-Hahn A, Higgins T. First therapeutic use of artesunate in treatment of human herpesvirus 6B myocarditis in a child. *J Clin Virol* 2013;**57**(2):157–60.
101. De Clercq E, Naesens L, De Bolle L, Schols D, Zhang Y, Neyts J. Antiviral agents active against human herpesviruses HHV-6, -7 and -8. *Rev Med Virol* 2001;**11**(6):381–95.
102. Yao K, Hoest C, Rashti F, Schott TC, Jacobson S. Effect of (r)-9-[4-hydroxy-2-(hydroxymethyl)butyl]guanine (H2G) and AZT-lipid-PFA on human herpesvirus-6B infected cells. *J Clin Virol* 2009;**46**(1):10–14.
103. Tyring SK, Plunkett S, Scribner AR, et al. Valomaciclovir versus valacyclovir for the treatment of acute herpes zoster in immunocompetent adults: a randomized, double-blind, active-controlled trial. *J Med Virol* 2012;**84**(8):1224–32.
104. Andrei G, Snoeck R. Emerging drugs for varicella-zoster virus infections. *Expert Opin Emerg Drugs* 2011;**16**(3):507–35.
105. Gerdemann U, Keirnan JM, Katari UL, et al. Rapidly generated multivirus-specific cytotoxic T lymphocytes for the prophylaxis and treatment of viral infections. *Mol Ther* 2012;**20**(8):1622–32.
106. Gerdemann U, Keukens L, Keirnan JM, et al. Immunotherapeutic strategies to prevent and treat human herpesvirus 6 reactivation after allogeneic stem cell transplantation. *Blood* 2013;**121**(1):207–18.

107. Bollard CM, Rooney CM, Heslop HE. T-cell therapy in the treatment of post-transplant lymphoproliferative disease. *Nat Rev Clin Oncol* 2012;**9**(9):510–9.
108. Ablashi DV, Berneman ZN, Williams M, Strayer DR, Kramarsky B, Suhadolnik RJ, et al. Ampligen inhibits human herpesvirus-6 in vitro. *In Vivo* 1994;**8**(4):587–91.
109. James SH, Hartline CB, Harden EA, Driebe EM, Schupp JM, Engelthaler DM, et al. Cyclopropavir inhibits the normal function of the human cytomegalovirus UL97 kinase. *Antimicrob Agents Chemother* 2011;**55**(10):4682–91.
110. Neyts J, Jahne G, Andrei G, Snoeck R, Winkler I, De Clercq E. In vivo antiherpesvirus activity of N-7-substituted acyclic nucleoside analog 2-amino-7-[(1,3-dihydroxy-2-propoxy)methyl]purine. *Antimicrob Agents Chemother* 1995;**39**(1):56–60.
111. De Clercq E, Naesens L, De Bolle L, Schols D, Zhang Y, Neyts J. Antiviral agents active against human herpesviruses HHV-6, HHV-7 and HHV-8. *Rev Med Virol* 2001;**11**(6):381–95.
112. Naesens L, Stephens CE, Andrei G, Loregian A, De Bolle L, Snoeck R, et al. Antiviral properties of new arylsulfone derivatives with activity against human betaherpesviruses. *Antiviral Res* 2006;**72**(1):60–7.
113. Ablashi DV, Berneman ZN, Lawyer C, Kramarsky B, Ferguson DM, Komaroff AL. Antiviral activity in vitro of Kutapressin against human herpesvirus-6. *In Vivo* 1994;**8**(4):581–6.
114. Rosenfeld E, Salimi B, O'Gorman MR, Lawyer C, Katz BZ. Potential in vitro activity of Kutapressin against Epstein-Barr virus. *In Vivo* 1996;**10**(3):313–8.
115. Chen JJ, Wei Y, Drach JC, Townsend LB. Synthesis and antiviral evaluation of trisubstituted indole N-nucleosides as analogues of 2,5,6-trichloro-1-(beta-D-ribofuranosyl)benzimidazole (TCRB). *J Med Chem* 2000;**43**(12):2449–56.
116. Yoshida M, Yamada M, Tsukazaki T, Chatterjee S, Lakeman FD, Nii S, et al. Comparison of antiviral compounds against human herpesvirus 6 and 7. *Antiviral Res* 1998;**40**(1-2):73–84.
117. De Clercq E, Naesens L. In search of effective anti-HHV-6 agents. *J Clin Virol* 2006; **37**(Suppl 1):S82–6.

Index

Note: Page numbers followed by *"f" and "t"* refer to figures and tables, respectively.

E

F

T

U

V

W

Z

Printed and bound by CPI Group (UK) Ltd, Croydon, CR0 4YY
08/05/2025
01865010-0001